Cold-adapted Microorganisms

Edited by

Isao Yumoto

National Institute of Advanced Industrial Science and Technology (AIST)
Sapporo
Japan

Caister Academic Press

Caister Academic Press
Norfolk, UK

www.caister.com

British Library Cataloguing-in-Publication Data
A catalogue record for this book is available from the British Library

ISBN: 978-1-908230-26-3

Description or mention of instrumentation, software, or other products in this book does not imply endorsement by the author or publisher. The author and publisher do not assume responsibility for the validity of any products or procedures mentioned or described in this book or for the consequences of their use.

Cover image courtesy of Yoshinobu Nodasaka.

Printed and bound in Great Britain

Contents

Contributors

Rifat Zubair Ahmed
Centre for Molecular Genetics
University of Karachi
Karachi
Pakistan

rifzak@yahoo.com

Pierre Amato
Clermont University
Blaise Pascal University
Institute of Chemistry of Clermont-Ferrand (ICCF)
Clermont-Ferrand;
CNRS
UMR 6296
ICCF
Aubière
France

pierre.amato@univ-bpclermont.fr

Corien Bakermans
Altoona College
Pennsylvania State University
Altoona, PA
USA

cub21@psu.edu

Ahmad Iskandar Bin Haji Mohd Taha
Division of Biosphere Science
Graduate School of Environmental Science
Hokkaido University
Sapporo
Japan

radiance163@yahoo.com

Shinsuke Fukata
Graduate School of Science and Technology
Ehime University
Matsuyama
Japan

v855004c@mails.cc.ehime-u.ac.jp

Charles Gerday
Laboratory of Biochemistry
University of Liege
Liege
Belgium

ch.gerday@ulg.ac.be

Isao Hara
Technology Research Laboratory
Shimadzu Corporation
Seika-cho
Soraku-gun
Kyoto
Japan

i-hara@shimadzu.co.jp

Hidenori Hayashi
Cell-Free Science and Technology Research Center
Ehime University
Matsuyama
Japan

hayashi.hidenori.mj@ehime-u.ac.jp

Shannon Hinsa-Leasure
Grinnell College
Grinnell, IA
USA

hinsa@grinnell.edu

Tamotsu Hoshino
National Institute of Advanced Industrial Science and Technology (AIST)
Toyohira-ku
and
Graduate School of Life Sciences
Hokkaido University
Sapporo
Hokkaido
Japan

tamotsu.hoshino@aist.go.jp

Taiki Katayama
Institute for Geo-Resources and Environment
National Institute of Advanced Industrial Science and Technology
Tsukuba
Ibaraki
Japan

katayama.t@aist.go.jp

Jun Kawamoto
Institute for Chemical Research
Kyoto University
Uji
Kyoto
Japan

jun_k@mbc.kuicr.kyoto-u.ac.jp

Tatsuo Kurihara
Institute for Chemical Research
Kyoto University
Uji
Kyoto
Japan

kurihara@scl.kyoto-u.ac.jp

Norio Kurosawa
Department of Environmental Engineering for Symbiosis
Soka University
Hachioji
Tokyo
Japan

kurosawa@soka.ac.jp

Taro Motoigi
Division of Biosphere Science
Graduate School of Environmental Science
Hokkaido University
Sapporo
Japan

t.motoigi@ees.hokudai.ac.jp

Yuichi Nogi
Extremobiosphere Research Program Institute of Biogeosciences
Japan Agency for Marine-Earth Science and Technology
Yokosuka
Japan

nogiy@jamstec.go.jp

Hidetoshi Okuyama
Division of Biosphere Science
Graduate School of Environmental Science
and
Laboratory of Environmental Molecular Biology
Faculty of Environmental Earth Science
Hokkaido University
Sapporo
Japan

hoku@ees.hokudai.ac.jp

Michiko Tanaka
Laboratory of Novel Microbial Function Research
Graduate School of Agriculture
Hokkaido University
Sapporo
Hokkaido
Japan

mtanaka@chem.agr.hokudai.ac.jp

Oleg B. Tkachenko
Main Botanical Garden named after N.V. Tsitsin, Russian Academy of Sciences
Moscow
Russia

otkach@postman.ru

Kentaro Watanabe
Bioscience Group
National Institute of Polar Research
Tachikawa
Tokyo
Japan

kentaro@nipr.ac.jp

Nan Xiao
Kochi Core Center
Japan Agency for Marine-Earth Science and Technology (JAMSTEC)
Nakoku
Kochi
Japan

nanxiao@jamstec.go.jp

Yuka Yajima
National Institute of Advanced Industrial Science and Technology (AIST)
Toyohira-ku
Sapporo
Hokkaido
Japan

yuka44@frontier.hokudai.ac.jp

Seiji Yamauchi
Cell-Free Science and Technology Research Center
Ehime University
Matsuyama
Japan

yamauchi@ccr.ehime-u.ac.jp

Koji Yamazaki
Reaearch Faculty of Fisheries
Hokkaido University
Hakodate
Japan

yamasaki@fish.hokudai.ac.jp

Kazuaki Yoshimune
Department of Applied Molecular Chemistry
Nihon University
Narashino
Chiba
Japan

yoshimune.kazuaki@nihon-u.ac.jp

Isao Yumoto
Bioproduction Research Institute
National Institute of Advanced Industrial Science and Technology (AIST)
Toyohira-ku
Sapporo
Japan

i.yumoto@aist.go.jp

Current books of interest

Myxobacteria: Genomics, Cellular and Molecular Biology	2014
Next Generation Sequencing: Current Technologies and Applications	2014
Omics in Soil Science	2014
Mollicutes: Molecular Biology and Pathogenesis	2014
Genome Analysis: Current Procedures and Applications	2014
Bacterial Membranes: Structural and Molecular Biology	2014
Bacterial Toxins: Genetics, Cellular Biology and Practical Applications	2013
Fusarium: Genomics, Molecular and Cellular Biology	2013
Prions: Current Progress in Advanced Research	2013
RNA Editing: Current Research and Future Trends	2013
Real-Time PCR: Advanced Technologies and Applications	2013
Microbial Efflux Pumps: Current Research	2013
Cytomegaloviruses: From Molecular Pathogenesis to Intervention	2013
Oral Microbial Ecology: Current Research and New Perspectives	2013
Bionanotechnology: Biological Self-assembly and its Applications	2013
Real-Time PCR in Food Science: Current Technology and Applications	2012
Bacterial Gene Regulation and Transcriptional Networks	2012
Bioremediation of Mercury: Current Research and Industrial Applications	2012
Neurospora: Genomics and Molecular Biology	2012
Rhabdoviruses	2012
Horizontal Gene Transfer in Microorganisms	2012
Microbial Ecological Theory: Current Perspectives	2012
Two-Component Systems in Bacteria	2012
Foodborne and Waterborne Bacterial Pathogens	2012
Yersinia: Systems Biology and Control	2012
Stress Response in Microbiology	2012
Bacterial Regulatory Networks	2012
Systems Microbiology: Current Topics and Applications	2012
Quantitative Real-time PCR in Applied Microbiology	2012
Bacterial Spores: Current Research and Applications	2012
Small DNA Tumour Viruses	2012
Extremophiles: Microbiology and Biotechnology	2012

Full details at www.caister.com

Preface

If we define environments lower than 10°C as cold environments, most of the earth would consist of cold environments. These include the conventional cold environments such as the subarctic regions in winter, most of the sea environments, and specific environments such as the Arctic regions, soils and ice wedges in permafrost, deep sea, sea ice and glaciers, and soils in high alpine regions. Some of these specific cold environments are far from human activities and exhibit totally natural ecosystems. Analyses of microbiota in such cold environments reveal that certain microbiota adaptable to certain environments are composed of various cold-adapted microorganisms, although most of the mesophilic microorganisms living in the moderate temperature environments would find it difficult to propagate in such cold environments. Studies of cold ecosystems need to be carried out to further our understanding of the interaction between microorganisms and surrounding organisms such as plants or other microorganisms, or between microorganisms and environmental factors such as oxidative stress, hydrostatic pressure or desiccation in cold environments. There are specific relationships between microorganisms and other organisms in cold environments. In addition, microorganisms adapted to such cold ecosystems are suitable for the investigation of the various strategies they use to adapt to cold environments only or to cold environments concomitant with other environmental factors such as those mentioned above. Although microorganisms are able to thrive at a temperature of approximately 5°C, all the activities of microorganisms stop under freezing conditions (e.g. −20°C). The analyses of microbiota in such environments may provide a clue to the microorganisms that existed more than 10,000 years ago. On the other hand, if microorganisms have evolved by the accumulation of subtle mutational changes during generational changes, in cold environments where microorganisms can barely metabolize these microorganisms will exhibit different evolutional strategies because of their longer generation times compared with microorganisms living in mesophilic or thermophilic environments. Thus, studies of microorganisms in cold environments may provide the most comprehensive information on the life of microorganisms during the course of their evolution over 10,000 years. Furthermore, bacterial adaptation mechanisms in such extreme environments will provide new findings regarding bacterial environmental adaptation and evolution.

Certain cold-adapted microorganisms are adapted to temperature fluctuating environments. Such microorganisms are sensitive to ambient temperature changes and demonstrate exquisite gene regulation mechanisms by which proteins corresponding to the changes in ambient temperature are expressed. The microorganisms are able to grow more effectively than the surrounding organisms when the temperature rises in spring after a long period of low temperatures in winter. On the other hand, a stable appropriately low temperature (e.g. 5–10°C) is suitable for promoting certain biological activitics because exhaustion of nutrients at low temperatures is slower than that at ordinary ambient temperature. Some psychrophiles adapt well only to cold environments and their functions degenerate in respond to fluctuations of the ambient temperature. In cold sea environments, the temperature is stable, the decomposition of ambient nutrients is slow, and the concentration of dissolved oxygen is high. Indeed, microbial activities flourish in cold sea water and cold sea ice. On the other hand, very low temperatures such as subzero temperatures provide dormant conditions for microorganisms. Under such conditions, although the biological activities are low, the microorganisms have a chance of having an extended lifetime. Thus, as described above, cold environments do not only lower metabolic activities and are barren but they also provide microorganisms with beneficial niches from several aspects.

Although cold environments are quite common on earth, biological activities and biological mass are not always abundant in such environments especially at temperatures lower than −2°C. This is because of the fact that biological activities are based on chemical reactions, and such reactions need a certain activation energy depending on temperature. Therefore, cold-adapted microorganisms are not able to exhibit energy production rates equivalent to those in low to moderate temperature adapted microorganisms. Certain cold adaptation strategies at the molecular level are necessary for organisms to live in cold environments. Indeed, ecosystems in moderate cold (e.g. around 5°C) embrace various microorganisms. They possess exquisite molecular mechanisms for adaptation to cold environments such as the production of cold active enzymes via the metabolic pathways or enzymes for hydrolysing substrates and cold shock proteins to accelerate the transcription and translation of genes at low temperatures. In addition, the maintenance of appropriate membrane fluidity at low temperatures is indispensable for survival because membrane-binding proteins have important roles in energy production in most aerobic microorganisms. These enzymes are able to react with each other only in the presence of the appropriate membrane fluidity.

As described above, cold environments do not simply provide negative factors for microorganisms that delay their metabolism, they also give niches for microorganisms owing to stable environments, less competition with organisms living in their surroundings and less heat induced exhaustion due a decreased metabolic turnover rate. In this book, prominent authors present basic and new knowledge and concepts concerning cold-adapted microorganisms from the aspects of each area of expertise.

The editor would like sincerely to thank all the authors for their excellent contributions. I believe that each chapter will reveal to the readers quite fascinating research horizons of cold-adapted microorganisms. I also thank Horizon Press, particularly Dr Hugh Griffin, for giving me the opportunity to edit this book and his continuous support and encouragement to hasten the publication of this book.

Isao Yumoto
Sapporo, Japan

Part I

Biodiversity in Cold Ecosystems

Diversity of Bacteria in Permafrost

1

Shannon Hinsa-Leasure and Corien Bakermans

Abstract

In the cold, challenging environment of the permafrost, bacteria have found a way to survive and grow for thousands to millions of years. In this chapter, we explore bacterial diversity in permafrost from around the world, identified through culture-dependent and -independent techniques. Members of the phylum *Actinobacteria*, *Firmicutes* and *Proteobacteria* have been found in every environment studied thus far, indicating that these bacteria are well suited for life at low temperatures with low water activity. Also, unique species specific to individual environments have been discovered at each site. Researchers are faced with the challenge of determining which bacteria are active in the permafrost and which are in a dormant state. The ability of bacteria to reside in a dormant state further complicates culture-independent experimental results, as DNA from both dormant and active cells has been analysed. We are only beginning to understand the metabolic capabilities of permafrost bacteria, many discoveries are still to come.

Introduction

We live on a planet that is dominated by low-temperature environments. These low-temperature environments, including 90% of oceans and 26% of terrestrial soil ecosystems, once thought too cold for life have been shown to support diverse microbial communities (Gilichinsky *et al.*, 1995; Zhang *et al.*, 2008b). In this chapter we focus on bacterial diversity in permafrost, soil that has been at or below 0°C for at least two years. Permafrost is an extreme type of environment due to stable low temperatures, low water activity, and radiation exposure (Gilichinsky, 2002a,b). Omelyansky first described viable permafrost microorganisms in 1911 (Omelyansky, 1911). In 1942, James and Sutherland isolated aerobic and anaerobic bacteria from Canadian permafrost (James and Sutherland, 1942). These studies were followed by investigations in the permafrost of Alaska and Antarctica (Becker and Volkmann, 1961; Boyd and Boyd, 1964; Cameron and Morelli, 1974) and laid the groundwork for the now thriving study of microbial diversity in permafrost environments. Researchers have found the permafrost to be full of bacteria, with counts ranging from 10^5 to 10^8 cells per gram dry weight (gdw) permafrost (Bai *et al.*, 2006; Gilichinsky, 2002b; Hansen *et al.*, 2007; Kochkina *et al.*, 2001; Steven *et al.*, 2008b; Wilhelm *et al.*, 2011). Questions arose about these bacterial cells- were they metabolically active or dormant in a spore form or cyst-like state? Soina and colleagues characterized cell structure in Arctic and Antarctic permafrost by microscopy and found no spores or sporulating cells, instead they observed cells with thickened cell walls, compact nucleoids, and altered cytoplasm indicative of dormant cells (Soina *et al.*, 2004). In contrast, metabolic activity has been measured in the Siberian permafrost from −10°C to −20°C (Panikov and Sizova, 2007; Rivkina *et al.*, 2000, 2004). Work now focuses on identifying which bacteria are metabolically active in the permafrost.

Both culture-dependent and -independent techniques have been employed to characterize bacterial diversity in permafrost environments around the world. Typically bacterial viability decreases with increasing age of the permafrost (Gilichinsky *et al.*, 2007; Shi *et al.*, 1997; Zhang *et al.*, 2007b). Few if any isolates recovered from permafrost are psychrophiles; instead isolates are psychrotolerant, often with optimal growth at temperatures much higher than *in situ* temperatures (Bakermans *et al.*, 2003; Hinsa-Leasure *et al.*, 2010; Shi *et al.*, 1997; Steven *et al.*, 2007; Wilhelm *et al.*, 2011). Cultivation on diluted media at low temperatures has been shown to result in the highest number of CFU per gram of permafrost, although bacteria grown on rich media and/or at higher temperatures presented more morphological diversity (Hansen *et al.*, 2007; Hinsa-Leasure *et al.*, 2010; Steven *et al.*, 2007; Vishnivetskaya *et al.*, 2000). Several types of media have been employed in bacterial isolations from the permafrost, with R2A, PYGV (peptone, yeast extract, glucose, and vitamin) and tryptone soy agar being the most common (Bai *et al.*, 2006; Bakermans *et al.*, 2003; Gilichinsky *et al.*, 2007; Hansen *et al.*, 2007; Hinsa-Leasure *et al.*, 2010; Reasoner and Geldreich, 1985; Shi *et al.*, 1997; Steven *et al.*, 2007, 2008b; Vishnivetskaya *et al.*, 2000, 2006; Wilhelm *et al.*, 2011; Zhang *et al.*, 2007b). Culture-independent techniques are increasingly applied to the study of bacterial diversity in permafrost (Gilichinsky *et al.*, 2007; Hansen *et al.*, 2007; Hinsa-Leasure *et al.*, 2010; Steven *et al.*, 2007, 2008a; Vishnivetskaya *et al.*, 2006; Wilhelm *et al.*, 2011; Yergeau *et al.*, 2010) but face significant challenges. Because low temperatures preserve cells and DNA (Lindahl, 1993; Poinar *et al.*, 1996; Soina *et al.*, 1995; Willerslev *et al.*, 2004) extracted DNA does not necessarily represent active populations of cells. In the following sections we will describe bacterial diversity found within permafrost from around the world (Fig. 1.1 and Table 1.1). Both culture-dependent and -independent studies are described; as is permafrost from Siberia, the high Arctic, Antarctica and Central Asia (Fig. 1.1). The active layer (the uppermost layer which does not remain frozen year-round), while an integral part of permafrost, is beyond the scope of this review.

Siberian permafrost

The Siberian permafrost is one of the best-studied permafrost areas on earth. Early work by Russian scientists determined that up to 10^8 bacterial cells could be found in each gram of permafrost (Gilichinsky *et al.*, 1995). At depth, Siberian permafrost maintains a constant temperature from −10°C to −12°C and has a high total organic content (1–1.3%). Diverse bacteria have been isolated from the Siberian permafrost including: *Actinobacteria*, Firmicutes, sulfate-reducing bacteria, aerobic and anaerobic bacteria, and *Proteobacteria*, including nitrite-oxidizing *Betaproteobacteria* (Table 1.1) (Alawi *et al.*, 2007; Bakermans *et al.*, 2006; Hinsa-Leasure *et al.*, 2010; Liebner *et al.*, 2009; Liebner and Wagner, 2007; Shi *et al.*, 1997; Vatsurina *et al.*, 2008; Vishnivetskaya *et al.*, 2000, 2006; Wagner *et al.*, 2009). Many isolates have been shown to tolerate high salt concentrations, a state bacterial cells are likely to encounter in the unfrozen water of the Siberian permafrost (Bakermans *et al.*, 2003; Hinsa-Leasure *et al.*, 2010; Vishnivetskaya *et al.*, 2000). Studies conducted on samples from the active and permafrost table layers have found a rich diversity of bacteria including type

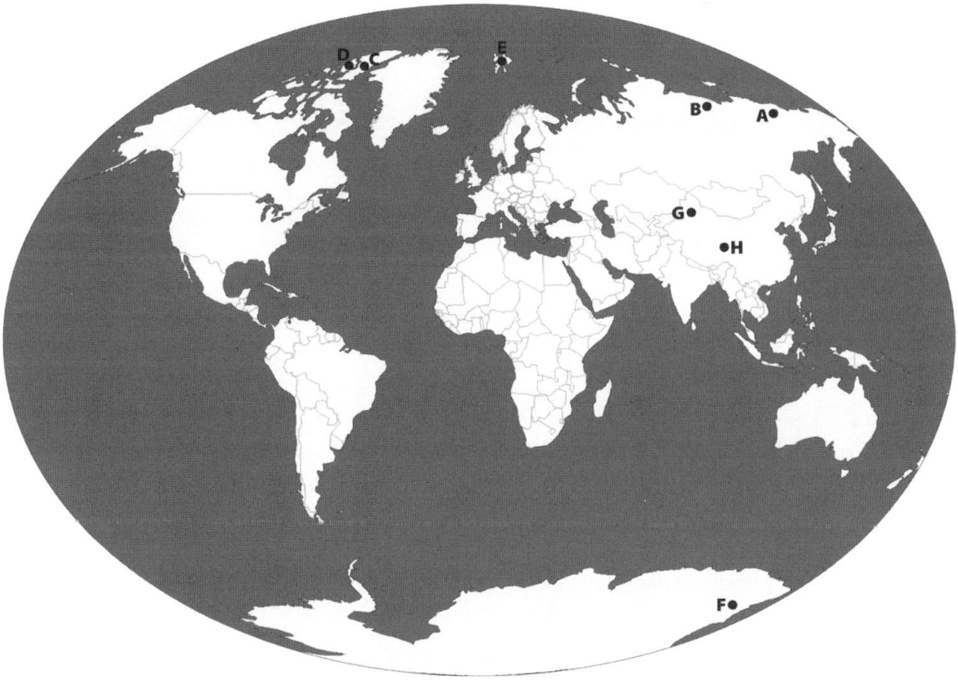

Figure 1.1 Location of permafrost bacterial diversity studies summarized in Table 1.1. Positions are marked at the approximate site where samples were collected. A, Kolyma lowland, Siberia; B, Lena Delta, Siberia; C, Ellesmere Island, Canada; D, Axel Heiberg Island, Canada; E, Spitsbergen, Norway; F, Dry Valleys, Antarctica; G, Tian Shan Mountains, China; H, Qinghai–Tibet Plateau, China.

I and type II methane-oxidizing bacteria, community members that will be of vital importance as temperatures rise and methane production increases (Liebner *et al.*, 2009; Liebner and Wagner, 2007).

Culture-dependent studies have commonly isolated *Arthrobacter, Planococcus, Exiguobacterium* and *Psychrobacter* species from several sites, while *Sporosarcina, Bacillus, Paenibacillus, Flavobacterium, Sphingobacterium, Pseudomonas, Serratia, Escherichia, Alcaligenes* and *Sphingomonas* have been identified at lower frequencies (Hinsa-Leasure *et al.*, 2010; Shi *et al.*, 1997; Vishnivetskaya *et al.*, 2006). However, when culture-independent techniques were employed, *Planomicrobium* and *Carnobacterium* rose to prominence in the study by Hinsa-Leasure and colleagues, while *Xanthomonadaceae* sequences were identified at high frequency from sites studied by Vishnivetskaya and colleagues (Hinsa-Leasure *et al.*, 2010; Vishnivetskaya *et al.*, 2006). The differences found between culture-dependent and -independent studies could reflect any combination of the following issues: the challenges faced in culturing permafrost bacteria in the laboratory, bias in DNA extraction and amplification, and the presence of non-viable cells.

Further examination of species from two genera commonly isolated from permafrost demonstrates their ubiquity in and adaptation to the permafrost environment. *Exiguobacterium* and *Psychrobacter* species have been detected in a majority of Siberian permafrost samples from cores ranging from 5000 to 3 million years old (Rodrigues *et al.*, 2009;

Table 1.1 Bacterial diversity in permafrost[a]

	Location[b]											
	Siberia				Canadian high Arctic				Norway	Antarctica	Central Asia	
	Kolyma lowland (A)			Lena Delta (B)	Ellesmere Island (C)			Axel Heiberg Island (D)	Spitsbergen (E)	Dry Valleys (F)	Tian Shan Mountains (G)	Qinghai–Tibet Plateau (H)
Phylum	Shi (1997)	Vishnivetskaya (2006)	Bakermans (2003)	Hinsa-Leasure (2010)	Steven (2007)	Steven (2008)	Yergeau[c] (2010)	Wilhelm (2011)	Hansen[d] (2007)	Gilichinsky (2007)	Bai (2007)	Zhang (2007)
Acidobacteria				C(1)		C(2)	Q(3.5)	C(20		C		
Actinobacteria	I(31)	C, I(47)	I(39)	C(19), I(51)	C(38),I(26)	C(13)	Q(38)	C(28), I(53)	C,I(81)	C, I	I(49)	II(36)
Bacteroides		I(10)					Q(11	C(2)	C,I(3)	C		I(12)
CFB[e]					C(11)	C(18)				C	I(16)	
Clostridia									C			
Chloroflexi										C		
Fibrobacteres										C		
Firmicutes	I(34)	C, I(29)	I(37)	C(72), I(45)	C(4), I(70)	I(100)	Q(1)	I(33)	I(4)	C,I	I(6)	I(6)
Gemmatinonadetes					C(3)	C(2)		C(15)				
Nitrospira								C(1)				
OD1									C			
OP10									C			
Planctomyces					C(1)			C(2)				
Proteobacteria						C(16)		C(15)				
Alpha		I(4)			C(12)		Q(6.5)	I(6.5)	I(1)	C	I(10)	I(18)
Beta	I(21)			C(1)	C(2)	C(41)	Q(40)	I(6.5)	C, I(6)	C	I(6)	I(12)
Delta					C(17)				C	I		
Gamma	I(14)	C, I(10)	I(24)	C(1), I(4)	C(8), I(4)				C, I(4)	C, I	I(14)	I(15)
TM7								C(1)		C		
Verrucomicrobia								C(2)				
Unclassified				C(6)	C(4)	C(8)		C(15)				

[a]Bacterial diversity as identified in a clone library (C), isolates (I), or through Q-PCR (Q). The percentage of sequences belonging to each phylum is given in parenthesis, if available from the study.
[b]The letters behind the location name identify the location in Fig. 1.1.
[c]Diversity shown from the sample that did not receive multiple displacement amplification.
[d]Diversity represents that found in original sample, enrichments not included.
[e]CFB, *Cytophaga–Flexibacter–Bacteroides*.

Rodrigues and Tiedje, 2007; Vishnivetskaya *et al.*, 2009). In the laboratory, two permafrost isolates, *Psychrobacter arcticus* and *Exiguobacterium sibiricum*, have been shown to be metabolically active when under low water activity conditions, conditions which simulate environmental conditions (Ponder *et al.*, 2008). *E. sibericum* and *P. arcticus* have also been shown to alter membrane composition by decreasing fatty acid saturation and chain length, in accordance with decreasing temperatures, which affect carbon utilization and antibiotic resistance at low temperatures (Ponder *et al.*, 2005). Further characterization of these two species has revealed additional adaptations to low temperatures which likely allow them to actively survive within permafrost (Ayala-del-Rio *et al.*, 2010; Bergholz *et al.*, 2009; Rodrigues *et al.*, 2008).

Owing to the age of Siberian permafrost and the ability of its low temperatures to preserve material, study of Siberian permafrost isolates can provide historical information that would not be available otherwise. For example, Petrova and colleagues isolated a novel integron, a mobile DNA element which confers antibiotic and chromate resistance, from a *Pseudomonas* isolate recovered from 15,000- to 40,000-year-old permafrost (Petrova *et al.*, 2011). This work demonstrated that the presence of antibiotic resistance genes predates human manufacturing and use of antibiotics. Additionally, Mindlin and colleagues identified mercury resistance transposons in *Pseudomonas* isolates from permafrost aged 15,000–40,000 years (Mindlin *et al.*, 2005). The mercury resistance transposons from permafrost are very similar to those found today, suggesting that mercury resistance gene content and distribution have not changed significantly over the last 40,000 years.

A distinct environment within the Siberian permafrost is the cryopeg, a pocket of highly saline (170–300 g/l) liquid water sandwiched between layers of permafrost. Cryopegs contain about 10^7 cells/ml (direct counts) and a diversity of prokaryotes (from heterotrophs to sulfate-reducers to methanogens) have been cultivated from them (Bakermans *et al.*, 2003; Gilichinsky *et al.*, 2003). Bakermans and colleagues characterized isolates from several cryopegs and determined that *Psychrobacter* and *Arthrobacter* species dominated the non-spore-forming isolates, while *Bacillus* and *Paenibacillus* species made up the spore-forming bacterial isolates (Bakermans *et al.*, 2003). Another study (Gilichinsky *et al.*, 2005) also found that isolates were dominated by spore-forming rod-shaped cells when incubated anaerobically in heterotrophic media. Two isolates from this study, *Psychrobacter murincola* 2pS and *Clostridia* sp. 14D, both reproduced at −5°C (Gilichinsky *et al.*, 2005). More spore-forming bacteria were isolated from cryopegs than have typically been isolated in the terrestrial permafrost. Further, two genera found in a majority of permafrost environments, *Planococcus* and *Exiguobacterium* were not isolated (Bakermans *et al.*, 2003; Gilichinsky *et al.*, 2005). These studies suggest that the high salinity and larger liquid volume found in cryopegs has selected for a different bacterial community than that found in the surrounding permafrost.

Canadian high Arctic permafrost

Low temperatures and high carbon content characterize high Arctic permafrost regions in Nunavut, Canada. Several studies conducted on unique environments in this region have found remarkably similar bacterial communities (Table 1.1) (Steven *et al.*, 2007, 2008a; Wilhelm *et al.*, 2011; Yergeau *et al.*, 2010). New species have been characterized including a psychrotolerant anaerobic bacterium, *Clostridium tagluense*, which was isolated from permafrost in the Mackenzie Delta, Nunavut, mainland Canada (Suetin *et al.*, 2009). This newly discovered species is similar to other cluster I psychrotolerant and psychrophilic *Clostridium* species. The bacterial diversity in the Canadian permafrost has only begun to be explored and it holds promise for the discovery of many new bacterial species.

The permafrost on Eureka, Ellesmere Island, Canada, is a unique permafrost environment because it is more similar in temperature to the Antarctic permafrost, yet it has a high carbon content (total carbon of 2.19% dry weight of permafrost), similar to the Siberian permafrost (Gilichinsky *et al.*, 1995, 2007; Steven *et al.*, 2007). Two studies conducted by Steven and colleagues on separate permafrost samples from Eureka isolated a bacterial community dominated by spore-forming *Bacillus, Paenibacillus* and *Sporosarcina*. The dominance of spore-forming bacteria is in contrast to Siberian permafrost studies that have found far fewer spore-forming bacteria and significantly more *Proteobacteria* (Table 1.1) (Hansen *et al.*, 2007; Hinsa-Leasure *et al.*, 2010; Shi *et al.*, 1997; Steven *et al.*, 2007, 2008b; Vishnivetskaya *et al.*, 2006). Three Gram-positive isolates from the Eureka permafrost sample were characterized as new species; *Virgibacillus arcticus, Tumebacillus permanentifrigoris* and *Planococcus halocryophilus* (Mykytczuk *et al.*, 2011; Niederberger *et al.*, 2009; Steven *et al.*, 2008a). Unlike the isolate diversity, the diversity found in the 16S rRNA clone libraries was highly similar to the diversity found in the 16S rRNA clone libraries generated from Siberian permafrost (Steven *et al.*, 2007, 2008b; Yergeau *et al.*, 2010). The drastic differences between the isolated and the 16S rRNA clone library communities from this permafrost could be due to DNA surviving outside of cells in the permafrost or the presence of dead or unculturable cells in the permafrost. Another possible explanation could be differences in PCR amplification of environmental DNA. Yergeau and colleagues determined that the whole-community genome amplification techniques, often employed to increase DNA concentrations, led to a misrepresentation of bacterial diversity by increasing *Bacteriodetes* while decreasing *Actinobacteria* (Yergeau *et al.*, 2010).

To examine the metabolic diversity found in Ellesmere Island permafrost and the surrounding active layer, Yergeau and colleagues employed culture-independent techniques (Yergeau *et al.*, 2010). Only type I methanotrophs were detected in metagenomic libraries from Eureka active-layer soil and permafrost; however, during qPCR experiments type II methanotrophs were found in these samples but at lower numbers than type I methanotrophs. This finding is in contrast to results from Siberian permafrost that found type I methanotrophs in the active-layer samples and type II methanotrophs distributed throughout the active layer and permafrost table (Liebner *et al.*, 2009; Liebner and Wagner, 2007; Martineau *et al.*, 2010). Genes known to be involved in ammonia oxidation and nitrogen fixation were also found in active-layer and permafrost (Yergeau *et al.*, 2010).

The bacterial diversity within an acidic wetland from Axel Heiberg Island, Nunavut, Canada, has been recently analysed (Wilhelm *et al.*, 2011). This area has an annual temperature of −15°C and a pH <5 (Wilhelm *et al.*, 2011). Similar to other permafrost environments, higher cell counts were found in the active-layer compared with the permafrost (Steven *et al.*, 2006), although a greater bacterial diversity was found in the permafrost (Wilhelm *et al.*, 2011). In the permafrost, *Actinobacteria* were the most numerous in the clone library, while the isolates contained members of the phyla *Firmicutes, Actinobacteria* and *Proteobacteria* (Table 1.1) (Wilhelm *et al.*, 2011). The bacterial diversity found in the clone library had overlap with that identified in other high Arctic sites, while it had no sequences in common with the Siberian permafrost (Table 1.1) (Hansen *et al.*, 2007; Steven *et al.*, 2008b;

Vishnivetskaya *et al.*, 2000; Wilhelm *et al.*, 2011). Interestingly a relatively large portion (15%) of sequences from the clone library could not be classified (Wilhelm *et al.*, 2011). This acidic high Arctic site has selected for a unique group of bacteria capable of growing at low temperatures and in acidic conditions.

The bacterial diversity in permafrost and from endostromatolites was studied in the Haughton Impact Structure Region on Devon Island. This impact crater formed approximately 39 million years ago and is considered a polar desert (Pellerin *et al.*, 2009). The permafrost had a greater bacterial diversity than that found in the endostromatolites and included members of various phyla with *Actinobacteria, Proteobacteria* and *Bacteriodetes* present in all samples (Pellerin *et al.*, 2009). Similar to other Canadian high Arctic sites, a large number of 16S rRNA sequences could not be classified, suggesting the presence of novel bacterial strains (Pellerin *et al.*, 2009).

Norwegian permafrost

Spitsbergen is also within the high Arctic and permafrost from there is similar (temperature and organic content, 1.6%) to other high Arctic permafrost. The aerobic and anaerobic microbial diversity from an acidic 2300 year old Arctic Spitsbergen permafrost sample was characterized utilizing culture-dependent and -independent techniques (Hansen *et al.*, 2007). Of the isolated aerobic bacteria, *Arthrobacter* and *Cellulomonas* were the most prevalent genera, together making up 55% of the isolates, while *Clostridium* species dominated the anoxic enrichments (Hansen *et al.*, 2007). Similar to Siberia and in contrast to the Canadian high Arctic, few spore-forming isolates were recovered (Hansen *et al.*, 2007; Hinsa-Leasure *et al.*, 2010; Shi *et al.*, 1997; Steven *et al.*, 2007, 2008a; Vishnivetskaya *et al.*, 2006). Hansen and colleagues used several enrichments to increase the diversity of bacteria and they successfully isolated sulfate-reducing *Desulfocapsa* and *Desulfobacterium* (Hansen *et al.*, 2007). Following growth in enrichments, 16S rRNA genes from *Actinobacteria* and candidate division TM7 were found in all nine enrichments (Hansen *et al.*, 2007). Furthermore, two new *Spirosoma* species, *S. spitsbergense* and *S. luteum*, and one *Actinobacteria, Demequina lutea*, from the Spitsbergen permafrost have been characterized (Finster *et al.*, 2009a; Finster *et al.*, 2009b). The 16S rRNA clone library generated from non-enriched permafrost shows similar diversity to that found in other Canadian and Antarctic permafrost environments, although several sequences unique to Spitsbergen were also identified (Table 1.1). The researchers determined that only 26% of the total cells in the permafrost were actually viable by epifluorescence microscopy (Hansen *et al.*, 2007). This finding implies that much of the DNA recovered from the permafrost is from non-viable cells.

Antarctic permafrost

Because Antarctica is primarily covered with glaciers and ice sheets, permafrost is limited to the McMurdo Dry Valleys and the Antarctic peninsula. The McMurdo Dry Valley permafrost is a cold and dry environment with mean annual temperatures of approximately −25°C (Gilichinsky *et al.*, 2007). Despite low total organic carbon levels (0 to 0.1%), Gilichinsky and colleagues have found 10^5 to 10^6 cells per gdw permafrost with *Proteobacteria* and *Actinobacteria* dominating the bacterial community (Table 1.1) (Gilichinsky *et al.*, 2007). Compared with the Siberian permafrost, the Antarctic permafrost has less microbial diversity and 100 to 1000 times fewer bacterial cells per gram of permafrost. Several genera including *Arthrobacter, Bacillus, Caulobacter, Comamonas, Flavobacterium, Nocardioides, Pseudomonas* and *Sphingobacterium* were identified in 16S rRNA clone libraries from every sample examined. Isolation experiments identified several genera of aerobic bacteria including *Arthrobacter, Micrococcus, Rhodococcus, Bacillus, Cellulomonas, Promicromonospora, Azotobacter, Aeromonas* and *Pseudomonas,* along with denitrifying bacteria and methanogens (Gilichinsky *et al.*, 2007). Owing to the extreme cold and dry conditions, the authors hypothesize that a majority of microorganisms reside in a dormant state in the Antarctic permafrost.

Central Asia

Permafrost can also occur at lower latitudes at high elevations. The largest area of extensive alpine permafrost is found in Central Asia. The permafrost in the Tian Shan Mountains in China is distinctive because it is warmer than most other studied permafrosts, with an average annual temperature of −5°C (Bai *et al.*, 2006). Isolates from Tian Shan permafrost fell into four major groups (Table 1.1), with the most prevalent and diverse being *Actinobacteria* isolates (Bai *et al.*, 2006). The isolate diversity is similar to the diversity found in other permafrost environments, particularly the high Arctic, and included isolates proposed to belong to *Arthrobacter, Exiguobacterium, Rhodococcus, Psychrobacter, Pseudomonas* and *Pedobacter* genera among others (Table 1.1) (Bai *et al.*, 2006). A new species, *Chryseobacterium xinjiangense,* was isolated from this permafrost area (Zhao *et al.*, 2011).

Another alpine permafrost, from the Qinghai–Tibet plateau, was studied to characterize overall bacterial diversity and specifically alkaliphiles present in this relatively warm (−1°C to −2.2°C), high elevation, alkaline area (Zhang *et al.*, 2007a; Zhang *et al.*, 2007b). Samples from this region yielded up to 10^7 CFU/gdw permafrost and were dominated by rod shaped bacterial cells, similar to reports from Tian Shan and Siberian permafrosts (Bai *et al.*, 2006; Shi *et al.*, 1997; Zhang *et al.*, 2007b). The bacterial diversity in this permafrost is similar to the diversity from the Tian Shan permafrost (Table 1.1), with the *Pseudomonas* and *Arthrobacter* as the dominant genera (Zhang *et al.*, 2007b). Active-layer and permafrost samples yielded large numbers of *Actinobacteria* with smaller numbers of Firmicutes and *Proteobacteria* alkaliphiles on low-nutrient medium at a pH of 9 (Zhang *et al.*, 2007a). A novel psychrotolerant *Hymenobacter* species, *H. psychrotolerans,* was isolated and characterized from this permafrost (Zhang *et al.*, 2008a). Although geographically separated from other permafrost regions, bacterial diversity in Central Asian permafrost is remarkably similar to high Arctic and Antarctic permafrost, suggesting that the permafrost environment selects for specific types of bacteria.

Conclusions and future trends

Over the last decade our understanding of bacterial diversity in permafrost environments has greatly expanded. Certain phyla such as *Actinobacteria, Firmicutes* and *Proteobacteria* are found in all permafrost examined thus far, which suggests they can adapt well to cold temperatures, low water availability, and varied soil composition. Certain genera, such as *Arthrobacter* and *Planococcus,* have been found in permafrost from Siberia, Antarctica, China and Canada, suggesting these genera are particularly well adapted to cold environments (Bai *et al.*, 2006; Gilichinsky *et al.*, 2007; Hinsa-Leasure *et al.*, 2010; Reddy *et al.*, 2002; Steven *et al.*, 2007; Vishnivetskaya *et al.*, 2006; Zhang *et al.*, 2007a). Not surprisingly, permafrost environments with additional constraints (acidic, alkaline, highly saline) support different populations (such as *Acidobacteria* or a higher proportion of spore-forming bacteria). Many new species have been identified from permafrost environments around the world, proving novel organisms can survive and thrive in this extreme environment. Indeed, a high number of unclassified sequences in clone libraries from the Canadian high Arctic point to novel, as yet undescribed bacterial diversity.

Although we have learned a great deal about bacteria present in permafrost, we are just beginning to understand the complexity of these environments. More work is needed to fully understand if the active-layer and permafrost communities interact, especially during this time of rising temperatures, when more permafrost will transition into the active-layer. The permafrost environment contains a vast carbon pool and understanding the metabolic capabilities of the microbes present will allow researchers to predict how permafrost affects and will affect global nutrient cycling. Based on the studies described in this chapter, we can begin to predict the type of microbes that will be present in a given environment, but more work is needed to overcome the biases introduced during isolation, DNA extraction, and DNA amplification. When analysing a community through culture-independent techniques- how can we know which DNA comes from metabolically active cells? The refinement of current techniques, such as stable isotope probing, and the development of new technologies will allow us to address this question.

Acknowledgements

We would like to thank Bruce Leasure for graphic design assistance.

References

Alawi, M., Lipski, A., Sanders, T., Pfeiffer, E.M., and Spieck, E. (2007). Cultivation of a novel cold-adapted nitrite oxidizing betaproteobacterium from the Siberian Arctic. ISME J. *1*, 256–264.

Ayala-del-Rio, H.L., Chain, P., Grzymski, J.J., Ponder, M., Ivanova, N., Bergholz, P.W., Bartolo, G.D., Hauser, L., Land, M., Bakermans, C., *et al.* (2010). The genome sequence of *Psychrobacter arcticus* 273–4, a psychroactive Siberian permafrost bacterium reveals mechanisms for adaptation to low temperature. Appl. Environ. Microbiol. *76*, 2304–2312.

Bai, Y., Yang, D., Wang, J., Xu, S., Wang, X., and An, L. (2006). Phylogenetic diversity of culturable bacteria from alpine permafrost in the Tianshan Mountains, northwestern China. Res. Microbiol. *157*, 741–751.

Bakermans, C., Ayala-del-Rio, H.L., Ponder, M.A., Vishnivetskaya, T., Gilichinsky, D., Thomashow, M.F., and Tiedje, J.M. (2006). *Psychrobacter cryohalolentis* sp. nov. and *Psychrobacter arcticus* sp. nov., isolated from Siberian permafrost. Int J. Syst. Evol. Microbiol.*56*, 1285–1291.

Bakermans, C., Tsapin, A.I., Souza-Egipsy, V., Gilichinsky, D.A., and Nealson, K.H. (2003). Reproduction and metabolism at −10 degrees C of bacteria isolated from Siberian permafrost. Environ. Microbiol. *5*, 321–326.

Becker, F.E., and Volkmann, C.M. (1961). A preliminary report on the bacteriology of permafrost in the Fairbanks area. Proc. Alaskan Sci. Conf. *12*, 188.

Bergholz, P.W., Bakermans, C., and Tiedje, J.M. (2009). *Psychrobacter arcticus* 273–4 uses resource efficiency and molecular motion adaptations for subzero temperature growth. J. Bacteriol. *191*, 2340–2352.

Boyd, W.L., and Boyd, J.W. (1964). The presence of bacteria in permafrost of the Alaskan Arctic. Can. J. Microbiol. *10*, 917–919.

Cameron, R.E., and Morelli, F.A. (1974). Viable microorganisms from ancient Ross Island and Taylor Valley drill core. Antarct J US *9*, 113–116.

Finster, K.W., Herbert, R.A., Kjeldsen, K.U., Schumann, P., and Lomstein, B.A. (2009a). *Demequina lutea* sp. nov., isolated from a high Arctic permafrost soil. Int. J. Syst. Evol. Microbiol. *59*, 649–653.

Finster, K.W., Herbert, R.A., and Lomstein, B.A. (2009b). *Spirosoma spitsbergense* sp. nov. and *Spirosoma luteum* sp. nov., isolated from a high Arctic permafrost soil, and emended description of the genus *Spirosoma*. Int. J. Syst. Evol. Microbiol. *59*, 839–844.

Gilichinsky, D. (2002a). Permafrost model of extraterrestrial habitat. In Astrobiology: The quest for the conditions of life, G. Horneck, and C. Baumstark-Khan, eds. (Berlin, Springer), pp. 125–142.

Gilichinsky, D., Rivkina, E., Bakermans, C., Shcherbakova, V., Petrovskaya, L., Ozerskaya, S., Ivanushkina, N., Kochkina, G., Laurinavichuis, K., Pecheritsina, S., *et al.* (2005). Biodiversity of cryopegs in permafrost. FEMS Microbiol. Ecol. *53*, 117–128.

Gilichinsky, D., Rivkina, E., Shcherbakova, V., Laurinavichuis, K., and Tiedje, J. (2003). Supercooled water brines within permafrost-an unknown ecological niche for microorganisms: a model for astrobiology. Astrobiology *3*, 331–341.

Gilichinsky, D., Wagener, S., and Vishnivetskaya, T. (1995). Permafrost microbiology. Permafrost Periglac Process *6*, 281–291.

Gilichinsky, D.A. (2002b). Permafrost as a microbial habitat. In Encyclopedia of Environmental Microbiology, G. Briton, ed. (New York, John Wiley & Sons), pp. 932–956.

Gilichinsky, D.A., Wilson, G.S., Friedmann, E.I., McKay, C.P., Sletten, R.S., Rivkina, E.M., Vishnivetskaya, T.A., Erokhina, L.G., Ivanushkina, N.E., Kochkina, G.A., *et al.* (2007). Microbial populations in Antarctic permafrost: biodiversity, state, age, and implication for astrobiology. Astrobiology *7*, 275–311.

Hansen, A.A., Herbert, R.A., Mikkelsen, K., Jensen, L.L., Kristoffersen, T., Tiedje, J.M., Lomstein, B.A., and Finster, K.W. (2007). Viability, diversity and composition of the bacterial community in a high Arctic permafrost soil from Spitsbergen, Northern Norway. Environ. Microbiol. *9*, 2870–2884.

Hinsa-Leasure, S.M., Bhavaraju, L., Rodrigues, J.L., Bakermans, C., Gilichinsky, D.A., and Tiedje, J.M. (2010). Characterization of a bacterial community from a Northeast Siberian seacoast permafrost sample. FEMS Microbiol. Ecol. *74*, 103–113.

James, N., and Sutherland, M.L. (1942). Are there living bacteria in permanently frozen subsoil? Can. J. Res. Sect. C Bot. Sci. *20*, 228–235.

Kochkina, G.A., Ivanushkina, N.E., Karasev, S.G., Gavrish, E.Y., Gurina, L.V., Evtushenko, L.I., Spirina, E.V., Vorob'eva, E.A., Gilichinskii, D.A., and Ozerskaya, S. (2001). Survival of micromycetes and actinobacteria under conditions of long-term natural cryopreservation. Microbiology *70*, 356–364.

Liebner, S., Rublack, K., Stuehrmann, T., and Wagner, D. (2009). Diversity of aerobic methanotrophic bacteria in a permafrost active layer soil of the Lena Delta, Siberia. Microb. Ecol. *57*, 25–35.

Liebner, S., and Wagner, D. (2007). Abundance, distribution and potential activity of methane oxidizing bacteria in permafrost soils from the Lena Delta, Siberia. Environ. Microbiol. *9*, 107–117.

Lindahl, T. (1993). Instability and decay of the primary structure of DNA. Nature *362*, 709–715.

Martineau, C., Whyte, L.G., and Greer, C.W. (2010). Stable isotope probing analysis of the diversity and activity of methanotrophic bacteria in soils from the Canadian high Arctic. Appl. Environ. Microbiol. *76*, 5773–5784.

Mindlin, S., Minakhin, L., Petrova, M., Kholodii, G., Minakhina, S., Gorlenko, Z., and Nikiforov, V. (2005). Present-day mercury resistance transposons are common in bacteria preserved in permafrost grounds since the Upper Pleistocene. Res. Microbiol. *156*, 994–1004.

Mykytczuk, N.C., Wilhelm, R.C., and Whyte, L.G. (2011). *Planococcus halocryophilus* sp. nov.: an extreme subzero species from high Arctic permafrost. Int. J. Syst. Evol. Microbiol..

Niederberger, T.D., Steven, B., Charvet, S., Barbier, B., and Whyte, L.G. (2009). *Virgibacillus arcticus* sp. nov., a moderately halophilic, endospore-forming bacterium from permafrost in the Canadian high Arctic. Int. J. Syst. Evol. Microbiol. *59*, 2219–2225.

Omelyansky, V.L. (1911). Bacteriological investigation of the Sanga mammoth and surrounding soil. Ark. Biol. Nauk. *16*, 335–340.

Panikov, N.S., and Sizova, M.V. (2007). Growth kinetics of microorganisms isolated from Alaskan soil and permafrost in solid media frozen down to −35 degrees C. FEMS Microbiol. Ecol. *59*, 500–512.

Pellerin, A., Lacelle, D., Fortin, D., Clark, I.D., and Lauriol, B. (2009). Microbial diversity in endostromatolites (cf. Fissure Calcretes) and in the surrounding permafrost landscape, Haughton impact structure region, Devon Island, Canada. Astrobiology *9*, 807–822.

Petrova, M., Gorlenko, Z., and Mindlin, S. (2011). Tn5045, a novel integron-containing antibiotic and chromate resistance transposon isolated from a permafrost bacterium. Res. Microbiol. *162*, 337–345.

Poinar, H.N., Hoss, M., Bada, J.L., and Paabo, S. (1996). Amino acid racemization and the preservation of ancient DNA. Science *272*, 864–866.

Ponder, M.A., Gilmour, S.J., Bergholz, P.W., Mindock, C.A., Hollingsworth, R., Thomashow, M.F., and Tiedje, J.M. (2005). Characterization of potential stress responses in ancient Siberian permafrost psychroactive bacteria. FEMS Microbiol. Ecol. *53*, 103–115.

Ponder, M.A., Thomashow, M.F., and Tiedje, J.M. (2008). Metabolic activity of Siberian permafrost isolates, *Psychrobacter arcticus* and *Exiguobacterium sibiricum*, at low water activities. Extremophiles *12*, 481–490.

Reasoner, D.J., and Geldreich, E.E. (1985). A new medium for the enumeration and subculture of bacteria from potable water. Appl. Environ. Microbiol. *49*, 1–7.

Reddy, G.S., Prakash, J.S., Vairamani, M., Prabhakar, S., Matsumoto, G.I., and Shivaji, S. (2002). *Planococcus antarcticus* and *Planococcus psychrophilus* spp. nov. isolated from cyanobacterial mat samples collected from ponds in Antarctica. Extremophiles *6*, 253–261.

Rivkina, E., Laurinavichius, K., McGrath, J., Tiedje, J., Shcherbakova, V., and Gilichinsky, D. (2004). Microbial life in permafrost. Adv. Space Res. *33*, 1215–1221.

Rivkina, E.M., Friedmann, E.I., McKay, C.P., and Gilichinsky, D.A. (2000). Metabolic activity of permafrost bacteria below the freezing point. Appl. Environ. Microbiol. *66*, 3230–3233.

Rodrigues, D.F., da, C.J.E., Ayala-Del-Rio, H.L., Pellizari, V.H., Gilichinsky, D., Sepulveda-Torres, L., and Tiedje, J.M. (2009). Biogeography of two cold-adapted genera: *Psychrobacter* and *Exiguobacterium*. ISME J. *3*, 658–665.

Rodrigues, D.F., Ivanova, N., He, Z., Huebner, M., Zhou, J., and Tiedje, J.M. (2008). Architecture of thermal adaptation in an *Exiguobacterium sibiricum* strain isolated from 3 million year old permafrost: a genome and transcriptome approach. BMC Genomics *9*, 547.

Rodrigues, D.F., and Tiedje, J.M. (2007). Multi-locus real-time PCR for quantitation of bacteria in the environment reveals *Exiguobacterium* to be prevalent in permafrost. FEMS Microbiol. Ecol. *59*, 489–499.

Shi, T., Reeves, R.H., Gilichinsky, D.A., and Friedmann, E.I. (1997). Characterization of viable bacteria from Siberian permafrost by 16S rDNA sequencing. Microb. Ecol. *33*, 169–179.

Soina, V.S., Mulyukin, A.L., Demkina, E.V., Vorobyova, E.A., and El-Registan, G.I. (2004). The structure of resting bacterial populations in soil and subsoil permafrost. Astrobiology *4*, 345–358.

Soina, V.S., Vorobiova, E.A., Zvyagintsev, D.G., and Gilichinsky, D.A. (1995). Preservation of cell structures in permafrost: a model for exobiology. Advanc. Space Res. *15*, 237–242.

Steven, B., Briggs, G., McKay, C.P., Pollard, W.H., Greer, C.W., and Whyte, L.G. (2007). Characterization of the microbial diversity in a permafrost sample from the Canadian high Arctic using culture-dependent and culture-independent methods. FEMS Microbiol. Ecol. *59*, 513–523.

Steven, B., Chen, M.Q., Greer, C.W., Whyte, L.G., and Niederberger, T.D. (2008a). *Tumebacillus permanentifrigoris* gen. nov., sp. nov., an aerobic, spore-forming bacterium isolated from Canadian high Arctic permafrost. Int. J. Syst. Evol. Microbiol. *58*, 1497–1501.

Steven, B., Leveille, R., Pollard, W.H., and Whyte, L.G. (2006). Microbial ecology and biodiversity in permafrost. Extremophiles *10*, 259–267.

Steven, B., Pollard, W.H., Greer, C.W., and Whyte, L.G. (2008b). Microbial diversity and activity through a permafrost/ground ice core profile from the Canadian high Arctic. Environ. Microbiol. *10*, 3388–3403.

Suetin, S.V., Shcherbakova, V.A., Chuvilskaya, N.A., Rivkina, E.M., Suzina, N.E., Lysenko, A.M., and Gilichinsky, D.A. (2009). *Clostridium tagluense* sp. nov., a psychrotolerant, anaerobic, spore-forming bacterium from permafrost. Int. J. Syst. Evol. Microbiol. *59*, 1421–1426.

Vatsurina, A., Badrutdinova, D., Schumann, P., Spring, S., and Vainshtein, M. (2008). *Desulfosporosinus hippei* sp. nov., a mesophilic sulfate-reducing bacterium isolated from permafrost. Int. J. Syst. Evol. Microbiol. *58*, 1228–1232.

Vishnivetskaya, T., Kathariou, S., McGrath, J., Gilichinsky, D., and Tiedje, J.M. (2000). Low-temperature recovery strategies for the isolation of bacteria from ancient permafrost sediments. Extremophiles *4*, 165–173.

Vishnivetskaya, T.A., Kathariou, S., and Tiedje, J.M. (2009). The *Exiguobacterium* genus: biodiversity and biogeography. Extremophiles *13*, 541–555.

Vishnivetskaya, T.A., Petrova, M.A., Urbance, J., Ponder, M., Moyer, C.L., Gilichinsky, D.A., and Tiedje, J.M. (2006). Bacterial community in ancient Siberian permafrost as characterized by culture and culture-independent methods. Astrobiology *6*, 400–414.

Wagner, D., Kobabe, S., and Liebner, S. (2009). Bacterial community structure and carbon turnover in permafrost-affected soils of the Lena Delta, northeastern Siberia. Can. J. Microbiol. *55*, 73–83.

Wilhelm, R.C., Niederberger, T.D., Greer, C., and Whyte, L.G. (2011). Microbial diversity of active layer and permafrost in an acidic wetland from the Canadian High Arctic. Can. J. Microbiol. *57*, 303–315.

Willerslev, E., Hansen, A.J., Ronn, R., Brand, T.B., Barnes, I., Wiuf, C., Gilichinsky, D., Mitchell, D., and Cooper, A. (2004). Long-term persistence of bacterial DNA. Curr. Biol. *14*, R9–10.

Yergeau, E., Hogues, H., Whyte, L.G., and Greer, C.W. (2010). The functional potential of high Arctic permafrost revealed by metagenomic sequencing, qPCR and microarray analyses. ISME J. *4*, 1206–1214.

Zhang, G., Ma, X., Niu, F., Dong, M., Feng, H., An, L., and Cheng, G. (2007a). Diversity and distribution of alkaliphilic psychrotolerant bacteria in the Qinghai–Tibet Plateau permafrost region. Extremophiles *11*, 415–424.

Zhang, G., Niu, F., Busse, H.J., Ma, X., Liu, W., Dong, M., Feng, H., An, L., and Cheng, G. (2008a). *Hymenobacter psychrotolerans* sp. nov., isolated from the Qinghai--Tibet Plateau permafrost region. Int. J. Syst. Evol. Microbiol. *58*, 1215–1220.

Zhang, G., Niu, F., Ma, X., Liu, W., Dong, M., Feng, H., An, L., and Cheng, G. (2007b). Phylogenetic diversity of bacteria isolates from the Qinghai–Tibet Plateau permafrost region. Can. J. Microbiol. *53*, 1000–1010.

Zhang, T., Barry, R.G., Knowles, K., Heginbottom, J.A., and Brown, J. (2008b). Statistics and characteristics of permafrost and ground-ice distribution in the Nothern Hemisphere. Polar Geog. *31*, 47–68.

Zhao, Q., Bai, Y., Zhang, G., Zhu, S., Sheng, H., Sun, Y., and An, L. (2011). *Chryseobacterium xinjiangense* sp. nov., isolated from alpine permafrost. Int. J. Syst. Evol. Microbiol. *61*, 1397–1401.

Ecology and Taxonomy of Psychrotolerant Bacteria in Artificial Cold Environments

2

Isao Yumoto and Koji Yamazaki

Abstract

A variety of artificial cold environmental conditions exist around human activities. These environments involve several factors that define the kinds of bacteria existing in niches. These environmental factors change with time owing to changes in the environment caused by existing microorganisms and their metabolisms. Namely, the inevitable selection of bacteria occurs in each artificial cold environment and the successive changes in microbiota occur according to changes in environmental factors. Some of the microorganisms in these microbial communities are useful for the degradation of organic matter in cold environments and for food preservation. On the other hand, several microorganisms adapted to artificial cold environments cause food spoilage. The frequently isolated psychrotolerant bacterial genera in artificial cold environments are *Pseudomonas*, *Psychrobacter*, *Staphylococcus* and *Photobacterium*. Lactic acid bacteria are also frequently detected under anaerobic conditions. To date, it has been reported that these genera are widely distributed in natural as well as artificial cold environments. It is intriguing why these bacterial genera are widely distributed in cold environments and why certain species belonging to these genera are able to survive in artificial cold environments. The frequent detection of members of above-mentioned genera in artificial cold environments suggests that the important factors that define the existence of these genera in cold artificial environments are (i) chances of invasion of such cold environments from ambient environments, (ii) chances of invasion through their own basic components, (iii) ability to rapidly propagate at low temperature and (iv) presence or absence of oxygen. It is also considered that these genera are genetically widely diverse, and cold artificial environment adaptation mechanisms of certain species belonging to these genera make their distribution wider not only in natural cold environments but also in artificial cold environments.

Introduction

Approximately 80% of earth occupies cold environments such as the deep sea, polar regions and high mountains. In addition, many areas on earth fall 5°C during winter. Certain microorganisms are able to grow at low temperature (lower than 5°C) and they are distributed in cold environments such as the polar regions and permafrost in natural environments. These cold-adapted microorganisms can be divided into two groups (Morita, 1975). The first category is psychrophilic bacteria. They are able to grow below 5°C and their optimum growth temperatures are below 20°C. The second category is psychrotolerant (pyschrotophic) bacteria. They can grow below 5°C and their optimum growth temperatures are higher than 20°C. The former microorganisms are distributed in permanently cold environments such as the polar regions and deep sea. The latter microorganisms are distributed not only in permanently cold environments but also in subarctic regions and temperate zones. The latter microorganisms are related to human activity such as wastewater treatment, food stored in the refrigerator and oil-contaminated soil in cold regions.

Considering the survival of microorganisms in artificial cold environments, adaptation to both cold and artificial environments is necessary. Firstly, it is considered that the physiologically important factors for growth in cold environments are as follows: (i) enzyme activities related to metabolism, (ii) acquisition of nutrients, (iii) production and maintenance of proteins and (iv) flexible membranes at low temperature and maintenance of membrane fluidity. Secondly, important factors for growth in artificial environments are as follows: (i) affinity for nutritional condition corresponding environment, (ii) rapidity of growth, (iii) presence or absence of oxygen, (iv) production of lactic acid or bacteriocins, (v) tolerance for reactive oxygen species, (vi) chances of invasion from ambient environments and (vii) spore formation.

Ecological, taxonomic and physiological understanding of microorganisms adapted to artificial cold environments adapted microorganism is very important for the utilization of useful microorganisms and the prevention of the proliferation of food spoilage microorganisms under regulated conditions. In this review, we focus on the ecological and taxonomic distribution of psychrotolerant microorganisms in artificial cold environments, cold storage foods, cold ripening foods and other artificially substrates rich conditions such as wastewater treatment systems. The relationships between the characteristics of microbiota and environmental factors in each artificial cold environment are discussed.

Foods in cold storage or cold ripening

Generally, it is difficult for mesophilic microorganisms to propagate rapidly at cold temperatures. To avoid spoilage of foods, we store foods in refrigerators. Storage in refrigerators of perishable foods such as fish, meat and dairy products is indispensable for keeping these items in good condition and avoiding propagation of microorganisms for a longer period. Even in cold storage, certain microorganisms propagate steadily over an extended period. We describe such microorganisms in each type of food. On the other hand, the cold ripening period is a very important step to produce a better taste and good flavour and prolong the edible state of fermented foods. The contribution of the activity of microorganisms is very important for a successful ripening period. Here, we describe microorganisms in various foods stored at low temperature.

Refrigerated raw marine products

Numerous studies to elucidate the mechanisms leading to the spoilage of cold-stored raw fish have been performed from the food biochemical and microbiological viewpoints (Suyama and Kounosu, 1987). Understanding of the process of spoilage is very important for determining shelf-life. Approximately 2 h after the death of a fish, muscle rigidity occurs. The velocity and period of rigidity depend on the ambient temperature and manner of death. After the rigidity period, fluid bleeds from fish meat. Domestic enzymes in the fluid in fish meat digest its own proteins and lipids contained in the fish meat. This is the self-digestion of fish meat and occurs in the fish itself. Then, microorganisms attached to the fish body propagate by utilizing the digested proteins and fatty acids resulting from the self-digestion. As microorganisms propagate, by-products of bacterial metabolisms, such as amines, biogenic amines, organic acids and sulfur compounds, result in intensity of fishy odour. The accumulation of these substances results in an intense putrid odour. Generally, resulting of such putrid order from fish is detected can be considered 'spoilage'. Thus, the spoilage of raw marine products consists of two steps, from death of fish to self-digestion and the propagation of bacteria. Although it is difficult to regulate the first stage to prevent the fish spoilage, the second stage can be prolonged by preventing the activity of domestic enzymes and the propagation of bacteria by keeping the temperature as low as possible. By contraries, even if the raw fish are kept at low temperature, reactions of domestic enzymes and the propagation of attached bacteria gradually proceed. Bacteria with a viable number of 10^2–10^5 colony-forming units (CFU)/cm^2 exist in ordinary landed fish, while this becomes more than 10^8 CFU/cm^2 in spoiled fish. Although the composition of microbiota in chilled fish differs depending on the origin of the fish and the landed season, bacteria frequently found in fish from the sea are *Photobacterium* spp., *Pseudomonas* spp., *Shewanella* spp., *Enterobacteriaceae,* lactic acid bacteria (LAB), *Pseudoalteromonas* spp., *Vibrio/Aliivibrio* spp., *Brochothrix thermosphacta* and *Psychrobacter* spp. (Hovda *et al.*, 2007a,b; Pantazi *et al.*, 2008; Broekaert *et al.*, 2011). *B. thermosphacta* is found in most samples at the end of the storage period when farmed Atlantic halibut is stored at 4°C (Hovda *et al.*, 2007b). It has been reported that sweet and fruity spoilage odours are due to the presence of *Pseudomonas* species and H_2S odours are due to the presence of *Shewanella putrefaciens* (Olafsdottir *et al.*, 2006). On the other hand, microbiota in chilled fish caught from fresh water, such as pearlspot (*Etroplus suratensis* Bloch), are somewhat different from those in fish from the sea (Ravi Sankar *et al.*, 2008). In addition to *Enterobacteriaceae,* pseudomonads and *Shewanella putrefaciens, Aeromonas* spp. including *Aeromonas hydrophila, Aeromonas veronii* biovar *sobria* and *Aeromonas veronii* biovar *veronii* and staphylococci are observed. The microbiota reflects microorganisms distributing in fresh water. They do not require sodium for their metabolisms.

In the case of iced squid, the initial microbiota mainly consists of *Pseudoalteromonas* spp. (31%), *Aeromonas* spp. (23%), *S. putrefaciens* (14%), *Pseudomonas* spp. (11%) and *Vibrio* species (11%) in whole squid (Paarup *et al.*, 2002). The microbiota of ice-stored whole squid until day 7 consists of *Pseudoalteromonas* spp. (70%), *Photobacterium phosphoreum* (15%) and *S. putrefaciens* (11%). On the other hand, the microbiota of ice-stored gutted squid until day 7 consists of *S. putrefaciens* (37%), *Pseudoalteromonas* spp. (29%) and *Pseudomonas* spp. (24%). From the above results, the presence of digestive glands defines the presence of *P. phosphoreum.* The bacterial species may contribute to the acceleration of spoilage of ice-stored squid.

The above findings indicate that those genera including bacterial species are able to grow rapidly at low temperature in the presence of self-digested materials. In addition, microbiota of chilled marine products is affected by where the products were obtained and by handling and processing procedures (Jay, 1986). Owing to the presence of Gram-positive bacteria and *Enterobacteriaceae,* it is considered that artificial factors are also involved in the formation of microbiota.

Refrigerated fabricated marine product

Cold-smoked salmon is a preserved fabricated fish product, which is typically vacuum-packed and stored at low temperature. The shelf-life of regular cold-smoked salmon is around 2–3 weeks at 7°C. The shelf-life fluctuates depending on the original microbiota. During the cold storage of smoked salmon, the microbiota consists of *S. putrefaciens, B. thermoshacta, Vibrio* sp., *Aeromonas* spp., *Lactobacillus alimentarius, Lactobacillus sakei, Lactobacillus farciminis, Carnobacterium maltaromaticum* (formerly *C. piscicola*), *P. phosphoreum* and *Serratia liquefaciens* (Leroi *et al.*, 1998; Paludan-Müller *et al.*, 1998; Joffraud *et al.*, 2001, 2006; Stohr *et al.*, 2001). Leroi *et al.* (1998) reported that the spoilage potential is associated with *S. putrefaciens, Aeromonas* spp. *and Brochothrix* spp. On the other hand, Stohr *et al.* (2001) reported that the bacteria mainly responsible for spoilage are *L. sake, L. farciminis* and *B. thermosphacta*. These constituents are different from those of non-fabricated fish owing to the fact that the commodity is vacuum packed. Therefore, a strictly aerobic bacterium is not able to propagate under this condition. The above findings are based on the isolated strains. The culture-independent approach for the analysis of microbiota in cold-smoked salmon stored at 7°C is also performed (Olofsson *et al.*, 2007). The indigenous microbiota of vacuum-packed fresh smoked salmon consists of, in order of abundance, *B. thermosphacta, Yersinia ruckeri, Photobacterium* spp. and *Carnobacterium* spp. On the other hand, the dominant constituents of the same sample stored at 7°C for 19 days are *Lactobacillus* spp. and *Photobacterium* spp. On the basis of the culture-independent approach, it was revealed that 10% of the total microbiota multiplying on chilled, vacuum-packed smoked salmon consisted of unknown species. These include at least three different unknown *Photobacterium* spp. From the above findings, it can be concluded that rapidly growing bacteria under relatively nutritionally rich, anaerobic and cold conditions become dominant members in the chilled and vacuum-packed smoked salmon during cold storage.

Refrigerated milk

The cold storage of milk produces selective conditions for the growth of psychrotolerant bacteria. Although psychrotolerant bacteria occupy only 10% of the initial raw milk bacterial population, they gradually become dominant during cold storage (Champagne *et al.*, 1994). Extracellular lipases or proteases produced by psychrotolerant bacteria cause spoilage of milk (Meer *et al.*, 1991; Ternström *et al.*, 1993). Lipases degrade the milk fat to produce short-chain fatty acids that cause rancidity and a soapy appearance. On the other hand, proteases degrade casein, causing the deterioration of colour and flavour.

To prevent milk spoilage, several pretreatments have been considered, such as thermization (TH), microfiltration (MF) and introduction of carbon dioxide (CO_2). Rasolofo *et al.* (2010) examined microbiota on the pretreated milk and compared with an untreated control (UT) stored at 4°C or 8°C using 16S rRNA gene clone libraries. The results showed that the dominant bacterial genera in UT, CO_2 and TH samples at day 3 at 4°C were *Staphylococcus, Clostridia, Streptococcus, Acinetobacter, Fackkamia, Aerococcus* and *Trichococcus*. In contrast, the dominant bacterial genera in the corresponding MF milk were *Pseudomonas, Stenotrophomonas* and *Delftia*. At day 7 at 4°C, genus *Pseudomonas* constituted 94% of the bacterial population in UT milk. In treated samples, the predominant genera were *Pseudomonas, Staphylococcus* and *Streptococcus* in the CO_2 sample, *Pseudomonas* and *Stenotrophomonas* in MF sample and *Staphylococcus, Delftia* and *Clostridia* in TH sample.

On the basis of the above results, *Pseudomonas* spp. were found to predominantly propagate during the cold storage of milk because these species can grow more rapidly than any other bacterial genera in chilled milk. However, they are susceptible to heat.

The above investigation was performed by a culture-dependent method. The effect of pretreatment on cold-stored milk under the same condition as described above was estimated by culture-independent methods, terminal restriction fragment length polymorphism (T-RFLP) and denaturing gradient gel electrophoresis (DGGE) (Rasolofo *et al.*, 2011). Although the genus *Staphylococcus* was predominant in the microbiota after day 3 in cold storage, this species was superseded by *Pseudomonas* species after day 7 in all samples except the TH sample, in which *Staphylococcus* remained predominant at the end of the storage period. The genus *Stenotrophomonas* was predominantly found in the MF sample, and *Pseudomonas* and *Trichococcus* species were the major constituents of the microbiota of the UT sample at day 7. *Streptococcus* was found in CO_2 sample, as in the culture-dependent experiment. It can be explained by the fact that the carbonation of milk decreased the pH of milk and *Streptococcus* species can tolerate such acidic conditions. *Trichococcus floculiformis* and *Prevotella* spp. were found to be the major species at day 7 in UT and CO_2 samples in the culture-independent experiment. The genus *Trichococcus* has been detected in raw milk and dairy waste. On the other hand, genus *Prevotella* was predominant in the rumen.

De Jonghe *et al.* (2011) investigated the influence of storage conditions on the growth of *Pseudomonas* species in refrigerated raw milk. The raw milk was firstly stored in a farm milk tank for 4 days and then, transferred (takes 8 h) to an industrial milk tank (stored for 24 h). During the storage in the farm tank (the beginning of the dairy chain), the outgrowth of *Pseudomonas* species occurs under both optimal (3.5°C) and suboptimal (6°C) conditions as determined by DGGE. Bacterial isolation of bacteria was performed at the beginning of the farm milk tank storage, at the end of the farm milk tank storage and at the end of the transportation. The isolates were identified as *Pseudomonas gessardii, Pseudomonas gessardii*-like, *Pseudomonas fluorescens*-like, *Pseudomonas lundensis, Pseudomonas fragi* and *Pseudomonas fragi*-like. These *Pseudomonas* spp. significantly increase (2 log CFU/ml) under the suboptimal conditions (6°C in farm milk tank; 10°C in transport; 10°C in industrial tank) in prolonged storage compared with optimal storage conditions (3.5°C in farm milk tank; 6°C in transport; 6°C in industrial tank). On the basis of the above facts, it is highly recommended that the storage temperature be as low as possible (3.5°C) and that the storage period be reduced to prevent milk spoilage. Psychrotolerant *Pseudomonas* spp. have little chance to propagate under 3.5°C during a very short storage period.

Recently, culture-independent bacterial community analyses, DGGE and clone library analyses have been carried out in farms and dairy plant tanks (Raats *et al.*, 2011). Although there are a few exceptions, *Gammaproteobacteria* (particularly *Pseudomonadales*) is predominant in both farms and dairy plant tanks. Farm samples exhibit the classes *Bacilli, Clostridia* and *Actinobacteria* in addition to *Gammaproteobacteria* (particularly *Pseudomonadales*). On the other hand, *Gamma proteobacteria* (particularly *Pseudomonadales*) is predominant in most dairy plant tanks. Thus, indigenous raw milk microbiota is already present before the milk arrives at the dairy plant tank. These bacteria have the potential to produce lipolytic and proteolytic enzymes. These enzymes have the potential to act in milk even after pasteurization. It is considered that *Pseudomonadales* can propagate to become a predominant member in dairy plant tanks because of the action of originally present microorganisms after the pasteurization.

The above reports indicated that *Pseudomonas* spp. are the most frequently detected microorganisms. This is probably because *Pseudomonas* spp. are ubiquitous around human activity and can grow rapidly at low temperature. The high water content of milk may also be suitable for the propagation of *Pseudomonas* spp.

Fermented salmon *Izushi*

Fermented fish, *Izushi* (Fig. 2.1), is a traditional preserved food prepared during winter in the Tohoku and Hokkaido areas of Japan. The main fish materials used are salmon, herring, flatfish, sandfish and Okhostk Atka mackerel. After slicing, the fish fillet is immersed in water and treated with vinegar. The main materials for making *Izushi* consist of the pre-treated fish, rice and rice malt (*Koji*) and vegetables. In addition, vinegar, rice wine (*Sake*), salt and sugar are added to adjust its taste. These three main materials are put alternately in a barrel, converted with a lid on top and pressurized by putting a stone weight on the lid. The preparation is fermented and ripened for more than a month at low temperature. The *Koji* contains amylase and it digests the starch in the rice. As a result, *Koji* acts as the accelerator for the fermentation at low temperature. Thin strips of vegetables such as carrot, Japanese radish, cabbage, cucumber, ginger, red pepper, Japanese pepper and lemon are added to prevent the smell of fish. Because *Izushi* contains a very small amount of salt, it can sometimes cause food poisoning. Changes in microbiota during fermentation and ripening of *Izushi* have been investigated (Sasaki *et al.*, 2004). Isolates are classified according to phenotype for aerobic bacteria (Shewan *et al.*, 1960; Komagata, 1985), lactic acid bacteria (Schleifer, 1986; Uchimura and Okada, 1992) and yeast (Kurtzman and Fell, 1998). Firstly, microbiota of each material is examined. In pickled fish, *Achromobacter* (*Alcaligenes*), and *Leuconostoc, Lactococcus* and *Streptococcus* are the predominant aerobic and LAB, respectively. In boiled

Figure 2.1 Salmon *Izushi*. *Izushi* is a traditional fermented food in the northern part of Japan. It is prepared from raw and salted sockeye salmon (*Oncorhynchus nerka*) fillets, rice and rice malt and vegetables (cabbage, carrot, ginger and red pepper). *Izushi* has been produced as a preserved food in the winter season. It is considered that the presence of lactic acid bacteria and the preservation at low temperature prevent its spoilage in a short period. Bar= 3cm. A colour version of this figure is located in the plate section at the back of the book.

rice and *Koji*, *Staphylococcus* and *Micrococccus* and *Leuconostoc* and *Streptococcus* are the predominant aerobic and lactic acid bacteria (LAB), respectively. In finely cut vegetables, *Achromobacter* and *Pseudomonas*, and *Lactococcus* and *Streptococcus* are the predominant aerobic and lactic acid bacteria, respectively. In aerobic microbiota, the major genera are *Micrococcus* (56–64%) and *Bacillus* (22–27%) at day 2 and *Bacillus* becomes the most predominant aerobes after day 14. On the other hand, the predominant lactic acid bacteria are *Streptococcus* from days 2 to 14 and *Leuconostoc mesenteroides* subsp. *cremoris* becomes predominant (88–100%) from days 35 to 44. On the other hand, *Debaryomyces hansenii* (100%) becomes predominant from days 35 to 44 in the yeasts. Thus, although microbiota in the early stage of fermentation and ripening are affected by those of the materials, it converges as fermentation and ripening proceed.

On the other hand, *Saccharomyces cerevisiae* has been detected instead of *D. hansenii* in experiments since 2005 (Sasaki *et al.*, 2005, 2006a,b). This is probably due to the difference in *Izushi* sample pH depending on the proportions of each constituent. *D. hansenii* is more sensitive than *S. cerevisiae* to acetic acid. Although the fermentation activity of *D. hansenii* is not as high as that of *S. cerevisiae*, it is a typical member, together with *Leuconostoc mesenteroides*, during the ripening period in such fermentation products. It has been reported that these microorganisms contribute to the production of the fermentation products' specific flavour.

The microbiota of processed *Izushi* is appropriately constructed without using sterilization or high salt concentration in open space. The fermentation and ripening of *Izushi* are carried out by traditional experience. It seems that the factors limiting the growth of microorganisms are low temperature, anaerobic condition and presence of appropriate microorganisms such as LAB (lactic acid production) and yeasts. Therefore, it is considered that the procedure for *Izushi* producing is quite sophisticated and can only be learned by experience.

Squid *Shiokara*

Fermented squid, *Shiokara* (Fig. 2.2), is a traditional Japanese food, and the original materials are shredded squid meat, squid liver and salt. Although the typical material of the main body is squid, instead of squid, octopus or scallop are occasionally used. The fermentation temperature differs depending on the procedure. A higher temperature (e.g. 25°C) makes the fermentation period shorter (5 days). Cold fermentation is performed at 4–10°C for 2–3 weeks. The salt concentration in the original procedure is more than 10%. However, because consumers in Japan prefer foods containing a low salt concentration these days, the salt concentration of *Shiokara* also tends to be lower. It has been reported that the major constituents of *Shiokara* ripening at 25°C are *Staphylococcus*, *Micrococcus* and yeast (Mori *et al.*, 1979; Takai *et al.*, 1992, 1993; Yamazaki *et al.*, 1992). In Ishikawa Prefecture, Japan, squid ink is added to *Shiokara* in addition to squid liver. Squid ink inhibits the propagation of a variety of bacteria, particularly *Staphylococcus epidermidis* and *Staphylococcus aureus* (Takai *et al.*, 1993; Kawai, 2009). The fermentation will proceed by the self-digestion of squid meat and the enzymes from squid liver and the products of the bacteria present during the fermentation. Because of the presence of a high salt concentration and squid liver, its microbiota are simpler than those of any other fermented foods (Takai *et al.*, 1993). However, without the addition of salt for the preparation, the squid meat spoils rapidly. It has been reported that the propagation of *Staphylococcus aureus* is prevented owing to the presence

Figure 2.2 Squid *Shiokara*. *Shiokara* is a traditional fermented food in Japan. The original materials are shredded squid meat, squid liver and salt. *Shiokara* has been produced as a preserved food in a wide area of Japan and in all seasons. The high salt concentration and squid liver prevent its spoilage. Bar= 2cm. A colour version of this figure is located in the plate section at the back of the book.

of salt and the high concentration of trimethylamine-*N*-oxide in the squid (Nishimura and Shinano, 1991; Suzuki *et al.*, 1992).

Recently, consumers have preferred a low salt concentration in foods. Therefore, commercially available *Shiokara* does not contain as much salt as the traditionally prepared *Shiokara*. Commercially available *Shiokara* commodities are normally fermented at low temperature for a short period and kept in the refrigerator. Morishita *et al.* (1994) examined the salt concentration and microbiota of commercial *Shiokara*. Isolates are classified according to phenotype by a simple identification scheme. In squid *Shiokara*, the NaCl concentration is 4.3–6.0% except for one sample (9.8%). The predominant genera in low-salt *Shiokara* are *Bacillus, Micrococcus, Staphylococcus,* coryneform bacteria and *Lactobacillus.* On the other hand, the most abundant genera in high-salt *Shiokara* are *Micrococcus* and *Staphylococcus*. It is considered that there is little chance of food poisoning with *Shiokara* prepared traditionally because the squid liver or squid ink prevents the growth of various bacteria in addition to the low temperature and high concentration of NaCl. However, most commercially available *Shiokara* commodities can be characterized as having a lower salt concentration and a shorter cold fermentation period than in the original procedure.

It is considered that the factors limiting the presence of microorganisms in low-salt and cold-fermented *Shiokara* are squid liver, low temperature and salt. The ubiquitously present microorganisms associated with human activities may mix in the food.

Cheese maturation

Various psychrotolerant bacteria belonging to the genera *Pseudomonas, Acinetobacter, Alcaligenes* and *Proteus* have been reported in terms of their association with cheese spoilage

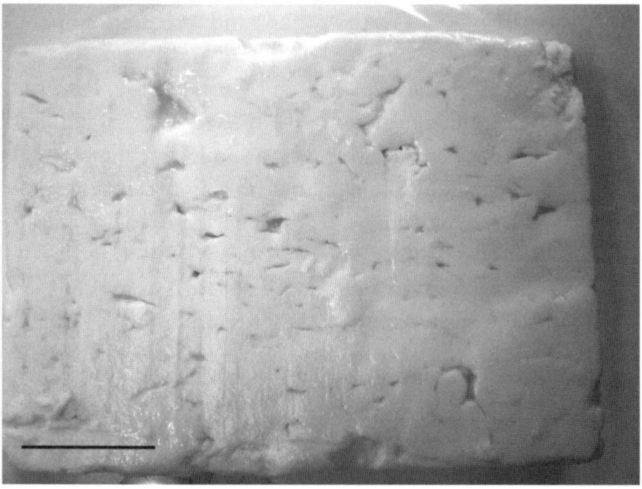

Figure 2.3 Feta cheese. Feta cheese is one of the most important and popular dairy products in Greece. It tastes slightly acidic and salty and is usually ripened in brine. Its characteristic flavour and taste are developed by the action of natural lactic acid microbiota of milk. It is considered that the presence of lactic acid and ripening at low temperature prevent its spoilage. Bar= 3cm. A colour version of this figure is located in the plate section at the back of the book.

(Jay, 1992). Biopreservation using a certain starter culture is one of the attractive candidates for the maintenance of the appropriate shelf-life of cheese. To date, a variety of suitable starter cultures have proposed as follows: bifidobacteria (Boylston *et al.*, 2004), *Lactococcus* (Litopoulation-Tzanetaki *et•al.*, 1993; Ryan *et al.*, 1996; Michaelidou *et al.*, 2003; Khosrowshahi *et al.*, 2006; Monteagudo-Mere *et al.*, 2011), *Lactobacillus* (Litopoulou-Tzanetaki *et al.*, 1993; Kieronczyk *et al.*, 2003; Kourkouta *et al.*, 2006a; Slattery *et al.*, 2010; Mills *et al.*, 2011; Monteagudo-Mere *et al.*, 2011), *Leuconostoc* (Litopoulou-Tzanetaki *et al.*, 1993; Monteagudo-Mere *et al.*, 2011) and *Enterococcus* sp. (Litopoulou-Tzanetaki *et al.*, 1993; Monteagudo-Mere *et al.*, 2011).

Kourkoutas *et al.* (2006b) investigated the effect of the use of freeze-dried kefir as a starter culture in the production of feta cheese aiming at the improvement of cheese quality characteristics, acceleration of cheese ripening and prolonging the shelf-life. Cheese maturation was performed at 4°C. Feta cheese (Fig. 2.3) is a soft white cheese usually ripened in brine and is one of the most popular dairy products in Greece with its characteristic slightly acid, salty taste. For the production of this cheese, 2 months of ripening is involved in the last step. Appropriate sanitary management is necessary during the ripening period, particularly for the product made from raw milk. Kefir was originally a probiotic fermented milk drink made with kefir grains in north Caucasus. Kefir is a mixed culture that consists of various yeasts as follows: *Kluyveromyces, Candida, Saccharomyces* and *Pichia* spp. It also contains various lactic acid bacteria (*Lactobacillus, Lactococcus* and *Leuconostoc*) and acetic acid bacteria (Garrote *et al.*, 1997; Luis *et al.*, 1993; Witthuhn *et al.*, 2005). *Staphylococcus* and *Psychrobacter* are detected in cheese with no kefir, while these genera are not detected in cheese using kefir as the starter. This suggests that the starter has the potential to act as a species-specific inhibitor of those genera. The main genera in cheese using kefir are *Pseudomonas, Lactobacillus* and *Janthinobacterium* as determined by DNA analysis. The genus *Lactococcus* was detected by

RNA analysis in cheese using kefir. On the other hand, the predominant members in the no-kefir-added control are *Lactobacillus, Staphylococcus* and *Psychrobacter* as determined by DNA analysis and *Pseudomonas, Janthinobacterium, Lactobacillus, Psychrobacter* and *Oxalobacteraceae* by RNA analysis. Spoilage of kefir-added cheese occurs at the 9th and 20th days of preservation at 10°C and 5°C, respectively, while spoilage of the non-kefir-added control occurs at the 7th and 16th days. Microbial counts (CFU) of aerobes, coliforms, enterobacteria, yeasts, staphylococci, lactococci and lactobacilli using each corresponding medium increased steadily and reached similar levels in both kefir-added and non-kefir-added cheeses except staphylococci (lower in kefir-added cheese) during the preservation period. The kefir-added cheese was approved by a panel in a preliminary sensory evaluation that compared the control and the commercially available feta-type cheese. These results suggested that although kefir is generally produced at 20–30°C, it may be useful for adjusting the microbiota in cheese stored at low temperature with appropriate starters.

Chilled beef

It has been reported that the microorganisms most commonly associated with chilled meat spoilage are *Pseudomonas* spp., *Enterobacteriaceae* and LAB such as *B. thermosphacta, Leuconostoc* and *Lactobacillus* (Nortjé and Shaw, 1989; Borch *et al.*, 1996). The microbiota contributing to the spoilage largely depends on the oxygen availability (Labadie, 1999; Jones, 2004; Koutsoumanis *et al.*, 2008). The combination of vacuum packaging and storage under chilled condition is considered to be very effective for prolonging the shelf-life of perishable food. Vacuum packaging prevents the growth of some aerobic pathogens and food spoilage bacteria.

Pennacchia *et al.* (2011) investigated the succession of microbiota of chilled beef in order to clarify the spoilage-related microbiota by culture-independent (DGGE of directly extracted DNA) procedures in nine different samples stored at 4°C in the presence of air or under vacuum-packed condition. In the original meat (day 0), *Carnobacterium divergens, B. thermosphacta, Pseudomonas* spp. and *Psychrobacter* spp. are present in the majority of samples. After 7 days storage in the presence of air, *B. thermosphacta, Pseudomonas* spp. and *Photobacterium* spp. are predominant in the majority of samples, while in the vacuum-packed samples, *Photobacterium* spp. are most predominantly observed. At day 20 in the presence of air, *B. thermosphacta, C. divergens* and *Photobacterium* spp. are frequently observed, and *C. divergens, Photobacterium* spp. and *Lactobacillus algidus* are the major members in vacuum-packed samples. Although strains belonging to *Photobacterium* are found in the original sample, they are found in more than 70% of the samples at later storage times. Although *Photobacterium* spp. are not well recognized as playing a role in the spoilage of chilled meats, the psychrotolerant *P. phosphoreum* is recognized as having a major role in the spoilage of fish stored under vacuum or modified atmosphere (Dalgaard *et al.*, 1993; Jørgensen *et al.*, 2000; Gram *et al.*, 2002; Hovda *et al.*, 2007b). This suggests that *Photobacterium* spp. play a role in the spoilage of chilled meat.

In addition, Pennacchia *et al.* (2011) performed DGGE analyses of bulk cells collected from Violet Red Bile Glucose Agar (VRBGA) and deMan–Rogosa–Sharpe (MRS) agar. In the original meat (day 0), *Pseudomonas* spp. and *Rahnella* spp. are mainly detected in DNA extracted from VRBGA plates, while the following species are heterogeneously distributed in MRS agar: *C. divergens, Leuconostoc* spp., *Leuconostoc mesenteroides/pseudomesenteroides,*

Lactobacillus sakei/curvatus and *Staphylococcus* spp. On day 7, *Pseudomonas* spp. and *Rahnella* spp. are frequently observed both in air-exposing and in vacuum-packed samples, while *C. divergens* is most predominantly observed in MRS agar. On day 20, strains belonging to genus *Pseudomonas* are detected in most samples in air-exposing and in vacuum-packed samples, and *Rhanella* spp. or *Serratia* spp. are occasionally observed in VRBGA. On the other hand, *C. divergens* is the most frequently observed and *L. sakei* and *Lc. mesenteroides/pseudomesenteroides* are also frequently observed in MRS agar.

In the studies described above, *Serratia, Rhanella* and *Leuconostoc* spp. are detected only by culture-dependent methods. This is probably because these species are well adapted to propagate on artificial media and the amount of DNA of these species were not sufficient to be amplify by PCR in these samples. Although cold-stored beef has little chance to encounter marine microorganisms, the presence of *Photobacterium* spp., *C. divergens* and *B. thermosphacta* is similar to case of cold-stored marine products. This means that cold-stored marine products and meat have common physicochemical environmental parameters.

Wastewater treatment system and oil-contaminated soil in cold environment

As described above, cold ripening or cold storage foods are the niches in artificial cold environments for cold-adapted microorganisms. In addition, wastewater treatment systems and cold environments artificially contaminated with organic matter also provide cold-adapted microorganisms with their niches. A common factor in between environments of cold stored foods and other artificial cold environments is the abundance of substrates for microorganisms. On the other hand, different factors between the two environments include the size of niches, qualities of substrates, fluctuation of temperature depending on the season and openness to ambient environments.

Wastewater treatment system in cold environment

The temperature of a wastewater treatment system of a fish processing plant in Hokkaido, Japan (sub-Arctic region), is mostly lower than 10°C throughout the year. The plant uses H_2O_2 as the bleaching agent to produce brightly coloured fish (herring) eggs. In addition to the low temperature, the particular characteristics of the wastewater are relatively high biological oxygen demand (BOD), presence of H_2O_2 and high NaCl concentration. Therefore, although the microbiota in the environment fluctuates depending on the operating rate of the plant, it is considered that the constituents of the microbiota of the wastewater treatment system are much simpler than those of an ordinary wastewater treatment system in the temperate zone. Several psychrotolerant bacterial strains have been isolated from the plant. They have been classified as novel species. These are *Vibrio rumoiensis* (Yumoto *et al.*, 1998, 1999), *Dietzia psychralcaliphila* (Yumoto *et al.*, 2002), *Exiguobacterium oxidotolerans* (Yumoto *et al.*, 2004), *Psychrobacter piscatorri* (Yumoto *et al.*, 2010; Kimoto *et al.*, 2012) and *Pseudomonas* sp. (I. Yumoto *et al.*, unpublished result). These genera are distributed in permanently cold environments in natural environments. *Vibrio* and *Dietiza* strains were isolated from a deep-sea environment (Uchimura *et al.*, 2010; Takami *et al.*, 1997). *Exiguobacterium* and *Psychrobacter* strains were isolated from permafrost soil, Antarctica soil, sea ice and Himalayan mountain soil (Rodrigues *et al.*, 2009). *Pseudomonas* strains were isolated from lake sediment and sea ice of Antarctica and a glacier in the Himalayan Mountain range (Bowman *et al.*, 1997; Shivaji *et al.*, 2011a,b). *Pseudomonas* and *Dietzia* strains were isolated from a permafrost ice wedge (Katayama *et al.*, 2007). In fact, the wastewater treatment system involves environmental pressure for the survival of microorganisms such as low temperature and high concentrations of H_2O_2 and NaCl. The constituent genera of isolates from the wastewater treatment system manifested that the environment is highly related to the microorganisms living in the cold environments. It is considered that these microorganisms are selected from the microbial population related to the marine products processed in the plant.

Oil-contaminated environments

Hydrocarbons are one of the most widespread contaminants in terrestrial as well as sea environments. Hydrocarbons are frequently required for electric power generation, heat production and operation of vehicles, aircraft and ships. Hydrocarbon spills occur by accident due to human activities, leakage from tanks in petroleum stands and underground leaching. Temperature is an important factor for the degradation of hydrocarbons by microorganisms, as well as the physicochemical characteristics of hydrocarbons. The viscosity of hydrocarbons increases, in cold temperature preventing the volatilization of toxic short-chain *n*-alkanes and long-chain *n*-alkanes are solidified. On the other hand, the metabolism and propagation of microorganisms are inhibited at low temperature. Although the degradation of hydrocarbons by microorganisms at low temperature is difficult, there are microorganisms that can degrade hydrocarbons even at low temperature.

Hydrocarbons contaminating polar soil are usually refined petroleum products such as diesel or aviation fuel comprising aliphatic and aromatic compounds. Therefore, *n*-alkane- and aromatic hydrocarbon-degrading bacteria were enriched in these contaminated sites. The observed *n*-alkane degraders are *Rhodococcus, Pseudomonas* and *Acinetobacter* (Aislabie *et al.*, 2006). *n*-Alkanes with short carbon chain lengths are relatively easier to degrade by microorganisms than those with long carbon chain lengths. Strains belonging to genus *Rhodococcus* isolated from Antarctic soil are able to utilize *n*-alkanes with carbon chain lengths of C_6–C_{20} (Bej *et al.*, 2000). *Pseudomonas* spp. isolated from Arctic soil utilizes *n*-alkanes with carbon chain lengths of C_5–C_{12} (Whyte *et al.*, 1997). Phylogenetic analysis based on 16S rRNA gene sequences from *n*-alkane-degrading *Rhodococcus* strains from cold soils indicates that they are similar to *Rhodococcus erythropolis* or *Rhodococcus fascians*. Gram-negative *n*-alkane-degrading isolates, *Pseudomonas* strains, are grouped with *P. fluorescens* or *Pseudomonas stutzeri*. Aromatic hydrocarbon-degrading isolates can be classified as *Pseudomonas* or *Sphingomonas*. Aromatic hydrocarbon-degrading *Pseudomonas* strains are grouped with *Pseudomonas syringae, P. fluorescens* or *P. stutzeri* on the phylogenetic tree on the basis of 16S rRNA gene sequences. *Sphingomonas* isolates cluster with *Sphingomonas chlorophenolica* or *Sphingomonas aerolata*. *Pseudomonas* stains tend to degrade a narrower range of aromatic hydrocarbons than *Sphingomonas* strains (Whyte *et al.*, 1997; Baraniecki *et al.*, 2002; Eriksson *et al.*, 2002).

Psychrotolerant hydrocarbon-degrading bacterial strains have been isolated from a sample of soil immersed in hot-spring water (cold mineral spring of 33°C when atmospheric temperature is at 13°C) containing hydrocarbons in Toyotomi, Hokkaido, Japan (subarctic region). Two strains were isolated and they identified as a different novel taxon, *Acinetobacter* sp. strain Ths (Yamahira *et al.*, 2008) and *Pseudomonas toyotomiensis* (Hirota *et al.*, 2011). Strain Ths grows at 0°C but not at temperatures higher than 45°C. It is an alkali-tolerant bacterium with the growth pH range of pH 7–9. The optimum temperature of the strain is 27°C. The strain utilizes *n*-alkanes of C_{13}–C_{30} chain length and aromatic hydrocarbon and fluorine at pH 9 and 4°C. On the other hand, *P. toyotomiensis* grows at 4–42°C with optimum growth at 35°C. It grows at pH 6.0–10.5, with an optimum pH range of 6–9. The strain utilizes *n*-alkanes of C_{12}–C_{30} chain length except C_{20} chain length. Although it utilizes the branched-chain hydrocarbon, pristane, it does not utilize aromatic hydrocarbons.

From the facts described above, the combination of low temperature and utilization of hydrocarbon limits the presence of microorganisms. Therefore, quite limited genera of bacteria utilize hydrocarbons in hydrocarbon contaminated soil in cold regions.

Conclusion

Cold-adapted microorganisms living in foods and artificial cold environments related to human activity are surveyed. The bacterial taxa detected in each artificial cold environment are listed in Table 2.1. All these are psychrotolerant microorganisms. Psychrophilic microorganisms have never been isolated from these environments. The reasons are that psychrophilic microorganisms do not ubiquitously exist in relation to human activity and that these artificial environments, except marine products, are not related to the environments where psychrophiles live. In addition, it is considered that psychrophiles may be eliminated in artificial environments if they invaded to artificial environments. On the other hand, psychrotolerant *Pseudomonas* spp. are always present under aerobic conditions. LAB tend to be present under anaerobic conditions. These are ubiquitously present in environments related to human activity. In addition to the invasion of ubiquitously present microorganisms from ambient environment, the affinity of microorganisms to foods or environments is also a very important factor that defines the existence of microorganisms. The detected members as described above suggested that the important factors that define the existence of certain genera in cold artificial environments are (i) chance of invasion from ambient environments, (ii) chance of invasion through basic ingredients, (iii) ability to rapidly propagate at low temperature, (iv) presence or absence of oxygen, (v) production of lactic acid or bacteriocins and (vi) tolerance for reactive oxygen species.

In addition to *Pseudomonas* spp. and LAB, strains belonging to genera *Psychrobacter, Shewanella* and *Photobacterium* have often been detected. In addition to their ability to grow at low temperatures and a wide range of distributions in various environments, it is considered that these genera involve a wide range of genetic diversity, and cold artificial environment adaptation mechanisms of certain species belonging to these genera make their distribution wider not only in natural cold environments but also in artificial cold environments.

Future trends

Analyses of microbiota have been carried out by a culture-dependent approach. This procedure is not appropriate enough for the analysis of microorganisms that do not adapt to culture media used for isolation. The identification of numerous isolates from fermented foods used to be performed according to the identification scheme based on the results of phenotypic characterization. In accordance with the development of molecular techniques,

Table 2.1 Bacterial taxa found in artificial cold environments

Raw marine product	
Brochothrix, *Enterobacteriaceae*, lactic acid bacteria, *Photobacterium*, *Pseudomonas*, *Pseudoalteromonas*, *Psychrobacter*, *Vibrio*/*Aliivibrio*	
Raw freshwater fish	
Aeromonas, *Enterobacteriaceae*, pseudomonads, *Shewanella*, staphylococci	
Fabricated marine product	
Aeromonas, *Brochothrix*, *Lactobacillus*, *Carnobacterium*, *Photobacterium*, *Yersinia*	
Refrigerated milk	
Acinetobacter, *Aerococcus*, *Clostridia*, *Delfita*, *Fackkamia*, *Pseudomonas*, *Staphylococcus*, *Stenotrophomonas*, *Streptococcus*, *Trichococcus*	
Fermented salmon Izushi	
Achromobacter (*Alcaligenes*), *Bacillus*, *Pseudomonas*, *Lactococcus*, *Leuconostoc*, *Micrococcus*, *Staphylococcus*, *Streptococcus*	
Squid Shiokara	
Bacillus, coryneform, *Lactobacillus*, *Micrococcus*, *Staphylococcus*	
Feta cheese	
Janthinobacterium, *Lactobacillus*, *Lactococcus*, *Oxalobacteraceae*, *Pseudomonas*, *Psychrobacter*, *Staphylococcus*	
Chilled beef	
Brochothrix, *Carnobacterium*, *Enterobacteriaceae*, *Lactobacillus*, *Leuconostoc*, *Photobacterium*, *Pseudomonas*, *Rahnella*, *Serratia*, *Staphylococcus*	
Wastewater treatment (marine product)	
Dietzia, *Exiguobacterium*, *Pseudomonas*, *Psychrobacter*, *Vibrio*	
Oil-contaminated environment	
Acinetobacter, *Pseudomonas*, *Rhodococcus*, *Sphingomonas*	

exhaustive culture-independent analysis of microbiota has become possible by analysing many clones in clone libraries based on DNA extracted from the environment. The culture-independent analysis of the succession of dominant microbiota in the environment is also possible by denaturation gradient gel electrophoresis (DGGE). In addition, a more precise identification of microorganisms has become possible using 16S rRNA gene analysis of isolated DNA and the enormous amount of sequence data in databases. More detailed and precise analyses may be possible with the application of recently advanced techniques. For example, next-generation sequencing approaches have already been performed in food microbial genomics (Solieri *et al.*, 2012). In addition, metagenomic approaches have been performed for analyses of natural environments, symbiotic systems and bioreactors and for obtaining useful genomic resources (Warnecke *at al.*, 2007; Simon and Daniel, 2009; Mackelprang *et al.*, 2011; Jaenicke *et al.*, 2011). However, few studies on the analysis of foods and other artificial cold environments are available. In the near future, the metagenomic approach will be applied to artificial cold environments.

In addition, we understand little about species–species or strain–strain interactions in artificial cold environments. Clarification of the interaction between microorganisms and substances produced during fermentation or spoilage is also important. Presently, no cold- environment-specific seed culture complex is available. However, acceleration of degradation in cold environments is very important for bioremediation and wastewater treatment systems. In fermented food, an effective seed culture complex is useful to construct appropriate microbiota effectively. Understanding of species–species interaction and affinities towards substances produced by bacteria will be useful for developing a suitable seed culture complex for starters of fermented foods or for bioremediation.

Although it is difficult to accelerate chemical reaction at low temperature, it is better to adjust more suitable microbiota in such an environment, because the exclusion of contaminants of mesophiles is easy and the microbial metabolisms of mesophiles are slower at low temperatures than under conventional conditions. Therefore, technologies for suitable adjustment of microbiota at low temperature are applicable to the food ripening process of fermentation products. It is considered that based on the information of suitable velocity of chemical reaction in food ripening and suitable selection of microorganisms corresponding to the reaction and niche, appropriate adjustment of microbiota will be possible in each case of food ripening.

Web resources

International Journal of Food Microbiology (http://www.journals.elsevier.com/international-journal-of-food-microbiology/)
Food Microbiology (http://www.journals.elsevier.com/food-microbiology/)
Frontiers in Food Microbiology (http://www.frontiersin.org/food_microbiology)

References

Aislabie, J., Saul, D.J., and Foght, J.M. (2006). Bioremediation of hydrocarbon-contaminated polar soils. Extremophiles *10*, 171–179.

Baraniecki, C.A., Aislabie, J., and Foght, J.M. (2002). Characterization of *Sphingomonas* sp. Ant 17, an aromatic hydrocarbon-degrading bacterium isolated from Antarctic soil. Microb. Ecol. *43*, 44–54.

Bej, A.K., Saul, D.J., and Aislabie, J. (2000). Cold tolerant alkane-degrading *Rhodococcus* species from Antarctica. Polar Biol. *23*, 100–105.

Borch, E., Kant-Muemans, M.L., and Blixt, Y. (1996). Bacterial spoilage of meat and cured meat products. Int. J. Food Microbiol. *33*, 103–120.

Bowman, J.P., McCammon, S.A., Brown, M.V., Nichols, D.S., and McMeekin, T.A. (1997). Diversity and association of psychrophilic bacteria in Antarctic sea ice. Appl. Environ. Microbiol. *63*, 3068–3078.

Boylston, T.D., Vinderola, C.G., Ghoddusi, H.B., and Reinheimer, J.A. (2004). Incorporation of bifidobacteria into cheeses: challenges and rewards. Int. Dairy J. *14*, 375–387.

Broekaert, K., Heyndrickz, M., Herman, L., Devilieghere, F., and Vlaemynck, G. (2011). Seafood quality analysis: molecular identification of dominant microbiota after ice storage on several general growth media. Food Microbiol. *28*, 1162–1169.

Champagne, C.P., Laing, R.R., Roy, D., Mafu, A.A., and Griffiths, M.W. (1994). Psychrotrophs in dairy products: their effects and their control. Crit. Rev. Food Sci. Nutr. *34*, 1–30.

Dalgaaed, P., Gram, L., and Huss, H.H. (1993). Spoilage and shelf life of cod fillets packed in vacuum or modified atmosphere. Int. J. Food Microbiol. *19*, 283–294.

De Jonghe, V., Coorevits, A., Van Hoorde, K., Messens, W., Van Landschoot, A., De Vos, P., and Heyndrickx, M. (2011). Influence of storage conditions on the growth of *Pseudomonas* species in refrigerated raw milk. Appl. Environ. Microbiol. *77*, 460–470.

Eriksson, M., Dalhammar, G., and Mohn, W.W. (2002). Bacterial growth and biofilm production on pyrene. FEMS Microbiol. Ecol. *40*, 21–27.

Garrote, G.L., Abraham, A.G., and De Antoni, G.L. (1997). Preservation of kefr grains, a comparative study. Lebens.-Wiss. Technol. *30*, 77–84.

Gram, L., Ravn, L., Rasch, M., Bruhn, J.B., Christensen, A.B., and Givskov, M. (2002). Food spoilage – interaction between food spoilage bacteria. Int. J. Food Mirobiol. *78*, 79–97.

Hirota, K., Yamahira, K., Nakajima, K., Nodasaka, Y., Okuyama, H., and Yumoto, I. (2011). *Pseudomonas toyotomiensis* sp. nov., a psychrotolerant facultative alkaliphile that utilizes hydrocarbons. Int. J. Syst. Evol. Microbiol. *61*, 1842–1848.

Hovda, M.B., Lunestad, B.T., Sivertsvik, M., and Rosnes, J.T. (2007a). Characterization of the bacteria flora of modified atmosphere packed famed Atlantic cod (*Gadus morhua*) by PCR-DGGE of conserved 16S rRNA gene regions. Int. J. Food Microbiol. *117*, 68–75.

Hovda, M.B., Sivertsvik, M., Lunestad, B.T., Lorentzen, G., and Rosnes, J.T. (2007b). Characterization of the dominant bacterial population in modified atmosphere packaged farmed halibut (*Hippoglossus hippoglossus*) based on 16S rDNA-DGGE. Food Microbiol. *24*, 362–371.

Jaenicke, S., Ander, C., Bekel, T., Bisdorf, R., Dröge, M., Gartemann, K.H., Jünemann, S., Kaiser, O., Tille, F., Zakrzewski, M., Pühler, A., Schlüter, A., and Goesmann, A. (2011). Comparative and joint analysis of two metagenomic datasets from a biogas fermenter obtained by 454-pyrosequencing. PLoS One *6*, e14519.

Jay, J.M. (1992). Microbial spoilage of foods. In: Modern food microbiology, 4th ed. Jay, J.M. ed., Van Nostrand Reinhold, New York, NY, pp. 187–248.

Jay, M. (1986). Modern Food Microbiology, 3th edn. Van Nostrand Reinhold Company, New York.

Joffraud, J.J., Leroi, F., Roy, C., and Berdague, J.L. (2001). Characterization of volatile compounds produced by bacteria isolated from the spoilage flora of cold-smoked salmon. Int. J. Food Microbiol. *66*, 175–184.

Joffraud, J.J., Cardinal, M., Cornet, J., Chasles, J.-S., Léon, S., Gigout, F., and Leroi, F. (2006). Effect of bacterial interaction on the spoilage of cold-smoked salmon. Int. J. Food Microbiol. *112*, 51–61.

Jones, R.J. (2004). Observations on the succession dynamics of lactic acid bacteria population in chill-stored vacuum-packaged beef. Int. J. Food Microbiol. 190: 273–282.

Jørgensen, L.V., Huss, H.H., and Dalgaard, P. (2000). The effect biogenic amine production by single bacterial culture and metabiosis on cold-smoked salmon. J. Appl. Microbiol. *89*, 920–934.

Katayama, T., Tanaka, M., Moriizumi, J., Nakamura, T., Brouchkov, A., Douglas, T.A., Fukuda, M., Tomita, F., and Asano, K. (2007). Phylogenetic analysis of bacteria preserved in a permafrost ice wadge for 25,000 years. Appl. Environ. Microbiol. 73, 2360–2363.

Kawai, Y. (2009). Research on chemical and microbiological factors relating to preservation and processing of seafood. Nippon Suisan Gakkaishi 75, 357–360 (in Japanease).

Khosrowshahi, A., Madadlou, A., Ebrahim zadeh Mousavi, M., and Emam-Djomeh, Z. (2006). Monitoring the chemical and textural changes during ripening of Iranian White cheese made different concentration of starter. J. Dairy Sci. *89*, 3318–3325.

Kieronczyk, A., Skeie, S., Langsrud, T., and Yvon, M. (2003). Cooperation between *Lactococcus lactis* and nonstarter lactobacillus in the formation of cheese aroma from amino acids. Appl. Environ. Microbiol. *69*, 734–739.

Kimoto, H., Yoshimune, K., Matsuyama, H., and Yumoto, I. (2012). Characterization of catalase from psychrotolerant *Psychrobacter piscatorii* T-3 exhibiting high catalase activity. Int. J. Mol. Sci. *13*, 1733–1746.

Komagata, K. 1985. Bacteria (1)–Aerobic bacteria. In: Classification and Identification of Bacteria. Hasegawa, T. ed., Gakkai Shuppan Center, Tokyo, pp. 99–161.

Kourkoutas, Y., Bosnea, L., Taboukos, S.,Baras, C., Lambrou, D., and Kanellaki, M. (2006a). Probiotic cheese production using *Lactobacillus casei* cells immobilized on fruit pieces. J. Dairy Sci. *89*, 1439–1451.

Kourkoutas, Y., Kandylis, P., Panas, P., Dooley, J.S.G., Nigam, P., and Koutinas, A.A. (2006b). Evaluation of freese-dried kefir culture as starter in feta-type cheese production. Appl Environ Microbiol. *72*, 6124–6135.

Koutsoumanis, K.P., Stamatiou, A.P., Drosinos, E.H., and Nycha, G.-J.E. (2008). Control of spoilage microorganisms in minced pork by a self-developed modified atomosphere induced by the respiratory activity of meat microflora. Food Microbiol. *25*, 915–921.

Kurtzman, C.P., Fell, J.W., eds. (1998). The Yeasts. A Taxonomic Study 4th edn, Elservier Science, Amsterdam, pp. 1–1055.

Labadie, J. (1999). Consequence of packing on bacterial growth. Meat is an ecological niche. Meat Sci. *52*, 299–305.

Leroi, F., Joffraud, J.-J., Chevalier, F., and Cardinal, M. (1998). Study of microbial ecology of colsmoked salmon during storage at 8°C. Int. J. Food Microbiol. 39, 111–121.

Litopoulou-Tzanetaki, E., Tzanetaki, N., and Vafopoulou-Mastrojiannaki, A. (1993). Effect of the type of lactic starter on microbiological, chemical and sencory characteristics of feta cheese. Food Microbiol. *10*, 31–41.

Luis, A., Lopez, E., and Lema, C. (1993). Microflora present in kefir grains of the Galican region. J. Dairy Res. *60*, 263–267.

Mackelprang, R., Waldrop, M.P., DeAngelis, K.M., David, M.M., Chavarria, K.L., Blazewicz, S.J., Rubin, E.M., and Jansson, J.K. (2011). Metagenomic analysis of a permafrost microbial community reveals a rapid response to thaw. Nature *480*, 368–371.

Meer, R.R., Bakker, J., Bodyfelt, F.W., and Griffiths, M.W. (1991). Phychrotrophilc *Bacillus* spp. in fluid milk product: a review. J. Food Prot. *54*, 969–979.

Michaelidou, A., Katsiari, M.C., Kondyli, E., Voutsinas, L.P., and Alichanidis, E. (2003). Effect of a commercial culture on proteolysis in low-fat feta-type cheese. Int. Dairy J. *13*, 179–189.

Mills, S., Serrano, L.M., Griffin, C., O'Connor, P.M., Schaad, G., Bruining, C., Hill, C., Ross, R.P., and Meijer, W.C. (2011). Inhibitory activity of *Lactobacillus plantarum* LMG P-26358 against *Listeria innocura* when used as an adjunct starter in the manufacture of cheese. Microb. Cell Fac. *10*, 1–11.

Monteagudo-Mere, A., Caro, I., Rodríguez-Aparicio, L.B., Rúa, J., Ferrero, M.A., and García-Armesto, M.R. (2011). Characterization of certain bacterial strains for potential use as starter or probiotic culture in dairy products. J. Food Prot. *74*, 1379–1386.

Mori, K., Shinano, H., and Akiba, M. (1979). The aerobic bacteria in the ripening process of 'Ika-Shiokara'. Bull. Jan. Soc. Sci. Fish. *45*, 771–779.

Morishita, K., Otakasaka, W., Yamazaki, K., Kawai, Y., Inoue, N., and Shinano, H. (1994). Chemical and microbiological characteristics of commercial '*Shiokara*'. Bull. Fac. Fish. Hokkaido. Univ. *45*, 100–107 (in Japanese).

Morita, R. Y. (1975). Psychrophilic bacteria. Bacteriol. Rev. *39*, 144–167.

Nishimura, M., and Shinano, H. (1991). The effect of trimethylamine-N-oxide on microbial flora in '*Ika-Shiokara*'. Nippon Suisan Gakkaishi *57*, 1141–1145 (in Japanese).

Nortjé G.L., and Shaw, B.G. (1989). The effect of ageing treatment on the microbiology and storage characteristics of beef in modified atmosphere packs containing 25% CO_2 plus 75% O_2. Meat Sci. *25*, 43–58.

Olofsson, T.C., Ahrné, S., and Molin, G. (2007). The bacterial flora of vacuum-packed cold-smoked salmon stored at 7°C, identified by direct 16S rRNA gene analysis and pure culture technique. J. Appl. Microbiol. *103*, 109–119.

Olafsdottir, G., Lauzon, H.L., Martindottir, E., and Krisbergesson, K. (2006). Influence of storage temperature on microbial spoilage characterics of haddock fillets (*Melanogrammus aeglefinus*) evaluated by multivariate quality prediction. Int. J. Food Microbiol. *111*, 112–125.

Paarup, T., Sanchez, J.A., Moral, A., Christensen, H., Bisgaard, M., and Gram, L. (2002). Sensory, chemical and bacteriological changes during storage of iced squid (*Todaropsis eblanae*). J. Appl. Microbiol. *92*, 941–950.

Paludan-Müller, C., Dalgaard, P., Huss, H.H., and Gram, L. (1998). Evaluation of the role of *Carnobacterium piscicola* in spoilage of vacuum- and modified-atomosphere-packed cold-smoked salmon stored at 5°C. Int. J. Food Microbiol. *39*, 155–166.

Pantazi, D., Papavergou, A., Pournis, N., Kontominas, M.G., and Savvaidis, I.N. (2008). Shelf-life of chilled fresh Mediterranean swordfish (*Xiphias gladius*) stored under various packaging conditions: micbiological, biochemical and sensory attributes. Food Microbiol. *25*, 136–143.

Pennacchia, C., Ercolini, D., and Villani, F. (2011). Spoilage-related microbiota associate with chilled beef stored in air or vacuum pack. Food Microbiol. *28*, 84–93.

Raats, D., Offek, M., Minz, D., and Halpern, M. (2011). Molecular analysis of bacterial communities in raw cow milk and the impact of refrigeration on its structure and dynamics. Food Microbiol. *28*, 465–471.

Rasolofo, E.A., St-Gelais, D., LaPointe, G., and Roy, D. (2010). Molecular analysis of bacterial population structure and dynamics during cold storage of untreated and treated milk. Int. J. Food Microbiol. *138*, 108–118.

Rasolofo, E.A., LaPointe, G., and Roy, D. (2011). Assessment of the bacterial divert of treated and untreated milk during cold storage by T-RFLP and PCR-DGGE methods. Dairy Sci. & Technol. *91*, 573–597.

Ravi Sankar, C.N., Lalitha, K.V., Jose, L., Manju, S., and Gopal, T.K.S. (2008). Effect of packing atmosphere on the microbial attributes of pearlspot (*Entoplus suratensis* Bloch) stored at 0–2°C. Food Microbiol. *25*, 518–528.

Rodrigues, D.F., da C Jesus, E., Ayala-Del-Río, H.L., Pellizari, V.H., Gilichinsky, D., Sepulveda-Torres, L., and Tiedje, J.M. (2009). Biogeography of two cold-adapted genera: *Psychrobacter* and *Exiguobacterium*. ISME J. 3, 658–665.

Ryan, M.P., Rea, M.C., Hill, C., and Ross, R.P. (1996). An application in cheddar cheese manufacture for a strain of *Lactococcus lactis* producing a novel broad-spectrum bacteriocin, lacticin 3147. Appl. Environ. Microbiol. *62*, 612–619.

Sasaki, M., Kawai, Y., Yoshimizu, M., and Shinano, H. (2004). Changes in chemical composition and microbial flora of salmon *Izushi* during the ripening process. Nippon Suisan Gakkaishi *70*, 928–937 (in Japanese).

Sasaki, M., Kawai, Y., Yoshimizu, M., and Shinano, H. (2005). Effect of salting of salmon on the chemical and microbiological characteristics of salmon *Izushi*. Nippon Susan Gakkaishi *71*, 618–627 (in Japanese).

Sasaki, M., Kawai, Y., Yoshimizu, M., and Shinano, H. (2006a). Effect of ripening temperature on the chemical and microbiological characteristics of salmon *Izushi*. Nippon Susan Gakkaishi 72, 223–230 (in Japanese).

Sasaki, M., Kawai, Y., Yoshimizu, M., and Shinano, H. (2006b). Effect of inoculation with starter cultures of yeast and lactic acid bacterium on the chemical and microbiological characteristics of salmon *Izushi*. Nippon Susan Gakkaishi 72, 231–238 (in Japanese).

Schleifer, K.H. (1986). Section 12 Gram-positive cocci. In: Bergey's Manual of Systematic Bacteriology Vol. 2, Sneath, P.H.A., Mar, N.S., Sharpe, M.E., and Holt, J.G. eds. Williams & Wilkins, Baltimore, MD, pp. 999–1103.

Shewan, J.M., Hobbs, G., and Hodgkiss, W. (1960). The *Pseudomonas* and *Achromobacter* groups of bacteria in the spoilage of marine white fish. J. Appl. Bacteriol. *23*, 463–468.

Shivaji, S., Kumari, K., Hara Kishore, K., Pindi, P.K., Rao, P.S., Srinivas, T.N.R., Astana, R., and Ravindra, R. (2011a). Vertical distribution of bacteria in a lake sediment from Antarctica by culture-independent and culture-dependent approaches. Res. Microbiol. *162*, 191–203.

Shivaji, S., Pratibha, M.S., Saliaja, B., Hara Kishore, K., Singh, A.K., Begum, Z., Anarasi, U., Prabagaran, S.R., Reddy, G.S.N., and Srinivas, T.N.R. (2011b). Bacterial diversity of soil in the vicinity of Pindari glacier, Himalayan mountain ranges, India, using cultureable bacteria and soil 16S rRNA gene clones. Extremophiles *15*, 1–22.

Simon, C., and Daniel, R. (2009). Achievements and new knowledge unraveled by matagenomic approaches Appl. Microbiol. Biotechnol. *85*, 265–276.

Slattery, L., O'Callaghan, J., Fitzgerald, G.F., Beresford, T., and Ross, R.P. (2010). *Lactobacillus helveticus* – a thermophilic dairy starter related to gut bacteria. J. Dairy Sci. *93*, 4435–4453.

Solieri, L., Dakal, T.C., and Giudici, P. (2012). Next-generation sequencing and its potential impact on food microbial genomics. Ann. Microbiol. Publish on-line May 2012.

Stohr, V., Joffraud, J.J., Cardinal, M., and Leroi, F. (2001). Spoilage potential and sensory profile associated with bacteria isolated from cold-smoked salmon. Food Res. Int. *34*, 797–806.

Suyama, M., and Kounosu, S. eds. (1987). Suisan Shokuhin Gaku (Aquatic Food Science), Kouseisha Kouseikaku, Tokyo.

Suzuki, S., Kubo, A., Shinano, H., and Takama, K. (1992). Inhibition of the electron transport system in *Staphylococcus* aureus by trimethylamine-N-oxide. Microbios *71*, 145–148.

Takai, M., Kawai, Y., Inoue, N., and Shinano, H. (1992). Comparative studies on microbiological and chemical characteristics of '*Ika-Shiokara Akazukuri*' and '*Ika-Shiokara Kurozukuri*'. Nippon Suisan Gakkaishi *58*, 2373–2378 (in Japanese).

Takai, M., Yamazaki, K., Kawai, Y., Inoue, N., and Shinano, H. (1993). Effect of squid liver, skin, and ink on microbiological characteristics of 'Ika-Shiokara' during ripening process. Nippon Suisan Gakkaishi *59*, 1617–1623 (in Japanese).

Takami, H., Inoue, A., Fuji, F., and Horikoshi, K. (1997). Microbial flora in the deepest sea mud of the Mariana Trench. FEMS Microbiol. Lett. *152*, 279–285.

Ternström, A., Lindberg, A.M., and Molin, G. (1993). Classification of the spoilage flora of raw and pasteurized bovine milk, with special reference to *Pseudomonas* and *Bacillus*. J. Appl. Bacteriol. *75*, 25–34.

Uchimura, K., Miyazaki, M., Nogi, Y., Kobayashi, T., and Horikoshi, K. (2010). Cloning and sequencing of alginate lyase genes from deep-sea strains of *Vibrio* and *Agarivorans* and characterization of a new *Vibrio* enzyme. Mar. Biotechnol. *12*, 526–533.

Uchimura, T., and Okada, S. (1992). Laboratory Manual for Lactic Acid Bacteria. Kozaki, M. ed. Asakura-shoten, Tokyo (in Japanese).

Warnecke, F., Luginbühl, P., Ivanova, N., Ghassemian, M., Richardson, T.H., Stege, J.T., Cayouette, M., McHardy, A.C., Djordjevic, G., Aboushadi, N., *et al.* (2007). Metagenomic and functional analysis of hindgut microbiota of a wood-feeding higher termite. Nature *450*, 560–565.
Whyte, L.G., Bourbonniere, L., and Greer, C.W. 1997. Biodegradation of petroleum hydrocarbons by psychrotrophic *Pseudomonas* strains possessing both alkane (alk) and naphthalene (nah) catabolic pathway. Appl. Environ. Microbiol. *63*, 3719–3723.
Witthuhn, R.C., Schoeman, T., and Britz, T.J. (2005). Characterization of the microbial population at different stages of kefir production and kefir grain mass cultivation. Int. Dairy J. *15*, 383–389.
Yamahira, K., Hirota, K., Nakajima, K., Morita, N., Nodasaka, Y., and Yumoto, I. (2008). *Acinetobacter* sp. strain Ths, a novel psychrotolerant and alkalitolerant bacterium that utilized hydrocarbon. Extremophiles *12*, 729–734.
Yamazaki, K., Kitamura, F., Inoue, N., and Shinano, H. (1992). Effect of squid liver of microflora of '*Ika-Shiokara*'. *58*, 1971–1976 (in Japanese).
Yumoto, I., Yamazaki, K., Kawasaki, K., Ichise, N., Morita, N., Hoshino, T., and Okuyama, H. (1998). Isolation of *Vibrio* sp. S-1 exhibiting extraordinarily high catalase activity. J. Ferment. Bioeng. *85*, 113–116.
Yumoto, I., Iwata, H., Sawabe, T., Ueno, K., Ichise, N., Matsuyama, H., Okuyama, H., and Kawasaki, K. 1999. Characterization of a facultatively psychrophilic bacterium, *Vibrio rumoiensis* sp. nov., that exhibits high catalase activity. Appl. Environ. Microbiol. *65*, 67–72.
Yumoto, I., Nakamura, A., Iwata, H., Kojima, K., Kusumoto, K., Nodasaka, Y., and Matsuyama, H. (2002). *Dietzia psychralcaliphila* sp. nov., a novel, facultative psychrophilic alkaliphile that grows on hydrocarbons. Int. J. Syst. Evol. Microbiol. *52*, 85–90.
Yumoto, I., Hishinuma-Narisawa, M., Hirota, K.; Shingyo, T., Takebe, F., Nodasaka, Y., Matsuyama, H., and Hara, I. (2004). *Exiguobacterium oxidotolerans* sp. nov., a novel alkaliphile exhibiting high catalase activity. Int. J. Syst. Evol. Microbiol. *54*, 2013–2017.
Yumoto, I., Hirota, K., Kimoto, H., Nodasaka, Y., Matsuyama, H., and Yoshimune, K. (2010). *Psychrobacter piscatorii* sp. nov., a psychrotolerant bacterium exhibiting high catalase activity isolated from an oxidative environment. Int. J. Syst. Evol. Microbiol. *60*, 205–208.

Psychrophilic Microorganisms In Marine Environments

3

Yuichi Nogi

Abstract

Psychrophilic microorganisms are extremophiles that are capable of growth and reproduction at low temperatures. They are present in marine environments, which occupy slightly more than 70% of earth's surface, especially in the Arctic, Antarctica and deep seas at temperatures lower than 15°C. Marine psychrophiles utilize a wide variety of metabolic pathways, including photosynthesis, chemoautotrophy, and heterotrophy. The deep-sea bacteria called psychropiezophiles 'love' both high pressure and low temperature. Marine psychrophiles are characterized by lipid cell membranes chemically resistant to hardening in response to the cold. Most psychrophiles are Bacteria, and psychrophily is present in widely diverse microbial lineages within the broad groups of *Alpha-*, *Beta-*, *Delta-*, and *Gammaproteobacteria* and the *Bacteroidetes* phylum.

Marine environments and psychrophilic bacteria

The surface area of the earth totals 5.1×10^8 km^2. The seas occupy slightly more than 70% (3.6×10^8 km^2) of that area, and land slightly less than 30%. The amount of land above sea level is much less than the volume of the seas. If the land mass were flattened and spread into the seas, the oceans would still be 3000 m deep. The highest point on land is Mt. Everest (8848 m), while the deepest underwater point is in the Mariana Trench (10,911 m). The seas have an average depth of about 3729 m, with average salinity of 3.5%. The warmest ocean temperatures are near the equator, and they cool gradually towards the poles, where the water temperatures are relatively cold. The seawater temperature surrounding the Arctic Circle and Antarctic Circle does not exceed 10°C year round (Fig. 3.1, upper panel). Therefore, these areas are the most suitable living environments for marine psychrophilic bacteria. The North–South thermal structures of the Atlantic and Pacific Oceans are shown in Fig. 3.1. The temperature tends to be highest at the surface and decreases with depth. In some places, the sea surface temperature may exceed 25°C, although in most it is lower than 10°C at depths of 500 m (Fig. 3.1, lower panel). In addition, at depths of 1000 m or more, the seawater temperature is lower than 4°C, which includes most of the marine environment.

The deep sea is divided into zones by depth (Fig. 3.2). The epipelagic zone includes all open ocean areas and can be further subdivided based on both depth and the abundance of penetrating light. The photic zone covers the oceans from the surface to a depth of approximately 200 m. This is the area where photosynthesis can occur and therefore is the most biodiverse. Although plants are capable of photosynthesis, it cannot occur in deep water

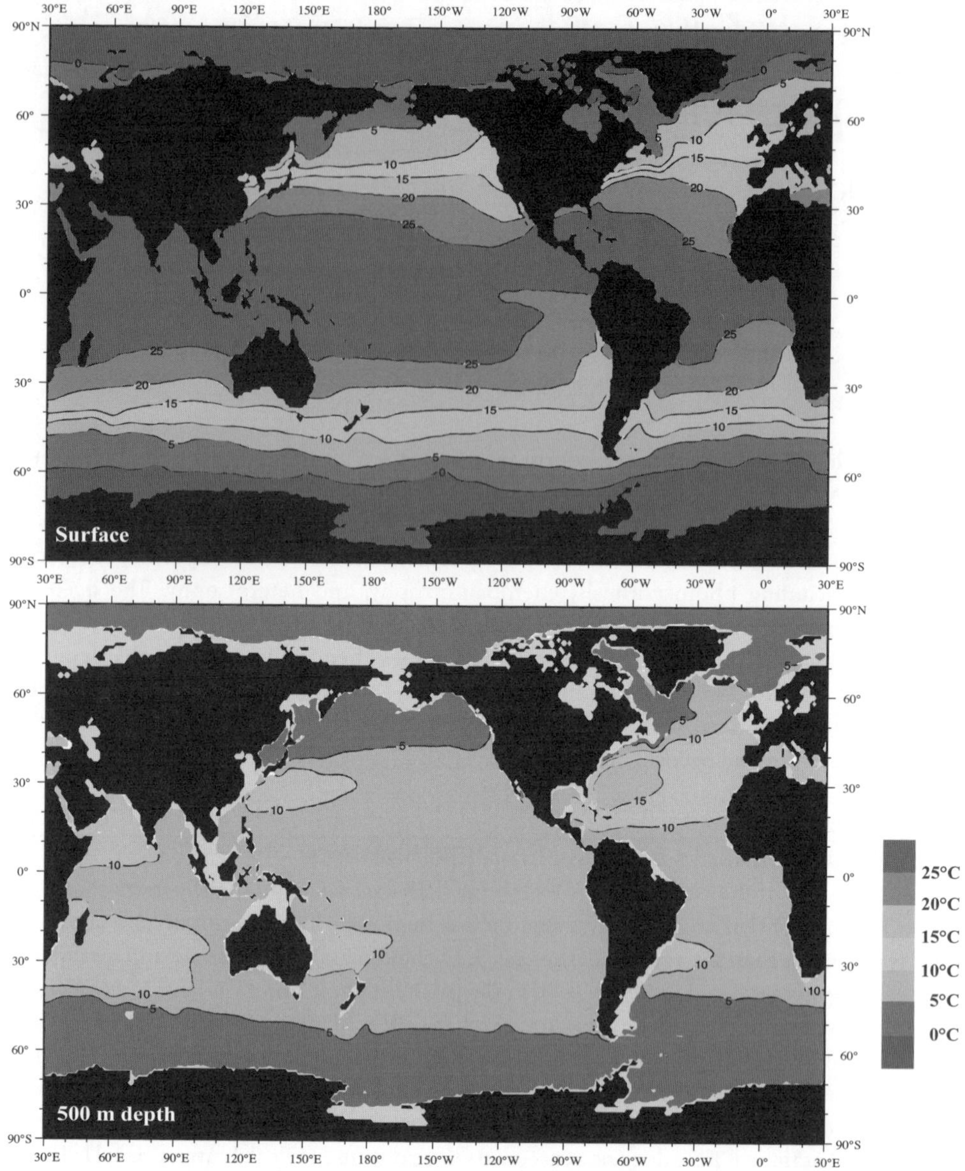

Figure 3.1 Annual ocean temperature observations at the surface (upper) and at 500m (lower). The map is based on data from the *World Ocean Atlas 2005* of the US National Oceanic and Atmospheric Administration.

where light cannot penetrate. Therefore, bacteria living in deeper water depend on sedimentation from the upper water layer or on organisms and compounds present in cold seeps and hydrothermal vents as energy sources. Methane, carbon dioxide, hydrogen sulfide, etc. are found in cold seeps and hydrothermal vents in the deep sea, which can support microbial life. Such energy sources are insufficient to support abundant life in the deep sea, although sufficient dissolved oxygen to support bacterial life is present in the deep sea (Fig. 3.3).

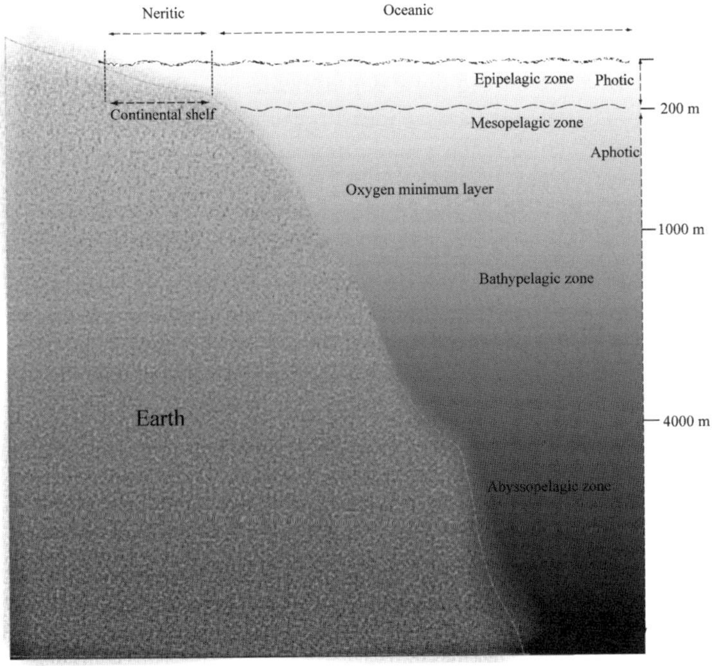

Figure 3.2 Layers of the ocean. The oceans are divided into two broad realms, the pelagic and the benthic. These layers, known as 'zones,' extend from the surface to the most extreme depths where light can no longer penetrate. Biologists divide the pelagic zone into the epipelagic (less than 200 m, where photosynthesis can occur), the mesopelagic (200–1000 m, the 'twilight' zone with faint sunlight but no photosynthesis), the bathypelagic (1000–4000 m), the abyssopelagic (4000–6000 m), and the deepest zone, the hadopelagic (trenches more than 6000m deep).

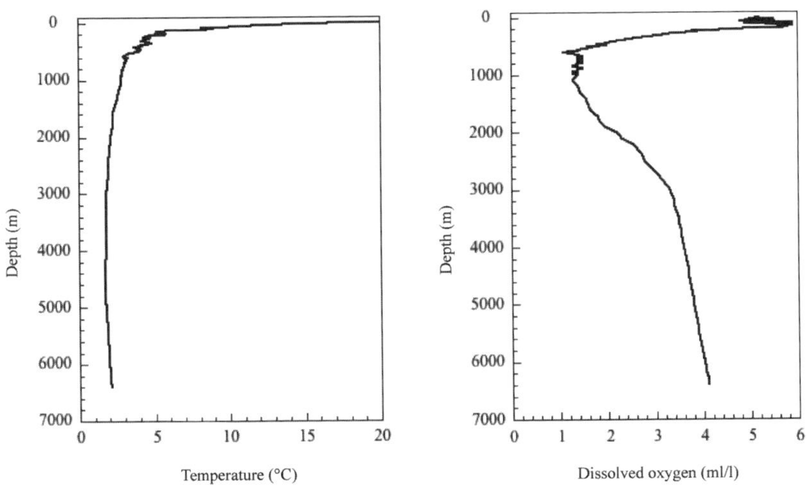

Figure 3.3 Depth relation of temperature and dissolved oxygen in the Japan Trench (in August 1995) as measured by the conductivity temperature depth (left)/dissolved oxygen profiler (right) of the submersible *Kaiko* system operated by JAMSTEC.

Surface water generally has an oxygen concentration close to that of the atmosphere. This oxygen-rich ocean water flows out of the mixed layer into the thermocline. However, the oxygen concentration decreases rapidly at depths below 50–100 m with the decline in photosynthesis rates, and living organisms require large amounts of oxygen. At depths between 500 and 1000 m, oxygen levels reach a minimum, and this area is referred to as the oxygen minimum zone. However, the still deeper sea is more abundant in oxygen than the oxygen minimum zone. The dark, cold waters of much of the deep sea contain oxygen levels adequate to support life because colder water retains more oxygen than warmer water, and the deepest waters generally originate from shallow polar seas. In certain places in the northern and southern seas, oxygen-rich water cools to such an extent that it becomes sufficiently dense to sink to the sea bottom. Such oxygen-rich cold water can flow at depth over the ocean floor by thermohaline convection, and sufficient oxygen remains to support bacterial life because there is not enough biomass to use it all.

Pressure increases by 0.1 megapascal [MPa; $1\ Pa = 1\ kg/(m\ s^2) = 9.87 \times 10^{-6}$ atm] for each 10 m in depth. The deep sea varies in depth from 200 m to about 11,000 m, and therefore pressure ranges from 2 MPa to more than 110 MPa. Pressure is the most significant environmental factor acting on deep-sea organisms. Although pressure tolerance differs by strain, if bacteria that are not piezophilic are subjected to pressure of approximately 20–30 MPa, they are still capable of growth (Nogi *et al.*, 1998b; Kato and Nogi, 2001). Therefore, the deep sea is the largest region inhabitable by various strains of psychrophiles.

Isolation and preservation of marine psychrophiles

Marine psychrophilic bacteria are generally found only in permanently cold environments, such as the Arctic Ocean, Antarctic Ocean and deep sea, with constant annual temperatures of less than 10°C and not subject to warming to more than 20°C even for short periods. Marine psychrophilic species have been isolated from seawater, ice, sediment, and marine organisms (Table 3.1).

The isolation of marine psychrophiles requires maintenance of samples at low temperature (0–4°C). Furthermore, the culture medium used for isolation must also be cooled in advance. It is preferable to use collected samples for isolation immediately, although they must not be frozen even if immediate isolation is not possible. If a sample is frozen without cyroprotectants, many marine psychrophiles are easily lysed in the freeze–thawing process.

In the polar zone where the temperature of surface seawater is low, deep-sea samples can be collected relatively easily at the optimum temperature. However, higher surface seawater temperatures areas, a special sampler is required. A temperature change affects deep-sea microbes more than a pressure change. Japan Agency for Marine-Earth Science and Technology (JAMSTEC) researchers use the sampler shown in Fig. 3.4 to collect deep-sea microbes (Ikemoto and Kyo, 1993; Kato *et al.*, 1995), which ensures only a minimal change in the sample temperature.

The preferred method for long-term preservation and storage of marine psychrophilic species is freezing in the vapour phase of liquid nitrogen (–130°C and below). This method yields good survival and recovery rates. Cultures to be stored in liquid nitrogen are usually grown to the late-log growth phase and mixed with a cryopreservative (10–20% glycerol or DMSO). Sample cultures can also be preserved at –70°C to –80°C. However, since the survival and recovery rates decrease with this method, it is necessary to use a preservative for a dense suspension of cells. Since almost all deep-sea microbe strains die after freeze-drying, this method cannot be used. For short periods of time, i.e. 1–2 months, many species remain viable on agar plates or slants if stored at 2–4°C.

Psychropiezophilic bacteria

Bacteria living in the deep sea have several unusual characteristics that allow them to grow in their extreme environment. In 1949, research on the effects of hydrostatic pressure on microbial activities was initiated (ZoBell and Johnson, 1949). The word 'barophilic' was first used to refer to those microbes, and barophily is defined today as optimal growth at pressure greater than 0.1 MPa or as a requirement for increased pressure for growth. However, 'piezophilic' is more suitable than barophilic to describe bacteria that grow better at high pressure than at atmospheric pressure. Presently, piezophilic is used to refer to high-pressure-loving bacteria. The growth patterns of piezotolerant and piezophilic bacteria are shown in Fig. 3.5. Piezophiles display maximum growth at high pressure. They may or may not be capable of growth at atmospheric pressure and are referred to as obligatory piezophiles in the latter case. Piezotolerant bacteria grow best at atmospheric pressure but can also grow at high pressure.

In 1979, the first pure culture isolate of a psychropiezophilic bacterium, a decaying amphipod from the central North Pacific, was reported (Yayanos *et al.*, 1979). The spirillum-like bacterium strain CNPT-3 had a rapid doubling rate at 50 MPa but did not grow at atmospheric pressure. However, no sample was deposited in public culture collections and no name was given to strain CNPT-3. Subsequent isolation of psychropiezophiles was reported, although they were not deposited in public collections and no names were added. The isolated deep-sea piezophilic bacterial strains were characterized in an effort to understand the interactions between the deep-sea environment and its microbial inhabitants (Kato *et al.*, 1998; Margesin and Nogi, 2004; Yayanos *et al.*, 1979).

The first psychropiezophile to be named was *Shewanella benthica* (Table 3.1). *S. benthica* strain W I45 was isolated from the holothurian, at depth of 4575 m in the South Atlantic Ocean (Deming *et al.*, 1984). Thereafter, JAMSTEC groups isolated and characterized numerous psychropiezophilic and psychropiezotolerant bacteria from deep-sea sediments at depths ranging from 2500 to 11,000 m collected using sterilized mud samplers on the submersibles *Shinkai 6500* and *Kaiko* (Kato *et al.*, 1995; Nogi and Kato, 1999; Nogi *et al.*, 1998a–c, 2002, 2004, 2007). Thus far, all psychropiezophilic bacterial isolates are members of the *Proteobacteria* according to phylogenetic classifications based on 16S rRNA gene sequence information (DeLong *et al.*, 1997; Margesin and Nogi, 2004; Eloe *et al.*, 2011). Based upon several taxonomic properties, JAMSTEC groups found that cultivated psychropiezophilic deep-sea bacteria are affiliated with one of the five genera within the *Gammaproteobacteria* subgroup: *Shewanella, Photobacterium, Colwellia, Moritella* and *Psychromonas* (Table 3.1). Recently, Scripps Institution of Oceanography researchers have reported a novel psychropiezophilic bacteria belonging to *Alphaproteobacteria* (Eloe *et al.*, 2011). Fig. 3.6 shows the phylogenetic relations among the taxonomically identified psychropiezophilic species (shown in boxed text) and psychrophilic bacteria.

The fatty acid composition of psychropiezophiles is characteristic. The psychropiezophilic *Shewanella* and *Photobacterium* strains produce eicosapentaenoic acid (EPA) (Nogi *et al.*, 1998b,c), *Moritella* strains produce docosahexaenoic acid (DHA) (Nogi *et al.*, 1998a;

Table 3.1 Well-characterized marine psychrophilic bacteria, habitat, optimum growth temperature, and pressure

Species		Type strain	T_{opt} (°C)	P_{opt} (MPa)	Habitat	Reference
Gammaproteobacteria						
Colwellia	*demingiae*	ACAM 459^T	10	0.1	Antarctic Sea, ice	Bowman *et al*. (1998a)
Colwellia	*hadaliensis*	BNL-1	10	90	Puerto Rico Trench (7410m depth)	Deming *et al*. (1988)
Colwellia	*hornerae*	ACAM 607^T	12	0.1	Antarctic Sea, ice	Bowman *et al*. (1998a)
Colwellia	*maris*	JCM 10085^T	15	0.1	Hokkaido, Japan, seawater	Yumoto *et al*. (1998)
Colwellia	*piezophila*	JCM 11831^T	10	60	Japan Trench, sediment (6278m depth)	Nogi *et al*. (2004)
Colwellia	*psychrerythraea*	ATCC 27364^T	10–12	0.1	Fish eggs	Deming *et al*. (1988)
Colwellia	*rossensis*	ACAM 608^T	8–10	0.1	Antarctic Sea, ice	Bowman *et al*. (1998a)
Glaciecola	*pallidula*	ACAM 615^T	13–14	–	Antarctic Sea, ice	Bowman *et al*. (1998b)
Glaciecola	*psychrophila*	JCM 13954^T	12	–	Arctic Sea, ice	Zhang *et al*. (2006)
Glaciecola	*punicea*	ACAM 611^T	16–18	–	Antarctic Sea, ice	Bowman *et al*. (1998b)
Marinobacter	*psychrophilus*	JCM 14643^T	16–18	–	Arctic Sea, ice	Zhang *et al*. (2008a)
Moritella	*abyssi*	JCM 11436^T	10	30	Atlantic Ocean, sediments (2815m depth)	Xu *et al*. (2003a)
Moritella	*dasanensis*	JCM 14759^T	9	–	Arctic, seawater	Kim *et al*. (2008)
Moritella	*japonica*	JCM 10249^T	15	50	Japan Trench, sediment (6356m depth)	Nogi *et al*. (1998a)
Moritella	*marina*	ATCC 15381^T	15–18	0.1	North-western Pacific Ocean (1200m depth)	Urakawa *et al*. (1999)
Moritella	*profunda*	JCM 11435^T	6	30	Atlantic Ocean, sediments (2815m depth)	Xu *et al*. (2003a)
Moritella	*viscosa*	NCIMB 13584^T	15	–	Fish in Iceland and Norway	Benediktsdottir *et al*., 2000)
Moritella	*yayanosii*	JCM 10263^T	10	80	Mariana Trench, sediments (10,898m depth)	Nogi and Kato (1999)
Photobacterium	*frigidiphilum*	JCM 12947^T	14	–	West Pacific Ocean, sediment (1450m depth)	Seo *et al*. (2005)
Photobacterium	*profundum*	JCM 10084^T	10	10	Rykyu Trench, sediment (5110m depth)	Nogi *et al*. (1998c)
Psychrobacter	*glacincola*	ACAM 483^T	15	–	Antarctic Sea, ice	Bowman *et al*. (1997c)
Psychromonas	*hadalis*	JCM 11830^T	6	60	Japan Trench, sediment (7542m depth)	Nogi *et al*. (2007)
Psychromonas	*ingrahamii*	CIP 108865^T	~10	–	Arctic Sea, ice	Auman *et al*. (2006)
Psychromonas	*kaikoae*	JCM 11054^T	10	50	Japan Trench, sediment (7434m depth)	Nogi *et al*. (2002)
Psychromonas	*marina*	JCM 10501^T	14–16	0.1	Okhotsk Sea, seawater	Kawasaki *et al*. (2002)
Psychromonas	*profunda*	JCM 11437^T	10	25	Atlantic Ocean, sediments (2770m depth)	Xu *et al*. (2003b)

Shewanella	*abyssi*	JCM 13041^{T}	10–14	-	Suruga Bay, sediment (2406m depth)	Miyazaki *et al*. (2006)
Shewanella	*benthica*	ATCC 43992^{T}	2	40	South Atlantic Ocean, holothurian (4575m depth)	Deming *et al*. (1984)
Shewanella	*donghaensis*	JCM 12524^{T}	17	10	Sea of Japan, sediment (3300 m depth)	Yang *et al*. (2007)
Shewanella	*gelidimarina*	ACAM 456^{T}	15–17	0.1	Antarctic Sea, sediment	Bowman *et al*. (1997b)
Shewanella	*hanedai*	ATCC 33224^{T}	14	0.1	Polar marine sediment	MacDonell and Colwell (1985)
Shewanella	*livingstonensis*	LMG 19866	15		Antarctic Sea, ice	Bozal N *et al*. (2002)
Shewanella	*psychrophila*	JCM 13876^{T}	10–15	20	West Pacific Ocean, sediment (1912 m depth)	Xiao *et al*. (2007)
Shewanella	*surugensis*	JCM 11835^{T}	13	–	Suruga Bay, sediment (2406m depth)	Miyazaki *et al*. (2006)
Shewanella	*violacea*	JCM 10179^{T}	10	30	Rykyu Trench, sediment (5110m depth)	Nogi *et al*. (1998b)
Alphaproteobacteria						
Octadecabacter	*antarcticus*	CIP 106731^{T}	10–12	–	Antarctic Sea, ice	Gosink *et al*. (1997)
Octadecabacter	*arcticus*	CIP 106732^{T}	10–12	–	Arctic Sea, ice	Gosink *et al*. (1997
Rhodobacterales bacterium		PRT1	10	80	Puerto Rico Trench, seawater (8350m depth)	Eloe *et al*. (2011)
Betaproteobacteria						
Polaromonas	*vacuolata*	ATCC 51984^{T}	4	–	Antarctic Sea, seawater	Irgens *et al*. (1996)
Deltaproteobacteria						
Desulfofaba	*gelida*	DSM 12344^{T}	7	–	Arctic Ocean, sediments	Knoblauch *et al*. (1999)
Desulfofrigus	*oceanense*	DSM 12341^{T}	10	–	Arctic Ocean, sediments	Knoblauch *et al*. (1999)
Desulfotalea	*psychrophila*	DSM 12343^{T}	10	–	Arctic Ocean, sediments	Knoblauch *et al*. (1999)
Desulfuromonas	*svalbardensis*	JCM 12927^{T}	14	–	Arctic Ocean, sediments (212m depth)	Vandieken *et al*. (2006)
Desulfuromusa	*ferrireducens*	JCM 12926^{T}	14–17	–	Arctic Ocean, sediments (64m depth)	Vandieken *et al*. (2006)
Bacteroidetes						
Gelidibacter	*algens*	ACAM 536^{T}	14–18	–	Antarctic Sea, ice	Bowman *et al*. (1997a)
Polaribacter	*filamentus*	ATCC 700397^{T}	10	–	Arctic Ocean, seawater	Gosink *et al*. (1998)
Polaribacter	*franzmannii*	ATCC 700399^{T}	10	–	Antarctic Sea, ice	Gosink *et al*. (1998)
Polaribacter	*glomeratus*	ACAM 171^{T}	15–17	–	Antarctic marine environments	Gosink *et al*. (1998)
Polaribacter	*irgensii*	ATCC 700398^{T}	10	–	Antarctic Sea, seawater	Gosink *et al*. (1998)
Psychroflexus	*torquis*	ACAM 623^{T}	8–10	–	Antarctic Sea, ice	Bowman *et al*. (1998c)

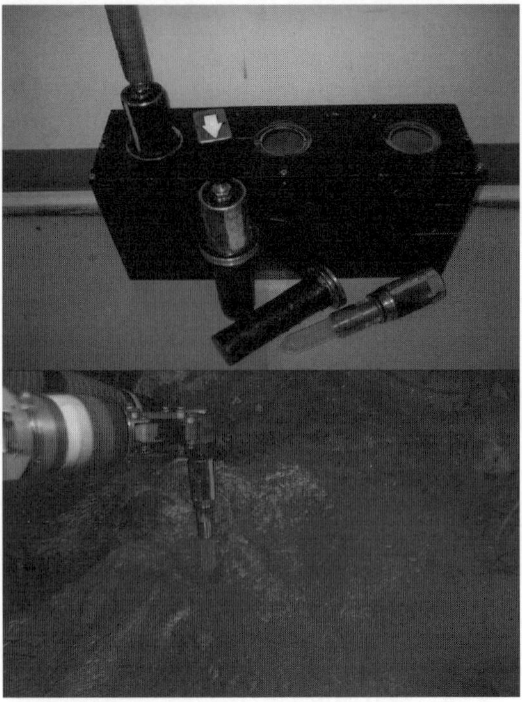

Figure 3.4 Sterilized sediment sampler (upper photograph) and sediment sampling (lower photograph).

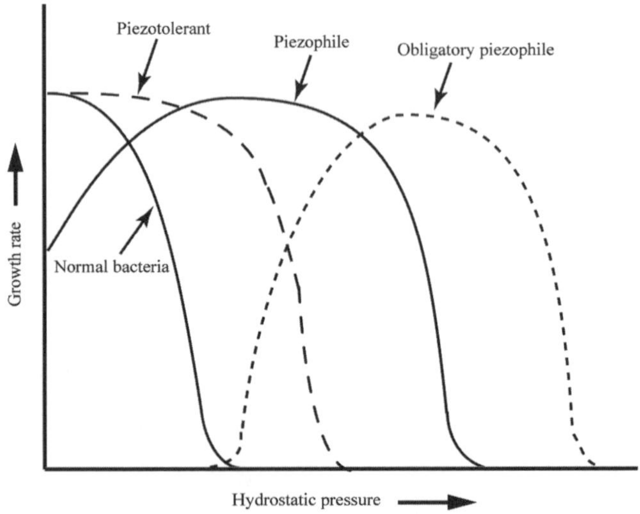

Figure 3.5 Definition of piezophilic growth properties. This figure is a conceptual chart, and x-axis value does not indicate specific pressure. Piezophiles display maximum growth at high pressure. They may or may not grow at atmospheric pressure, and in the latter case are obligatory piezophiles. Piezotolerant bacteria grow optimally at atmospheric pressure, but can sustain growth at high pressure (about 30–50MPa), whereas normal bacteria (piezosensitive) stop growing at about 30–50MPa.

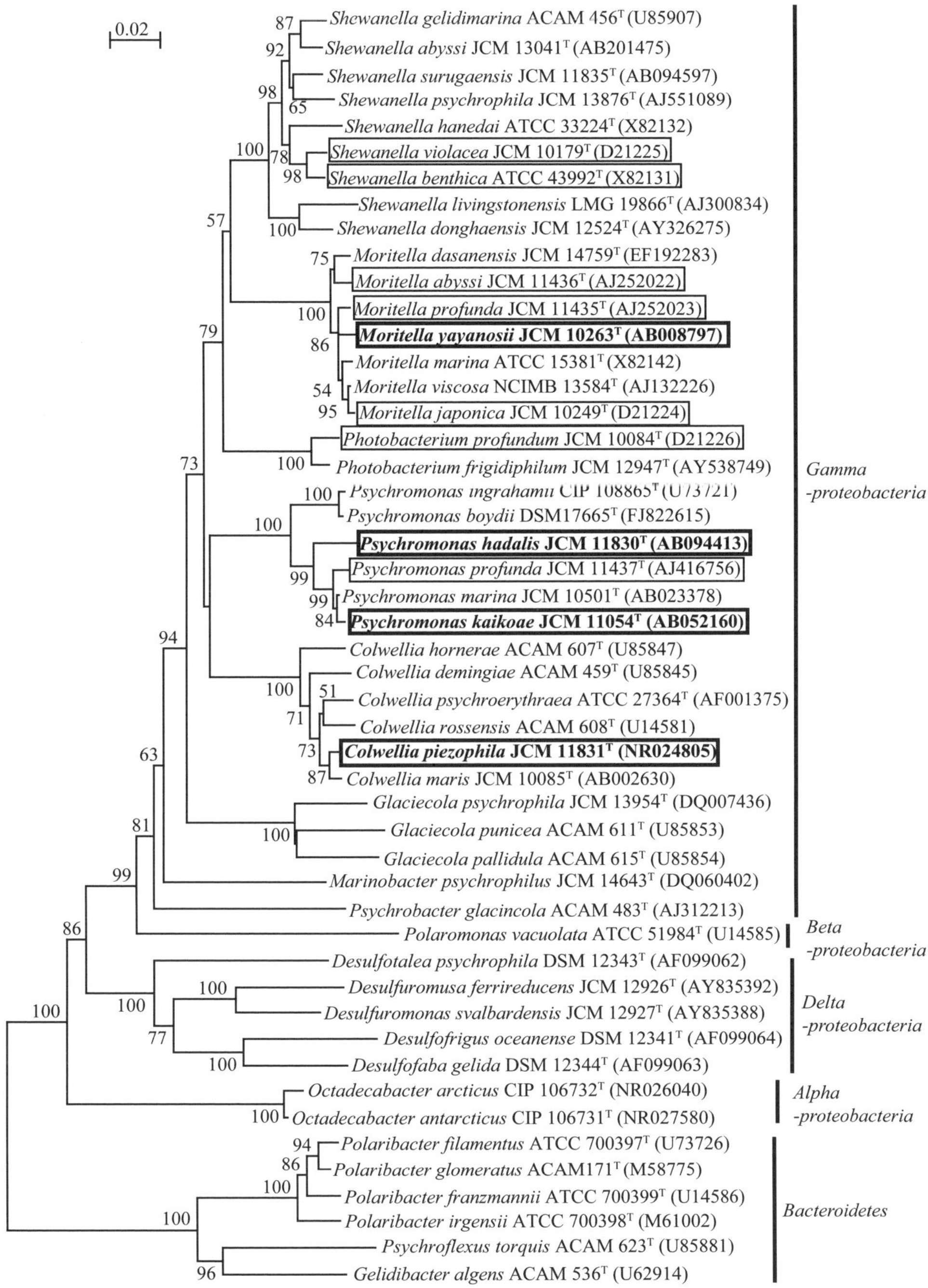

Figure 3.6 Phylogenetic relationships of marine psychrophilic bacteria isolated from several environments. The tree was produced by comparing 16S rDNA sequences using the neighbour-joining method. The scale represents the average number of nucleotide substitutions per site. Bootstrap values (%) are calculated from 1,000 trees and shown for frequencies above the threshold of 50%. Non-boxed text indicates psychrophilic bacteria; boxed text indicates psychropiezophilic bacteria; boxed and bold text indicates obligate psychropiezophilic bacteria.

Nogi and Kato, 1999), and *Psychromonas kaikoae* produces both EPA and DHA (Nogi *et al.*, 2002), but *Colwellia piezophila* does not produce polyunsaturated fatty acids (PUFAs) (Nogi *et al.*, 2004). The fatty acid compositions of those psychropiezophilic strains are distinct depending on their genus, and high levels of unsaturated fatty acids (about 50–70%), including EPA or DHA, are commonly involved in their membrane layer. The fatty acid composition of psychropiezophilic strains changes as a function of pressure, and in general greater amounts of PUFAs are synthesized under higher-pressure conditions for growth (DeLong and Yayanos, 1985, 1986). Psychropiezophilic bacteria were believed to produce one of the long-chain PUFAs, either EPA or DHA, but this does not appear to be obligatory. For example, Allen *et al.* (1999) reported that monounsaturated fatty acids, but not PUFAs, are required for the growth of the psychropiezophilic bacterium *Photobacterium profundum* SS9 based on the analysis of pressure-sensitive mutants. In their mutant experiments, 18:1 fatty acids proved to be necessary for growth under low-temperature and/or high-pressure conditions. *C. piezophila* has no 18:1 fatty acids but produces a large amount of 16:1 fatty acids in its cell membrane layer (Nogi *et al.*, 2004). All psychropiezophilic bacteria analysed so far have 16:1 fatty acids, and thus this fatty acid appears to be one of important components required for high-pressure growth.

Taxonomy of marine psychrophilic bacteria

Numerous marine psychrophilic bacterial strains have been isolated and characterized in an effort to understand the interactions between the polar-sea or deep-sea environment and its microbial inhabitants (DeLong *et al.*, 1997; Bowman *et al.*, 1998a; Margesin and Nogi, 2004). Thus far, marine psychrophilic bacterial isolates fall into diverse bacterial genera according to the phylogenetic classifications based on 16S rRNA sequence information. However, the only genera composed solely of marine psychrophilic bacteria are *Moritella* and *Octadecabacter*. Other bacterial genera include both mesophilic and psychrophilic species. No genus identified so far consists solely of psychropiezophilic bacteria (Fig. 3.6). The most commonly reported microorganisms in deep-sea or polar areas are the Gram-negative *Alpha-*, *Beta-*, *Delta-* and *Gammaproteobacteria* and the *Bacteroidetes* phylum (Table 3.1). The taxonomic features of marine psychrophilic genera were determined as described below.

Gammaproteobacteria

The genus *Colwellia*

The genus *Colwellia* was originally proposed by Deming *et al.* (1988) with the description of two facultatively anaerobic bacteria, *Colwellia psychrerythraea* and *Colwellia hadaliensis*. Species of the genus *Colwellia* are characterized as Gram-negative and motile (except for *Colwellia rossensis*), chemoorganotrophic, curved or straight rods, psychrophilic, halophilic, and containing ubiquinone-8 (Q-8) as the predominant isoprenoid quinone (Nogi *et al.*, 2004; Ivanova *et al.*, 2004; Zhang *et al.*, 2008b). Most species produce PUFAs such as 16:1 and DHA (Bowman *et al.*, 1998a; Nogi *et al.*, 2004). The genus *Colwellia* comprises 12 species isolated from marine environments. Two psychropiezophilic species, *C. hadaliensis* (Deming *et al.*, 1988) and *C. piezophila* (Nogi *et al.*, 2004) were isolated from deep-sea sediment. Five psychrophilic species, *C. psychrerythraea*, *Colwellia demingiae*, *Colwellia hornerae*, *C. rossensis* and *Colwellia maris*, were isolated from fish eggs, the Antarctic, and Japanese coastal ice or water (Deming *et al.*, 1988; Bowman *et al.*, 1998a; Yumoto *et al.*, 1998). Five non-psychrophilic species, *Colwellia asteriadis*, *Colwellia chukchiensis*, *Colwellia psychrotropica*, *Colwellia aestuarii* and *Colwellia polaris*, were isolated from starfish, the Arctic Ocean, Antarctic lake pycnocline, Antarctic tidal flats, and Arctic Sea ice, respectively (Choi *et al.*, 2010; Yu *et al.*, 2011; Bowman *et al.*, 1998a; Jung *et al.*, 2006; Zhang *et al.*, 2008b).

The genus *Glaciecola*

The genus *Glaciecola* was originally described as Gram-negative, aerobic, psychrophilic, slightly halophilic bacteria and initially comprised two species, *Glaciecola punicea* and *Glaciecola pallidula* (Bowman *et al.*, 1998b). Both species were isolated from sea-ice diatom assemblage samples collected from coastal areas of eastern Antarctica. Since then, further species of the genus *Glaciecola* have been isolated from marine environments: marine invertebrates, Arctic Ocean seawater, coastal surface seawater, deep-sea sediment, and East Sea sediments. They show diverse biochemical properties, such as agar digestion, nitrate reduction, and polysaccharide formation (Romanenko *et al.*, 2003; Van Trappen *et al.*, 2004; Baik *et al.*, 2006; Matsuyama *et al.*, 2006; Zhang *et al.*, 2006; Chen *et al.*, 2009; Zhang *et al.*, 2011). The genus *Glaciecola* comprises 10 recognized species. However, the range of growth temperature for *Glaciecola nitratireducens* (Baik *et al.*, 2006) is 15–30°C (optimum temperature 25°C), and thus it is not a psychrophile. All other species grow at temperatures below 7°C. There are three species of true psychrophiles with growth at temperatures lower than 15°C: *G. punicea*, *G. pallidula*, and *Glaciecola psychrophila* (Zhang *et al.*, 2006).

The genus *Marinobacter*

The genus *Marinobacter* was established with the description of *Marinobacter hydrocarbonoclasticus* in 1992 (Gauthier *et al.*, 1992). *Marinobacter* species are characterized as Gram-negative, chemoheterotrophic, straight or curved rods, aerobic, and moderately halophilic. The genus *Marinobacter* comprises 29 recognized species. Only *Marinobacter psychrophilus* is a marine psychrophilic bacteria isolated from sea-ice samples from the Arctic (Zhang *et al.*, 2008a).

The genus *Moritella*

The genus *Moritella* is characterized by Gram-negative, curved or straight rods, motile, chemoorganotrophic, psychrophilic, halophilic, and aerobic or facultatively anaerobic. Most species produce PUFAs such as DHA and 16:1 fatty acid (Urakawa *et al.*, 1998; Nogi and Kato, 1999; Xu *et al.*, 2003a). The genus *Moritella* consists of seven species. All members of this genus are true psychrophilic bacteria isolated from marine environments. Non-piezophilic *Moritella marina* and *Moritella dasanensis* were isolated from seawater of the North Pacific Ocean and near a glacier in Norway, respectively (Urakawa *et al.*, 1998; Kim *et al.*, 2008). *Moritella viscosa* was isolated from fish farms and is pathogenic for salmon species (Lunder *et al.*, 2000; Benediktsdottir *et al.*, 2000). Four psychropiezophilic species, *Moritella japonica*, *Moritella yayanosii*, *Moritella profunda* and *Moritella abyssi* were isolated from the Japan Trench, Mariana Trench, and the deep sea of the eastern tropical Atlantic Ocean (Nogi *et al.*, 1998a; Nogi and Kato, 1999; Xu *et al.*, 2003a).

The genus *Psychrobacter*

The genus *Psychrobacter* was first described by Juni and Heym (1986) as comprising psychrophilic or psychrotolerant and halotolerant, Gram-negative coccobacilli that are aerobic, non-motile, non-pigmented and oxidase positive. It was first described as a genus separate from *Acinetobacter* in 1986 (Juni, 2005; Juni and Heym, 1986). Members of the genus *Psychrobacter* are capable of growing at temperatures ranging from −10°C to 37°C and they vary from stenothermal psychrophiles, e.g. *Psychrobacter frigidicola* (0–22°C), to eurythermal psychrophiles, e.g. *Psychrobacter fulvigenes* (−5°C to 37°C). At the time of writing, the genus *Psychrobacter* comprised 34 recognized species. Fifteen species of this genus were isolated from a variety of low-temperature marine environments including Antarctic Sea ice, ornithogenic soil and sediments, the stomach contents of the Antarctic krill *Euphausia*, seawater (north-western Pacific Ocean, 300 m in depth), the deep sea, and the internal tissues of a marine ascidian (Bowman *et al.*, 1997c; Maruyama *et al.*, 2000; Romanenko *et al.*, 2002, 2009). However, the optimum growth temperature of most species is higher than 20°C, and only *Psychrobacter glacincola* isolated from Antarctic Sea ice has an optimum growth temperature lower than 15°C (Bowman *et al.*, 1997c).

The genus *Psychromonas*

The genus *Psychromonas* was originally proposed by Mountfort *et al.* (1988) with the description of the anaerobic bacteria *Psychromonas antarctica*. However, species of the genus *Psychromonas* are characterized as Gram-negative, curved or straight rod-shaped bacteria, motile or non-motile, chemoorganotrophic, psychrophilic, aerobic or facultatively anaerobic, and halophilic but some species do not require Na^+ ions for growth. The major isoprenoid quinone is Q-8, and the major fatty acids are 16:0 and 16:1ω7. Beyond these similarities, members of the genus *Psychromonas* described to date exhibit wide variation in their other physiological characteristics including cell size, temperature range for growth, piezophily, presence of gas vacuoles, and carbon source utilization (Bowman and McMeekin, 2005; Nogi *et al.*, 2007; Hosoya *et al.*, 2008; Miyazaki *et al.*, 2008; Auman *et al.*, 2006, 2010). The genus *Psychromonas* comprises 14 recognized species isolated from marine environments. Half of them are true psychrophiles. Three psychropiezophilic species, *P. kaikoae*, *Psychromonas hadalis* (Nogi *et al.*, 2002, 2007), and *Psychromonas profunda* (Xu *et al.*, 2003b) were isolated from deep-sea sediment. Three psychrophilic species, *Psychromonas ingrahamii*, *Psychromonas boydii* (Auman *et al.*, 2006, 2010), and *Psychromonas marina* (Kawasaki *et al.*, 2002) were isolated from the Antarctic, Arctic Sea ice, and Japanese coastal waters, respectively.

The genus *Shewanella*

The genus *Shewanella* was created from the reclassification of two species previously assigned to the genus *Alteromonas*, [*Alteromonas*] *putrefaciens* and [*Alteromonas*] *hanedai*. *Shewanella* species comprise Gram-negative, motile, straight or curved rod-shaped, aerobic or facultatively anaerobic, and halophilic but some species do not require NaCl for growth, oxidase-positive, non-fermentative bacteria (MacDonell and Colwell, 1985; Bowman, 2005). There are 56 species of *Shewanella* with validly published names, mostly isolated from marine environments, with a few isolated from freshwater sediment, industrial wastewater, a clinical specimen, and spoiled food. In the past decade, the number of recognized species in this genus has increased; they have been studied extensively because of their capacity for dissimilatory reduction of manganese and iron oxides (Myers and Nealson, 1988; Bowman *et al.*, 1997b; Venkateswaran *et al.*, 1998), reducing alternative electron acceptors such as trimethylamine *N*-oxide and nitrate (Bowman, 2005), and the production of tetrodotoxin (Simidu *et al.*, 1990). Some species isolated from cold deep-sea environments, such as *Shewanella hanedai, Shewanella gelidimarina, Shewanella violacea* and *S. benthica,* are known for the production of PUFAs (Kato and Nogi, 2001; Russell and Nichols, 1999; Bowman *et al.*, 1997b). The nine marine psychrophilic *Shewanella* species isolated from the deep sea are *Shewanella abyssi, Shewanella surugensis* (Miyazaki *et al.*, 2006), *S. benthica* (MacDonell and Colwell, 1985), *Shewanella donghaensis* (Yang *et al.*, 2007), *Shewanella psychrophila* (Xiao *et al.*, 2007), and *S. violacea* (Nogi *et al.*, 1998b). Those isolated from Antarctic Sea ice or coastal waters and sponges are *S. gelidimarina* (Bowman *et al.*, 1997b), and *Shewanella livingstonensis* (Bozal *et al.*, 2002). They all require Na^+ and produce PUFAs such as 16:1, 18:1 and/or EPA.

Alphaproteobacteria

The genus *Octadecabacter*

The genus *Octadecabacter* was originally proposed by Gosink *et al.* (1997). It was described as having octadecenoic acid (18:1) in excess of 70% of the total fatty acid content. *Octadecabacter arcticus* and *Octadecabacter antarcticus* are characterized as straight or pleomorphic rods, Gram-negative, non-motile, psychrophilic, moderately halophilic, produce gas vesicles, are catalase positive, oxidase negative, aerobic or microaerophilic, and utilize a strictly respiratory type of metabolism, with oxygen as the terminal electron acceptor. Phylogenetic analysis of the 16S rRNA showed that *Octadecabacter* is most closely related to the genus *Roseobacter* (Gosink *et al.*, 1997). However, this genus lacks bacteriochlorophyll *a*. The genus *Octadecabacter* comprises these two species isolated from the Antarctic Sea and Arctic Sea ice, respectively.

Deltaproteobacteria

Some marine psychrophilic species occur in the class *Deltaproteobacteria*. The order *Desulfuromonadales,* family *Desulfuromonadaceae* of the *Deltaproteobacteria* (Kuever *et al.*, 2005), includes the genera *Desulfuromonas* and *Desulfuromusa* (Pfennig and Biebl, 1976; Liesack and Finster, 1994). These genera are Gram-negative, rod-shaped, curved cells, strictly anaerobic chemolithoheterotrophs or chemoorganotrophs with a respiratory or fermentative type of metabolism. Marine strains require NaCl for growth. Sulfur or ferric iron can be used as an electron acceptor and is reduced to sulfide and ferrous iron. The genera *Desulfuromonas* and *Desulfuromusa* comprise seven and four species, respectively, isolated from anoxic environments. Two marine psychrophilic species, *Desulfuromonas svalbardensis* and *Desulfuromusa ferrireducens,* were isolated from Arctic sediments (Vandieken *et al.*, 2006).

The order *Desulfobacterales,* families *Desulfobacteraceae* and *Desulfobulbaceae,* which belong to the *Deltaproteobacteria* (Kuever *et al.*, 2005), include the genera *Desulfofrigus, Desulfofaba* and *Desulfotalea* (Knoblauch *et al.*, 1999). These genera are Gram-negative, strictly anaerobic chemoorganotrophs, chemolithoheterotrophs, or chemolithoautotrophs with a respiratory metabolism. Require NaCl for growth. Sulfate is used as an electron acceptor and reduced to sulfide. Fermentative growth on pyruvate or other carbon substrates is possible. Simple organic compounds such as long-chain fatty acids and alcohols serve as electron donors and carbon sources. The genera *Desulfofrigus, Desulfofaba,* and *Desulfotalea* comprise three, two, and two species, respectively, isolated from marine environments. Three marine psychrophilic species, *Desulfofrigus oceanense, Desulfofaba gelida,* and *Desulfotalea psychrophila,* were isolated from Arctic marine sediments (Knoblauch *et al.*, 1999).

Bacteroidetes

The genus *Polaribacter*

The genus *Polaribacter* was proposed by Gosink *et al.* (1998) and belongs to the family *Flavobacteriaceae* (Bernardet *et al.*, 2002; Bernardet and Nakagawa, 2006) in the phylum *Bacteroidetes.* It currently accommodates Gram-negative, aerobic, psychrophilic or mesophilic, rod-shaped bacteria that are usually non-motile, gas vesicles are produced by some species, and the DNA G + C content is in the range of 30–36 mol%. All strains were isolated from marine habitats and grow well in marine media or media that have been supplemented with NaCl. The genus *Polaribacter* comprises seven recognized species. There are four species of true marine psychrophiles in this genus, *Polaribacter filamentus* (type species), *Polaribacter franzmannii, Polaribacter glomeratus* and *Polaribacter irgensii* (Gosink *et al.*, 1998), which were isolated from polar marine environments and cannot be cultivated at temperatures above 22°C (Gosink *et al.*, 1998).

The genus *Psychroflexus*

The genus *Psychroflexus,* in the family *Flavobacteriaceae,* was first proposed by Bowman *et al.* (1998c) with the description of *Psychroflexus torquis* as well as the reclassification of *Flavobacterium gondwanense* (Dobson *et al.*, 1993) as *Psychroflexus gondwanensis.* The genus was defined as Gram negative. Cellular morphology ranges from rod-like cells to coiled filaments and filaments of indeterminate length, non-motile or motile via gliding, catalase and oxidase positive, strictly aerobic, heterotrophic, slightly or moderately halophilic, psychrophilic or psychrotrophic, with the major respiratory quinone MK-6 (Bowman *et al.*, 1998c). Members of the genus *Psychroflexus* have been isolated from a variety of salty or marine environments including Antarctic Sea ice and lakes, salt lakes, and marine solar salterns (Bowman *et al.*, 1998c; Donachie *et al.*, 2004; Yoon *et al.*, 2009). The genus *Psychrobacter* comprises five recognized species. However, most species grow at 4°C, and the optimum growth temperatures are higher than 20°C. Only *P. torquis,* isolated from Antarctic sea ice, is a true psychrophilic species.

References

Allen, E.E., Facciotti, D., and Bartlett, D.H. (1999). Monounsaturated but not polyunsaturated fatty acids are required for growth of the deep-sea bacterium *Photobacterium profundum* SS9 at high pressure and low temperature. Appl. Environ. Microbiol. *65*, 1710–1720.

Auman, A.J., Breezee, J.L., Gosink, J.J., Kampfer, P., and Staley, J.T. (2006). *Psychromonas ingrahamii* sp. nov., a novel gas vacuolate, psychrophilic bacterium isolated from Arctic polar sea ice. Int. J. Syst. Evol. Microbiol. *56*, 1001–1007.

Auman, A.J., Breezee, J.L., Gosink, J.J., Schumann, P., Barnes, C.R., Kampfer, P., and Staley, J.T. (2010). *Psychromonas boydii* sp. nov., a novel gas vacuolate, psychrophilic bacterium isolated from an Arctic Sea ice core from Point Barrow, Alaska. Int. J. Syst. Evol. Microbiol. *60*, 84–92.

Baik, K.S., Park, Y.D., Seong, C.N., Kim, E.M., Bae, K.S., and Chun, J. (2006). *Glaciecola nitratireducens* sp. nov., isolated from seawater. Int. J. Syst. Evol. Microbiol. *56*, 2185–2188.

Benediktsdottir, E., Verdonck, L., Sproer, C., Helgason, S., and Swings, J. (2000). Characterization of *Vibrio viscosus* and *Vibrio wodanis* isolated at different geographical locations: a proposal for reclassification of *Vibrio viscosus* as *Moritella viscosa* comb. nov. Int. J. Syst. Evol. Microbiol. *50*, 479–488.

Bernardet, J.F., and Nakagawa, Y. (2006). An introduction to the family Flavobacteriaceae. In The Prokaryotes: A Handbook on the Biology of Bacteria, 3rd edn., vol. 7, M. Dworkin, S. Falkow, E. Rosenberg, K.H. Schleifer, and E. Stackebrandt, eds. (New York, New York: Springer), pp. 455–480.

Bernardet, J.F., Nakagawa, Y., and Holmes, B. (2002). Proposed minimal standards for describing new taxa of the family Flavobacteriaceae and emended description of the family. Int. J. Syst. Evol. Microbiol. *52*, 1049–1007.

Bowman, J.P. (2005). Genus XIII. *Shewanella* MacDonell and Colwell 1986, 355^{VP}. In Bergey's Manual of Systematic Bacteriology, 2nd edn., vol. 2, part B, D.J. Brenner, N.R. Krieg, J.T. Staley, and G.M. Garrity, eds. (New York, New York: Springer), pp. 480–491.

Bowman, J.P., Gosink, J.J., McCammon, S.A., Lewis, T.E., Nichols, D.S., Nichols, P.D., Skerratt, J.H., Staley, J.T., and McMeekin, T.A. (1998a). *Colwellia demingiae* sp. nov., *Colwellia hornerae* sp. nov., *Colwellia rossensis* sp. nov. and *Colwellia psychrotropica* sp. nov.: psychrophilic Antarctic species with the ability to synthesize docosahexaenoic acid (22:6ω3). Int. J. Syst. Bacteriol. *48*, 1171–1180.

Bowman, J.P., McCammon, S.A., Brown, J.L., and McMeekin, T.A. (1998b). *Glaciecola punicea* gen. nov., sp. nov. and *Glaciecola pallidula* gen. nov., sp. nov.: psychrophilic bacteria from Antarctic sea-ice habitats. Int. J. Syst. Bacteriol. *48*, 1213–1222.

Bowman, J.P., McCammon, S.A., Brown, J.L., Nichols, P.D., and McMeekin, T.A. (1997a). *Psychroserpens burtonensis* gen. nov., sp. nov., and *Gelidibacter algens* gen. nov., sp. nov., psychrophilic bacteria isolated from Antarctic lacustrine and sea ice habitats. Int. J. Syst. Bacteriol. *47*, 670–677.

Bowman, J.P., McCammon, S.A., Lewis, T., Skerratt, J.H., Brown, J.L., Nichols, D.S., and McMeekin, T.A. (1998c). *Psychroflexus torquis* gen. nov., sp. nov., a psychrophilic species from Antarctic sea ice, and reclassification of *Flavobacterium gondwanense* (Dobson *et al.* 1993) as *Psychroflexus gondwanense* gen. nov., comb. nov. Microbiology *144*, 1601–1609.

Bowman, J.P., McCammon, S.A., Nichols, D.S., Skerratt, J.H., Rea, S.M., Nichols, P.D., and McMeekin, T.A. (1997b). *Shewanella gelidimarina* sp. nov. and *Shewanella frigidimarina* sp. nov., novel Antarctic species with the ability to produce eicosapentaenoic acid (20:5 omega 3) and grow anaerobically by dissimilatory Fe(III) reduction. Int. J. Syst. Bacteriol. *47*, 1040–1047.

Bowman, J.P., and McMeekin, T.A. (2005). Order X. *Alteromonadales* ord. nov. In Bergey's Manual of Systematic Bacteriology, 2nd edn., vol. 2, part B, D.J. Brenner, N.R. Krieg, J.T. Staley, and G.M. Garrity, eds. (New York, New York: Springer), p. 443.

Bowman, J.P., Nichols, D.S., and McMeekin, T.A. (1997c). *Psychrobacter glacincola* sp. nov, a halotolerant, psychrophilic bacterium isolated from Antarctic sea ice. Syst. Appl. Microbiol. *20*, 209–215.

Bozal, N., Montes, J., Tudela, E., Jimenez, F., and Guinea, J. (2002). *Shewanella frigidimarina* and *Shewanella livingstonensis* sp. nov. isolated from Antarctic coastal areas. Int. J. Syst. Evol. Microbiol. *52*, 195–205.

Chen, L.P., Xu, H.Y., Fu, S.Z., Fan, H.X., Liu, Y.H., Liu, S.J., and Liu, Z.P. (2009). *Glaciecola lipolytica* sp. nov., isolated from seawater near Tianjin city, China. Int. J. Syst. Evol. Microbiol. *59*, 73–76.

Choi, E.J., Kwon, H.C., Koh, H.Y., Kim, Y.S., and Yang, H.O. (2010). *Colwellia asteriadis* sp. nov., a marine bacterium isolated from the starfish *Asterias amurensis.* Int. J. Syst. Evol. Microbiol. *60*, 1952–1957.

DeLong, E.F., Franks, D.G., and Yayanos, A.A. (1997). Evolutionary relationship of cultivated psychrophilic and barophilic deep-sea bacteria. Appl. Environ. Microbiol. *63*, 2105–2108.

DeLong, E.F., and Yayanos, A.A. (1985). Adaptation of the membrane lipids of a deep-sea bacterium to changes in hydrostatic pressure. Science *228*, 1101–1103.

DeLong, E.F., and Yayanos, A.A. (1986). Biochemical function and ecological significance of novel bacterial lipids in deep-sea prokaryotes. Appl. Environ. Microbiol. *51*, 730–737.

Deming J.W., Hada, H., Colwell, R.R., Luehrsen, K.R., and Fox, G.E. (1984). The ribonucleotide sequence of 5s rRNA from two strains of deep-sea barophilic bacteria. J. Gen. Microbiol. *130*, 1911–1920.

Deming, J.W., Somers, L.K., Straube, W.L., Swartz, D.G., and MacDonell, M.T. (1988). Isolation of an obligately barophilic bacterium and description of a new genus, *Colwellia* gen. nov. Syst. Appl. Microbiol. *10*, 152–160.

Dobson, S.J., Colwell, R.R., McMeekin, T.A., and Franzmann, P.D. (1993). Direct sequencing of the polymerase chain reaction-amplified 16S rRNA gene of *Flavobacterium gondwanense* sp. nov. and *Flavobacterium salegens* sp. nov., two new species from a hypersaline Antarctic lake. Int. J. Syst. Bacteriol. *43*, 77–83.

Donachie, S.P., Bowman, J.P., and Alam, M. (2004). *Psychroflexus tropicus* sp. nov., an obligately halophilic *Cytophaga–Flavobacterium–Bacteroides* group bacterium from an Hawaiian hypersaline lake. Int. J. Syst. Evol. Microbiol. *54*, 935–940.

Eloe, E.A., Malfatti, F., Gutierrez, J., Hardy, K., Schmidt, W.E., Pogliano, K., Pogliano, J., Azam, F., and Bartlett, D.H. (2011). Isolation and characterization of a psychropiezophilic alphaproteobacterium. Appl. Environ. Microbiol. *77*, 8145–8153.

Gauthier, M.J., Lafay, B., Christen, R., Fernandez, L., Acquaviva, M., Bonin, P., and Bertrand, J.C. (1992). *Marinobacter hydrocarbonoclasticus* gen. nov., sp. nov., a new, extremely halotolerant, hydrocarbon-degrading marine bacterium. Int. J. Syst. Bacteriol. *42*, 568–576.

Gosink, J.J., Herwig, R.P., and Staley, J.T. (1997). *Octadecabacter arcticus* gen. nov., sp. nov., and *O. antarcticus*, sp. nov., nonpigmented, psychrophilic gas vacuolate bacteria from Polar sea ice and water. Syst. Appl. Microbiol. *20*, 356–365.

Gosink, J.J., Woese, C.R., and Staley, J.T. (1998). *Polaribacter* gen. nov., with three new species, *P. irgensii* sp. nov., *P. franzmannii* sp. nov. and *P. filamentus* sp. nov., gas vacuolate polar marine bacteria of the *Cytophaga–Flavobacterium–Bacteroides* group and reclassification of *'Flectobacillus glomeratus'* as *Polaribacter glomeratus* comb. nov. Int. J. Syst. Bacteriol. *48*, 223–235.

Hosoya, S., Yasumoto-Hirose, M., Adachi, K., Katsuta, A., and Kasai, H. (2008). *Psychromonas heitensis* sp. nov., a psychrotolerant bacterium isolated from seawater in Japan. Int. J. Syst. Evol. Microbiol. *58*, 2253–2257.

Ikemoto, E., and Kyo, M. (1993). Development of microbiological compact mud sampler. Jpn. Mar. Sci. Technol. Res. *30*, 1–16.

Irgens, R.L., Gosink, J.J., and Staley, J.T. (1996). *Polaromonas vacuolata* gen. nov., sp. nov., a psychrophilic, marine, gas vacuolated bacterium from Antarctica. Int. J. Syst. Bacteriol. *46*, 822–826.

Ivanova, E.P., Flavier, S., and Christen, R. (2004). Phylogenetic relationships among marine *Alteromonas*-like proteobacteria: emended description of the family *Alteromonadaceae* and proposal of *Pseudoalteromonadaceae* fam. nov., *Colwelliaceae* fam. nov., *Shewanellaceae* fam. nov., *Moritellaceae* fam. nov., *Ferrimonadaceae* fam. nov., *Idiomarinaceae* fam. nov. and *Psychromonadaceae* fam. nov. Int. J. Syst. Evol. Microbiol. *54*, 1773–1788.

Jung, S.Y., Oh, T.K., and Yoon, J.H. (2006). *Colwellia aestuarii* sp. nov., isolated from a tidal flat sediment in Korea. Int. J. Syst. Evol. Microbiol. *56*, 33–37.

Juni, E. (2005). Family II. Moraxellaceae, Genus III. Psychrobacter. In Bergey's Manual of Systematic Bacteriology, vol. 2, part B, G.M. Garrity, ed. (New York, New York: Springer), pp. 437–441.

Juni, E., and Heym, G.A. (1986). *Psychrobacter immobilis* gen. nov., sp. nov.: genospecies composed of gram-negative, aerobic, oxidase-positive coccobacilli. Int. J. Syst. Bacteriol. *36*, 388–391.

Kato, C., and Nogi, Y. (2001). Correlation between phylogenetic structure and function: examples from deep-sea *Shewanella*. FEMS Microbiol. Ecol. *35*, 223–230.

Kato, C., Li, L., Nakamura, Y., Nogi, Y., Tamaoka, J., and Horikoshi, K. (1998). Extremely barophilic bacteria isolated from the Mariana Trench, Challenger Deep, at a depth of 11,000 meters. Appl. Environ. Microbiol. *64*, 1510–1513.

Kato, C., Sato, T., and Horikoshi, K. (1995). Isolation and properties of barophilic and barotolerant bacteria from deep-sea mud samples. Biodiv. Conserv. *4*, 1–9.

Kawasaki, K., Nogi, Y., Hishinuma, M., Nodasaka, Y., Matsuyama, H., and Yumoto, I. (2002). *Psychromonas marina* sp. nov., a novel halophilic, facultatively psychrophilic bacterium isolated from the coast of the Okhotsk Sea. Int. J. Syst. Evol. Microbiol. *52*, 1455–1459.

Kim, H.J., Park, S., Lee, J.M., Park, S., Jung, W., Kang, J.S., Joo, H.M., Seo, K.-W., and Kang, S.-H. (2008). *Moritella dasanensis* sp. nov., a psychrophilic bacterium isolated from the Arctic ocean. Int. J. Syst. Evol. Microbiol. *58*, 817–820.

Knoblauch, C., Sahm, K., and Jorgensen, B.B. (1999). Psychrophilic sulfate-reducing bacteria isolated from permanently cold Arctic marine sediments: description of *Desulfofrigus oceanense* gen. nov., sp. nov., *Desulfofrigus fragile* sp. nov., *Desulfofaba gelida* gen. nov., sp. nov., *Desulfotalea psychrophila* gen. nov., sp. nov. and *Desulfotalea arctica* sp. nov. Int. J. Syst. Bacteriol. *49*, 1631–1643.

Kuever, J., Rainey, F.A., and Widdel, F. (2005). Class IV. Deltaproteobacteria class nov. In Bergey's Manual of Systematic Bacteriology, 2nd edn., vol. 2, part C, D.J. Brenner, N.R. Krieg, J.T. Staley, and G.M. Garrity, eds. (New York, New York: Springer), *p*. 922.

Liesack, W., and Finster, K. (1994). Phylogenetic analysis of five strains of gram-negative, obligately anaerobic, sulfur-reducing bacteria and description of *Desulfuromusa* gen. nov., including *Desulfuromusa kysingii* sp. nov., *Desulfuromusa bakii* sp. nov., and *Desulfuromusa succinoxidans* sp. nov. Int. J. Syst. Bacteriol. *44*, 753–758.

Lunder, T., Sørum, H., Holstad, G., Steigerwalt, A.G., Mowinckel, P., and Brenner, D.J. (2000). Phenotypic and genotypic characterization of *Vibrio viscosus* sp. nov. and *Vibrio wodanis* sp. nov. isolated from Atlantic salmon (*Salmo salar*) with 'winter ulcer'. Int. J. Syst. Evol. Microbiol. *50*, 427–450.

MacDonell, M.T., and Colwell, R.R. (1985). Phylogeny of the Vibrionaceae, and recommendation for two new genera, *Listonella* and *Shewanella*. Syst. Appl. Microbiol. *6*, 171–182.

Margesin, R., and Nogi, Y. (2004). Psychropiezophilic microorganisms. Cell. Mol. Biol. *50*, 429–436.

Maruyama, A., Honda, D., Yamamoto, H., Kitamura, K., and Higashihara, T. (2000). Phylogenetic analysis of psychrophilic bacteria isolated from the Japan Trench, including a description of the deep-sea species *Psychrobacter pacificensis* sp. nov. Int. J. Syst. Evol. Microbiol. *50*, 835–846.

Matsuyama, H., Hirabayashi, T., Kasahara, H., Minami, H., Hoshino, T., and Yumoto, I. (2006). *Glaciecola chathamensis* sp. nov., a novel marine polysaccharide-producing bacterium. Int. J. Syst. Evol. Microbiol. *56*, 2883–2886.

Miyazaki, M., Nogi, Y., Fujiwara, Y., and Horikoshi, K. (2008). *Psychromonas japonica* sp. nov., *Psychromonas aquimarina* sp. nov., *Psychromonas macrocephali* sp. nov. and *Psychromonas ossibalaenae* sp. nov.: psychrotrophic bacteria isolated from sediment adjacent to sperm whale carcasses off Kagoshima, Japan. Int. J. Syst. Evol. Microbiol. *58*, 1709–1714.

Miyazaki, M., Nogi, Y., Usami, R., and Horikoshi, K. (2006). *Shewanella surugensis* sp. nov., *Shewanella kaireitica* sp. nov. and *Shewanella abyssi* sp. nov., isolated from deep-sea sediments of Suruga Bay, Japan. Int. J. Syst. Evol. Microbiol. *56*, 1607–1613.

Mountfort, D.O., Rainey, F.A., Burghardt, J., Kaspar, H.F., and Stackebrandt, E. (1998). *Psychromonas antarcticus* gen. nov., sp. nov., a new aerotolerant anaerobic, halophilic psychrophile isolated from pond sediment of the McMurdo Ice Shelf, Antarctica. Arch. Microbiol. *169*, 231–238.

Myers, C.R., and Nealson, K.H. (1988). Bacterial manganese reduction and growth with manganese oxide as the sole electron acceptor. Science *240*, 1319–1321.

Nogi, Y., Hosoya, S., Kato, C., and Horikoshi, K. (2004). *Colwellia piezophila* sp. nov., a novel piezophilic species from deep-sea sediments of the Japan Trench. Int. J. Syst. Evol. Microbiol. *54*, 1627–1631.

Nogi, Y., Hosoya, S., Kato, C., and Horikoshi, K. (2007). *Psychromonas hadalis* sp. nov., a novel piezophilic bacterium isolated from the bottom of the Japan Trench. Int. J. Syst. Evol. Microbiol. *57*, 1360–1364.

Nogi, Y., and Kato, C. (1999). Taxonomic studies of extremely barophilic bacteria isolated from the Mariana Trench and description of *Moritella yayanosii* sp. nov., a new barophilic bacterial isolate. Extremophiles *3*, 71–77.

Nogi, Y., Kato, C., and Horikoshi, K. (1998a). *Moritella japonica* sp. nov., a novel barophilic bacterium isolated from a Japan Trench sediment. J. Gen. Appl. Microbiol. *44*, 289–295.

Nogi, Y., Kato, C., and Horikoshi, K. (1998b). Taxonomic studies of deep-sea barophilic *Shewanella* strains and description of *Shewanella violacea* sp. nov. Arch. Microbiol. *170*, 331–338.

Nogi, Y., Kato, C., and Horikoshi, K. (2002). *Psychromonas kaikoae* sp. nov., a novel piezophilic bacterium from the deepest cold-seep sediments in the Japan Trench. Int. J. Syst. Evol. Microbiol. *52*, 1527–1532.

Nogi, Y., Masui, N., and Kato, C. (1998c). *Photobacterium profundum* sp. nov., a new, moderately barophilic species isolated from a deep-sea sediment. Extremophiles *2*, 1–7.

Pfennig, N., and Biebl, H. (1976). *Desulfuromonas acetoxidans* gen. nov. and sp. nov., a new anaerobic sulfur-reducing, acetate oxidizing bacterium. Arch. Microbiol. *110*, 1–10.

Romanenko, L.A., Schumann, P., Rhode, M., Lysenko, A.M., Mikhailov, V.V., and Stackebrandt, E. (2002). *Psychrobacter submarinus* sp. nov. and *Psychrobacter marincola* sp. nov., psychrophilic halophiles isolated from marine environments. Int. J. Syst. Evol. Microbiol. *52*, 1291–1297.

Romanenko, L.A., Tanaka, N., Frolova, G.M., and Mikhailov, V.V. (2009). *Psychrobacter fulvigenes* sp. nov., isolated from a marine crustacean from the Sea of Japan. Int. J. Syst. Evol. Microbiol. *59*, 1480–1486.

Romanenko, L.A., Zhukova, N.V., Rohde, M., Lysenko, A.M., Mikhailov, V.V., and Stackebrandt, E. (2003). *Glaciecola mesophila* sp. nov., a novel marine agar-digesting bacterium. Int. J. Syst. Evol. Microbiol. *53*, 647–651.

Russell, N.J., and Nichols, D.S. (1999). Polyunsaturated fatty acids in marine bacteria – a dogma rewritten. Microbiology *145*, 767–779.

Seo, H.J., Bae, S.S., Lee, J.-H., and Kim, S.-J. (2005). *Photobacterium frigidiphilum* sp. nov., a psychrophilic, lipolytic bacterium isolated from deep-sea sediments of Edison Seamount. Int. J. Syst. Evol. Microbiol. *55*, 1661–1666.

Simidu, U., Kita-Tsukamoto, K., Yasumoto, T., and Yotsu, M. (1990). Taxonomy of four marine bacterial strains that produce tetrodotoxin. Int. J. Syst. Bacteriol. *40*, 331–336.

Urakawa, H., Kita-Tsukamoto, K., Steven, S.E., Ohwada, K., and Colwell, R.R. (1998). A proposal to transfer *Vibrio marinus* (Russell 1891) to a new genus *Moritella* gen. nov. as *Moritella marina* comb. nov. FEMS Microbiol. Lett. *165*, 373–378.

Van Trappen, S., Tan, T.L., Yang, J., Mergaert, J., and Swings, J. (2004). *Glaciecola polaris* sp. nov., a novel budding and prosthecate bacterium from the Arctic Ocean, and emended description of the genus *Glaciecola*. Int. J. Syst. Evol. Microbiol. *54*, 1765–1771.

Vandieken, V., Mussmann, M., Niemann, H., and Jorgensen, B.B. (2006). *Desulfuromonas svalbardensis* sp. nov. and *Desulfuromusa ferrireducens* sp. nov., psychrophilic, Fe(III)-reducing bacteria isolated from Arctic sediments, Svalbard. Int. J. Syst. Evol. Microbiol. *56*, 1133–1139.

Venkateswaran, K., Dollhopf, M.E., Aller, R., Stackebrandt, E., and Nealson, K.H. (1998). *Shewanella amazonensis* sp. nov., a novel metal-reducing facultative anaerobe from Amazonian shelf muds. Int. J. Syst. Bacteriol. *48*, 965–972.

Xiao, X., Wang, P., Zeng, X., Bartlett, D.H., and Wang, F. (2007). *Shewanella psychrophila* sp. nov. and *Shewanella piezotolerans* sp. nov., isolated from west Pacific deep-sea sediment. Int. J. Syst. Evol. Microbiol. *57*, 60–65.

Xu, Y., Nogi, Y., Kato, C., Liang, Z., Ruger, H.J., De Kegel, D., and Glansdorff, N. (2003a). *Moritella profunda* sp. nov. and *Moritella abyssi* sp. nov., two psychropiezophilic organisms isolated from deep Atlantic sediments. Int. J. Syst. Evol. Microbiol. *53*, 533–538.

Xu, Y., Nogi, Y., Kato, C., Liang, Z., Ruger, H.J., De Kegel, D., and Glansdorff, N. (2003b). *Psychromonas profunda* sp. nov., a psychropiezophilic bacterium from deep Atlantic sediments. Int. J. Syst. Evol. Microbiol. *53*, 527–532.

Yang, S.-H., Lee, J.-H., Ryu, J.-S., Kato, C., and Kim, S.-J. (2007). *Shewanella donghaensis* sp. nov., a psychrophilic, piezosensitive bacterium producing high levels of polyunsaturated fatty acid, isolated from deep-sea sediments. Int. J. Syst. Evol. Microbiol. *57*, 208–212.

Yayanos, A.A., Dietz, A.S., and Van Boxtel, R. (1979). Isolation of a deep-sea barophilic bacterium and some of its growth characteristics. Science *205*, 808–810.

Yoon, J.H., Kang, S.J., Jung, Y.T., and Oh, T.K. (2009). *Psychroflexus salinarum* sp. nov., isolated from a marine solar saltern. Int. J. Syst. Evol. Microbiol. *59*, 2404–2407.

Yu, Y., Li, H.R., and Zeng, Y.X. (2011). *Colwellia chukchiensis* sp. nov., a psychrotolerant bacterium isolated from the Arctic Ocean. Int. J. Syst. Evol. Microbiol. *61*, 850–853.

Yumoto I., Kawasaki, K., Iwata, H., Matsuyama, H., and Okuyama, H. (1998). Assignment of *Vibrio* sp. strain ABE-1 to *Colwellia maris* sp. nov., a new psychrophilic bacterium. Int. J. Syst. Bacteriol. *48*, 1357–1362.

Zhang, D.C., Li, H.R., Xin, Y.H., Chi, Z.M., Zhou, P.J., and Yu, Y. (2008a). *Marinobacter psychrophilus* sp. nov., a psychrophilic bacterium isolated from the Arctic. Int. J. Syst. Evol. Microbiol. *58*, 1463–1466.

Zhang, D.-C., Yu, Y., Chen, B., Wang, H.-X., Liu, H.-C., Dong, X.-Z., and Zhou, P.-J. (2006). *Glaciecola psychrophila* sp. nov., a novel psychrophilic bacterium isolated from the Arctic. Int. J. Syst. Evol. Microbiol. *56*, 2867–2869.

Zhang, D.C., Yu, Y., Xin, Y.H., Liu, H.C., Zhou, P.J., and Zhou, Y.G. (2008b). *Colwellia polaris* sp. nov., a psychrotolerant bacterium isolated from Arctic sea ice. Int. J. Syst. Evol. Microbiol. *58*, 1931–1934.

Zhang, Y.J., Zhang, X.Y., Mi, Z.H., Chen, C.X., Gao, Z.M., Chen, X.L., Yu, Y., Chen, B., and Zhang, Y.Z. (2011). *Glaciecola arctica* sp. nov., isolated from Arctic marine sediment. Int. J. Syst. Evol. Microbiol. *61*, 2338–2341.

Zobell, C.E., and Johnson, F.H. (1949). The influence of hydrostatic pressure on the growth and viability of terrestrial and marine bacteria. J. Bacteriol. *57*, 179–189.

Fungi in Cryosphere: Their Adaptations to Environments

4

Tamotsu Hoshino, Nan Xiao, Yuka Yajima and Oleg B. Tkachenko

Abstract

Cold-adapted fungi are widely distributed in the cryosphere where biosphere is constantly or seasonally covered with snow and/or ice. Fungi normally have different cells in their life cycle; therefore, thermal dependence of fungal life cycle stages is completely different from that of bacteria. We showed examples from fungi that the concept of 'psychrophile' by Moria (1975) was not applicable and propose a new term 'cryophilic fungi' for those that spend a certain life stage or whole life cycle (sexual and/or asexual reproductions) in cryosphere. Several groups of fungi associated with snow and/or ice, i.e. cryophilic fungi, are illustrated in terms of their ecology, and their ecophysiological adaptation mechanisms to freezing stress are reviewed, here.

Introduction: fungi and cryosphere

Fungi are eukaryotic organisms without plastids, nutrition absorptive, unicellular or filamentous and consisting of multicellular coenocytic hyphae (Kirk *et al.*, 2008). The true fungi form kingdom Fungi and the organisms studied by mycologist, fungi are mostly placed here, but others belong to the kingdoms Protozoa and Chromista (Cavalier-Smith, 1998; Kirk *et al.*, 2008). According to sexual reproduction, the true fungi were divided into four phyla (*Chytridiomycota, Zygomycota, Ascomycota* and *Basidiomycota*). Though phylogenetic analysis proposed fungi are currently composed of 7 phyla and uncertain four subphyla (Hibbet *et al.*, 2007), we will use the former four phyla in this review.

Approximately 80% of the biosphere are constantly and seasonally cold and have a temperature below 5°C, and polar regions occupies 14% of the biosphere (e.g. Gounot, 1999). The biosphere without deep sea are almost identical to the cryosphere. The term cryosphere was proposed by Dobrowolski (1923; Barry *et al.*, 2011) and collectively describes the portions of the earth's surface where water exists as the frozen state – snow cover, glaciers, ice sheets and shelves, freshwater ice, sea ice, icebergs, permafrost, and ground ice (Barry and Gan, 2011). Several kinds of microorganisms especially fungi were reported from the cryosphere (Friedmann, 1993; Robinson, 2001; Margesin *et al.*, 2007; Frisvad, 2008; Ludley and Robinson, 2008; Hoshino *et al.*, 2009; Ozerskaya *et al.*, 2009; Tojo and Newsham, 2012) and summarized in Table 4.1. Fungal species were less frequently recorded from the cryosphere than those of temperate zone despite that all fungal taxa have already been found in the cryosphere. These records suggest a possibility that various fungi are active under various cold environments. In this reviews, we introduced their niche and environmental adaptation strategies in the cryosphere.

Table 4.1 Fungal taxa and species numbers in cryosphere

	True fungi					Cromista	Protista	
Area	Chytridiomycetes	Zygomycetes	Ascomycetes	Basidiomycetes	Anamorphic fungi	Oomycetes	Myxomycetes	Total
World	ca. 900	ca. 1000	ca. 33,000	ca. 30,000	ca. 14,000	ca. 800	ca. 900	ca. 80,600[a]
Japan	140	24	7419	4160	941	361	454	13,739[b]
Norway	ND	68	2318	4046	1270	211	286	8199[c]
Arctic								
Russian Arctic	15	ND	733	633	289	26	8	1792[d]
Svalbard	3	28	867	200	112	2	3	1217[e]
Greenland	15	3	1534	837	200	ND	ND	ca. 2600[f]
Antarctica								
Antarctic Region	24	63	588	181	ND	7	14	877[g]
Continent	9	22	64	48	185	3	ND	331[h]
Alpine								
NW Himalaya	3	17	321	611	745	63	2	1852[i]

ND, not described. [a](Union of Japanese Societies for Systematic Biology, 2003); [b](Union of Japanese Societies for Systematic Biology, 2003); [c](Aarnæs, 2002); [d](Karatygin *et al.*, 1999); [e](Elvebakk and Prestrud, 1996); [f](Knudsen, 2006); [g]Bridge, 2010); [h](Onofri *et al.*, 2007); [i](Paul and Sharma, 2003).

History and terminology of fungi in the cryosphere

History

Fungi in the cryosphere were less studied than those in temperate zone; however, fungi and their activities were recorded before two centuries. The first observation was probably rotted winter barley under snow cover in Toyama, Japan, in 1788 (Hori, 1934), and Eriksson (1879) published a scientific report on winter injury of winter crops in Sweden. Rostrup (1888) was the first to publish an overview of the fungi of Greenland. In his review, he listed phytopathogenic fungi from *Flora Danica* (Hornemann, 1836). Fisher (1888) was probably the first to report that an unidentified fungus grew at 0°C. Fungi in Antarctica were surveyed since the beginning of the twentieth century (e.g. Bommer and Rousseau, 1900). Cooke (1944) reported mushrooms in melted snowbanks in alpine North America. Berry and Magoon (1934) and Morita (1966) summarized early studies of microorganisms, including fungi that grew at low temperatures.

Terminology

Schmidt-Nielsen (1902) reported the growth at 0°C of some bacteria and two yeasts (*Saccharomyces pastorianus* and red Torula) and coined 'psychrophile' (cold-loving organisms) that could grow at 0°C. Humphrey and Siggers (1933) divided wood rotting fungi into three groups, according to their optimal growth temperature in culture. That of 'low-temperature group' was 20–24°C, an intermediate-temperature group was 24–32°C, and a high temperature was above 32°C. Panassenko (1944, 1967) proposed distinguished three groups of microfungi based on mycelial growth temperature: 'psychrotolerant' that grew at subzero temperatures, mesotolerant at 5–30°C, and thermotolerant even at 40–50°C. Baxter and Gibbons (1962) proposed psychrophiles of yeast with an optimum growth temperature at 20°C or below. Stokes (1963) described the term of 'obligate psychrophile' including yeasts that could not grow at the usual room temperature from 20°C to 25°C. Deverall (1968) defined fungal psychrophiles as those with an optimum growth temperature near 10°C or below. He chose 10°C that was growth minima for most fungi.

Morita (1975) compared previous reports, and then proposed that psychrophiles were defined as bacteria having an optimal temperature for growth at about 15°C or lower and a maximum at about 20°C. He mentioned in his review that 'yeasts that meet the above definition are known to exist but they will not be discussed'. Morita's definition on psychrophile is widely received in modern microbiology, and the identical concept was adopted in fungi by Watson *et al.* (1978), Griffin (1981), Robinson (2001), and Maheshwari (2005). However, Vidal-Leiria *et al.* (1979) reported that maximal growth temperature of psychrophilic yeast group was below 24°C, and Margesin *et al.* (2007) used the concept of psychrophile of Schmidt-Nielsen (1902). These reports suggest that the psychrophilism of fungi is not accepted as a common concept.

Fungi growing at subzero temperatures are also termed as 'cold-adapted' fungi, 'cryophile' (cold, freezing-loving organism) or 'nivicolous species' (species living under the snow). Cold-adapted microorganisms included both psychrophile and 'psychrotolerants' that grew over a wide temperature range with the fastest growth rates above 20°C (Margesin, 2007). Ronikier and Ronikier (2009) defined the term 'nivicolous myxomycetes' as describing the distinct ecological groups of myxomycetes requiring for development of their sporangia specific microenvironmental conditions that can be found under long persisting snow cover, at margins of such snow patches and in place where the snow cover has just disappeared. These species probably have attractive nature, especially in terms of cold adaptation. However, almost nivicolous myxomycetes are unculturable, and few reports on their ecophysiological characteristics are available. Consequently, we do exclude this group in this review. Cryophiles have two kinds of meanings in modern mycology. One is synonymous with psychrophiles (Jennings and Lysek, 1999), and the other refers to fungi growing under snow (Eckblad, 1978).

Exceptions to thermal response in cold-adapted fungi

Biological life cycles of bacteria are basically simple, consisting of cell division in propagation. Therefore, measurements of bacterial growth temperature indicate their cardinal temperature range. On the other, fungi have various cell morphologies for sexual reproduction and survival stages in their life cycles. Mycelia and yeast cells represent the vegetative stages and are relatively easy to culture under artificial conditions. Consequently, growth temperature ranges of mycelia and yeast cells are not always equal to possible cardinal temperature ranges of fungi. We illustrated some inconsistence and contradiction with psychrophiles and psychrotolerants.

Snow moulds are fungal pathogens of overwintering plants. Their ecology will be discussed later. Mycelia of ascomycetous snow mould, *Sclerotinia borealis* develop under the snow cover, and the fungus forms sclerotia to survive from spring to autumn (e.g. Smith, 1986; Hsiang *et al.*, 1999; Hoshino *et al.*, 2009). Sclerotia germinate in autumn to produce cup-shaped ascocarps. Its optimum mycelial growth temperature is 4–10°C, with a maximum less than 20°C (Smith, 1986; Hsiang *et al.*, 1999; Hoshino *et al.*, 2009). This fungus is a rare psychrophile in the terrestrial ecosystem. However, *S. borealis* is not regard as psychrophilic, considering the conditions for sclerotial germination to produce ascocarps: optimal germination occurs when sclerotia were subjected to a daily thermal cycle at 20°C/5°C after incubation at 25°C/15°C for 4 weeks (Table 4.2; Istuki, 1984; Saito, 2001). The optimum mycelial growth temperature of oomycetous snow mould, *Pythium iwayamai*, ranged from 18°C to 20°C, and the maximum was 25–30°C (Hirane, 1960). However, germination of their oospores occurred exclusively below 15°C and never occurred over at 20°C or above (Takamatsu, 1989). *P. iwayamai* is psychrotolerant based on thermal dependence of mycelial growth, but this fungus is considered 'psychrophile' based on the biological importance of oospores. These facts suggest that fungal thermal sensitivity should be estimated at each

Table 4.2 Effect of incubation temperature on germination of sclerotinia of *Sclerotinia borealis* (revised from Itsuki, 1984)

Incubation condition*	Temperature (°C)	Germination rate (%)
Constant	9	0.0
Repetition	25/15**	53.3
	20/5**	16.7
	25/15 to 20/5***	76.7

All cultivation conditions were as as follows: Thirty sclerotina were used for each condition and were incubated for 66 days. *Light/dark cycle was 8/16h and light condition is 2000 lm. **Temperature A/B is light/dark. ***Sclerotia were first incubated at 25°C light/15°C dark for 28 days and then transferred to 20°C light/5°C dark for 38 days.

stage in their life cycle. Wolf and Wolf (1947) indicated that no single temperatures represented optimum for germination, mycelial development, and reproduction in all fungal species.

Most studies on the relationship between microorganisms and temperature focused on physiological characteristics for its own sake and did not consider biological interactions, despite that fungi in nature interact with various creatures and live in several environments. Mycelial growth rate of basidiomycetous snow mould, *Typhula ishikariensis* on potato dextrose agar (PDA) at 0°C was half that of its optimal growth at 10°C. However, when cultures were covered with unsterile soil to introduce natural microbial antagonists, mycelial growth was much less at 10°C than at 0°C (Matsumoto, 1989; Hsiang *et al.*, 1999). Although ascomycetous snow mould, *Sclerotinia nivalis* is psychrotolerant and its optimal mycelial growth temperature is 20°C, the fungus could infect santonica and carrot at 0°C or 3°C but not at 15°C (Saito, 2001). Sclerotia germinated rapidly on unsterilized soil at 0°C, producing abundant mycelia to invade carrot roots. In contrast, at 20°C, myceliogenic germination failed to occur, and carrot roots were not infected. Germination of sclerotia and mycelial growth were probably affected by the suppressive activities of soil microorganisms (Saito, 2001). This fungus may be considered as 'psychrophile' from its ecological characteristics. Wolf and Wolf (1949) also suggested that, conceivably, the temperature differential might be an important factor when two or more species were competing for substrata but that it might not necessarily constitute the critical controlling factor.

In combination with bacteriology, mycology derived from botany constitutes a major part of microbiology; however, fungi are completely different from bacteria in terms of ecological characteristics in the cryosphere, i.e. cold-adaptation and acclimation. The concept of psychrophile of Morita (1975) that described various bacteria from cold environment does not apply to fungi. Thus, we need to coin another term for fungi in the cryosphere.

Redefinition of cryophilic fungi

We propose the term 'cryophilic fungi' to denote the fungi adapted to the cryosphere. The concept of cryophilic fungi is defined as fungi that are present in the cryosphere, complete their life cycles (sexual and/or asexual reproductions), and grow under subzero temperature where water remains in the solid state such as snow and ice. The concept of cryophilic fungi also applies to uncultured fungi such as mycorrhizal fungi.

Diversity of cryophilic fungi

Snow fungi (fungi on the snow)

Cryoseston is defined as the community of organisms living on snow (Nedbalová, 2008). Snow algae are major organisms in cryoseston, and fungi that associate with snow algae are referred to as snow fungi. Kol (1968a) listed 82 species of snow fungi. Fungi incertae sedis *Chionaster nivalis, C. bicornis* and ascomycete *Selenotila nivalis* represent typical snow fungi that were distributed in the Arctic (Kol, 1974; Leya *et al.*, 2000), mountains in Europe (Kol, 1956, 1966, 1970; Light and Belcher, 2007; Nedbalová, 2008; Lukavský, 2009; Lukavský and Cepák, 2010; Cepák and Lukavský, 2012), Asia (Fukushima, 1963; Yamagishi, 1972), North America (Garric, 1965; Stein and Amundsen, 1967; Hoham *et al.*, 1993) and Australia (Marchant, 1982). *C. nivalis* was also isolated from samples in European lake

(Lukavský, 2009). Kol (1968a, 1974) listed that three fungi were pathogenic with snow algae, the chytridiomycetes *Chytridium chlamydococci* and *Rhizophydium sphaerocarpum* and the ascomycete *Oospora nivalis*. Other fungi, the oomycete cf. *Myzocytium* (Cepák and Lukavský, 2012), the chytridiomycetes *Rhizophydium* sp. (Stein and Amundsen, 1967) and unidentified species (Leya *et al.*, 2000), the zygomycete *Cochlonema verrucosum* (Cepák and Lukavský, 2012), the ascomycetes *Alternaria* sp. (Cepák and Lukavský, 2012), *Tricladium* sp. and *Varicosporium elodeae* (Yamagishi, 1972), the basidiomycete cf. *Rhodosporidium toruloides* (Cepák and Lukavský, 2012) also found on snow surface.

Chionaster nivalis and *S. nivalis* do not appear to enter into any type of symbiotic associations with snow algae, but probably receive their external carbon requirements from snow algae, bacteria or from coniferous leachates (Hoham, 1989). Geographical distribution of both fungi in North America is different; *C. nivalis* is found on residual snow in the Pacific Northwest, the northern Rocky Mountains, New York State, New Hampshire and Quebec, but absent from south-western USA (Hoham and Blinn, 1979), and *S. nivalis* is prevalent on residual snow throughout western North America (Hoham and Blinn, 1979; Kol, 1968b), but absent in eastern North America.

Fungi in cryoconite (fungi on glacier and icecap)

The cryoconite is a dark-coloured material covering glaciers and icecaps to greatly accelerate glacier-melting by reducing albedo. The cryoconite was revealed to be a stromatolite-like algal mat, a product of microbial activity on glacier (e.g. Takeuchi *et al.*, 2001). Various fungi were found in cryoconite in both polar regions (Gerdel and Drouet, 1960; Mueller *et al.*, 2001; Singh and Singh, 2011), and DNA clone of fungus was also obtained from Antarctica (Christner *et al.*, 2003).

Fungi from the cryoconite are as follows: zygomycete *Circinella* sp. (Gerdel and Drouet, 1960), ascomycetes *Altarnaria* sp., *Cephalothecium* sp., *Penicillium solitum* var. *crustosum* (syn. *P. terrestre*) (Gerdel and Drouet, 1960), *Hymenscyphus tetracladius* (syn. *Articulospora tetracladia*), *Phialophora alba* (Singh and Singh, 2011), basidiomycetes *Cryptococcus gilvescens*, *Mrakia* sp., *Rhodotorulla* sp. (Singh and Singh, 2011), unidentified fungus (Mueller *et al.*, 2001). Major fungal taxon in cryoconite was dark-coloured spore-forming ascomycetes. These fungal spores probably also influenced with glacier melting.

Snow moulds (pathogenic fungi under the snow)

Snow moulds attack dormant plants such as forage crops, winter cereals, and conifer seedlings at low temperatures under snow cover (e.g. Smith, 1986; Hsiang *et al.*, 1999; Hoshino *et al.*, 2009). Snow cover maintains darkness, humidity, and low temperature and protects plants from cold and desiccation (Matsumoto and Hoshino, 2008). 'Snow moulds' or 'snow mould fungi' is a generic name including diverse taxonomic groups of fungi.

Important pathogens of agricultural crops are oomycetes: *Pythium iwayamai*, *P. paddicum*, *P. okanoganense*, ascomycetes: *Microdochium nivale* (syn. *Fusarium nivale*), *Sclerotinia borealis* (syn. *Myriosclerotinia borealis*), *S. nivalis*, *S. trifoliorum*, basidiomycetes: *Coprinus psychromorbidus*, *Typhula incarnata*, *T. ishikariensis* (syn. *T. idahoensis*) and *T. trifolii*. Ascomycetes attacking conifer seedlings: *Phacidium infestans*, *P. abietis* and *Racodium therryanum*.

The snow moulds *S. borealis* and *T. ishikariensis* are widely distributed in the cryosphere in northern Hemisphere, the cool temperate zone to frigid zone and Arctic regions such as Alaska and the Yukon (Lebeau and Longston, 1958), Greenland (Hoshino *et al.*, 2006),

Finnmark (northern Norway: Årsvoll, 1975), Iceland (*S. borealis* not found: Kristinsson and Guðleifsson, 1976), Lapland (northern Finland and Sweden: Jamalainen, 1949, 1957; Ekstrand, 1955), Kola Peninsula in the European part of Russia (Petrov, 1983) and Svalbard (Hoshino *et al.*, 2003c). Their distribution pattern suggests that *S. borealis* and *T. ishikariensis* are highly adapted to the Arctic environment.

Other fungal species are recognized as snow moulds in both polar regions (Tojo and Newsham, 2012). In the Arctic, *Pythium* spp., morphologically similar to *P. ultimum* var. *ultimum* and other *Pythium* spp., were isolated from moribund tissues of polar mosses (Hoshino *et al.*, 1999, 2000, 2001b, 2006b). Though usually regard as a saprobe, an ascomycete *Trichoderma polysporum* caused asymptomatic infections on the moss *Sanionia uncinata* (Yamazaki *et al.*, 2011). Basidiomycetous *Microbotryum bistortarum*, the smut fungus, was shown to have deleterious effects on the growth and survival of Alpine Bistort *Polygonum vivparum* under snow cover (Tojo and Nishitani, 2005). In Antarctica, *Pythium* sp. from moribund mosses in Signy Island, South Orkney Islands, had a high degree of similarity of ITS regions of *P. iwayamai* (Bridge *et al.*, 2008). Similar symptoms were also found on mosses on King George Island, South Shetland Islands (Hoshino *et al.*, 2001a). Zygomycetous *Rhizopus* sp. was a pathogen on the moss *Bryum anatrcticum* at Cape Bird on Ross Island (Greenfield, 1983). Several ascomycetes *Thyronectria antarctica* var. *hyperantarctica, Coleroa turfosorum, Bryosphaeria megaspora, Epibryon chorisodontii* and an unidentified Plectomycete were recorded from ring infection and macroscopic section of mosses (Hawksworth, 1973; Longton, 1973; Fenton, 1983).

No sclerotium-forming snow moulds had been reported from Antarctica. However, *Sclerotinia antarctica* on Antarctic hair grass *Deschampsia antarctica* was found in Antarctic peninsula, and the fungus was morphological similar to *S. borealis* (Gamundi and Spinedi, 1987). We also found another fungal disease on the moss, *Polytrichum juniperium* on King George Island, Antarctica (Fig. 4.1). Sclerotia were not found on diseased plants; however, molecular data suggested that *Typhula* sp. was present there (T. Hoshino, B. Chen, and G. Laursen, unpublished results). These findings suggest that snow moulds and their relatives are widely distributed in both polar regions.

Snowbank fungi (mushrooms emerging after snowmelt)

Snowbanks linger until spring and summer in heavily timbered areas of the high Rocky Mountains. The snow bank fungi comprised ecologically diverse groups of fleshy fungi that belong to both ascomycetes and basidiomycetes. They are adapted to the unique microclimate provided by remnant snows in high elevation conifer forests (Cripps, 2009). This set of fungi was subsequently called the 'Snowbank flora' by Smith (1975). Moser (2004) described the snowbank fungi as a phenomenon unique to North American. However, Cha *et al.* (2010) found the snowbank fungi in Hokkaido, northern Japan.

The snowbank fungi are ubiquitous where certain conditions are met. They proliferate in regions of high elevation with short, cold summers where snowbanks remain until July. Sufficient elevation is necessary for a deep snowpack in mature forests suffused with downed logs and abundant litter as well as woody debris. Spring and summer nights must be cool enough to retain snowbanks, and days warm enough to provide melt water for the fungi which fruit as the soil warms and dries. The fungi can occur on steep slopes and level ground, but snowbanks persist longer on northern slopes and in the deep shade where fruiting is protracted. Fruiting can stretch into July and August at higher elevations (Cripps, 2009).

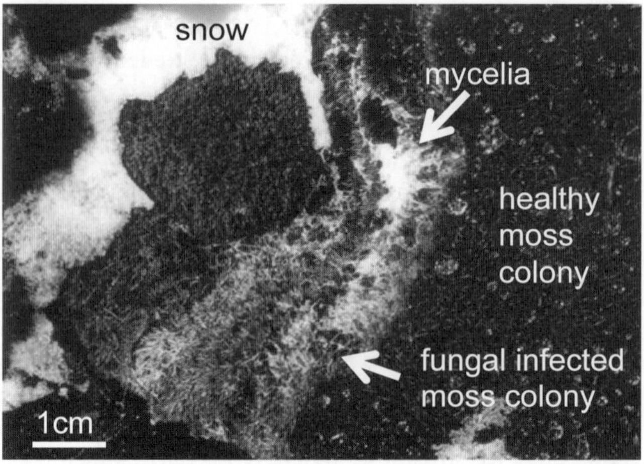

Figure 4.1 *Typhula* blight on Antarctic moss colony (T. Hoshino *et al.*, unpublished).

Following fungi are listed as the snowbank fungi; ascomycetes *Caloscypha fulgens, Discina ancillis* (syn. *D. perlata*), *Gelatinodiscus* spp. *Gyromitra montana, Sarcosoma latahense, Sarcosoma mexicanum*; basidiomycetes, gilled mushrooms *Clitocybe albirhiza, C. glacialis, Cortinarius ahsii, C. auchmerus, C. clandestinus, C. croceus, C. flavobasalis* nom. prov., *C. flavoroseus* nom. prov., *C. subalpinus* nom. prov., *Cortinarius* (*Phlegmacium*) spp., *Entoloma* sp. *Hygrophorus goetzii, H. marzuolus, H. subalpinus, Lentinellus montanus, Melanoleuca angelesiana, Mycena overholtsii, Neohygrophorus angelesianus, Nivatogastrium nubigenum, Strobilurus albipilatus, S. occidentalis*; non-gilled, *Pycnoporellus alboluteus, Tyromyces leucospongia*; jelly fungus, *Guepiniopsis alpine*. These fungi are mycorrhizal with conifers or wood decomposer without *C. fulgens* (seed pathogen on *Picea*) (Cripps, 2009).

Ecophysiological strategies of cryophilic fungi to adapt different environments

Temperature can influence the response of microorganisms either directly though the change in growth rate, enzyme activity, cell composition, and nutritional requirements, or indirectly, by affecting the solubility of solute molecules, ion transport and diffusion, osmotic effects on membranes, surface tension, and density (Herbert, 1986).

Cryophilic fungi and their relatives are taxonomically diverse and can grow at subzero temperatures and tolerate freeze–thaw event. Their strategies are different at the phylum level (Hoshino *et al.*, 2009).

General strategy

Ascomycetous yeast *Cyberlindnera jadinii* (syn. *Candida utilis*) increased cell size in culture incubated at low temperatures as compared to their optimal growth temperature (Rose, 1968; Beales, 2006). Rose (1968) reported that similar applied to various eukaryotic microorganisms.

Fungi in permafrost are characterized by both the presence of natural cryoprotectants in these ecotopes and the ability to utilize their inherent mechanisms of protection. For example, species belonging to the genera *Arthrinium, Aureobasidium, Botrytis, Fusarium, Geotrichum* and *Oidiodendron*, as well as many others, are usually isolated from plant material and/or present as phytopathogens. It is conceivable that plant substrates or derivatives thereof are natural cryoprotectants, which enables them to provide advantageous conditions to microorganisms under freezing conditions (Ozerskaya *et al.*, 2009). Stakhov *et al.* (2008) demonstrated that ancient seeds of higher plants constituted a specific habitat for microorganisms in frozen ground, which favoured their preservation for millennia. The presence of such natural protectants made it possible to preserve certain microbial species such as the genus *Phoma* specific to plants (Ozerskaya *et al.*, 2009).

Oomycetous snow moulds, *Pythium* spp. were reported to be less tolerant to chilling and freezing temperatures than other snow moulds taxa (Takamatsu, 1989; Hoshino *et al.*, 2002). However, *Pythium* species are often found pathogenic to mosses in polar regions (Hoshino *et al.*, 1999, 2000, 2001a, 2001b, 2006b; Tojo and Newsham, 2012). Tokura (2009) and Murakami (2012) investigated the freezing resistance of *Pythium* spp. from temperate and polar regions. Free mycelia and hyphal swellings, structures for survival, of *P. iwayamai* and *P. paddium* lost viability after 3 cycles of freezing-thawing; however, mycelia in host plants survived the treatment (Fig. 4.2; Tokura, 2009; Murakami, 2012). These results supported the above hypothesis of Ozerskaya *et al.* (2009).

Various fungi other than oomycetes have freeze tolerance, but mycelia growth of snow moulds was inhibited on frozen media with an exception of ascomycete *S. borealis* (for detail, see the next section, Hoshino *et al.*, 2009, 2010). Basidiomycetes produce extracellular antifreeze proteins (AFPs) which the surface of ice crystal to inhibit ice crystal growth (Hoshino *et al.*, 2009). Extracellular antifreeze activity is the mechanism unique to basidiomycetous snow moulds but not in other snow moulds belonging to oomycetes and ascomycetes (Snider *et al.*, 2000; Hoshino *et al.*, 2003a, 2003b, 2009). AFP of *T. ishikariensis* (TisAFP) exhibited a bursting ice growth normal to the c-axis of the ice crystal and high thermal hysteresis activity (approximately 2°C), as in the case of insect hyperactive antifreeze proteins (Xiao *et al.*, 2010b). Thus, diffusion of AFPs secreted in the extracellular environment does not support mycelial growth under subzero temperatures. *T. incarnata* and *T. ishikariensis* produce extracellular polysaccharides (Fig. 4.3; T. Hoshino *et al.*, 2009, T. Hoshino, C. Kuwabara, and R. Imai, unpublished), and they probably bind the AFP molecules they secreted. *M. nivale* does not produce AFPs but also produces extracellular polysaccharides such as cellulose (Schweiger-Hufnagel *et al.*, 2001) and fructan (Cairns *et al.*, 1995). These extracellular polysaccharides of *M. nivale* are considered to bind plant antifungal peptides and reduce activities of these peptides (Koike *et al.*, 2003; Hoshino *et al.*, 2009).

Duman and Olsen (1993) first reported antifreeze activities from fruit bodies of four kinds of mushrooms, i.e. *Trametes versicolor, Flammulina velutipes, Pleurotus ostreatus*, and *Stereum* sp. We purified and cloned AFPs from basidiomycetous snow moulds *Coprinus psychromorbidus* and *T. ishikariensis* (Hoshino *et al.*, 2003a,b). Similar kinds of AFP genes were found from other edible mushrooms (*Lentinula edodes* and *F. velutipes*) (Raymond and Janech, 2009) and Arctic basidiomycetous yeast *Leucosporidium* sp. AY30 strain (Lee *et al.*, 2010). Kawahara *et al.* (2006) reported that an ascomycete *Penicillum bilaiae* from Antarctic soil produced ionic and heat-stable antifreeze substance having polysaccharide. Antifreeze activities were detected in various taxa including oomycete, blastocladiomycete, ascomycetes and basidiomycetous yeasts from Antarctica (Xiao *et al.*, 2010a).

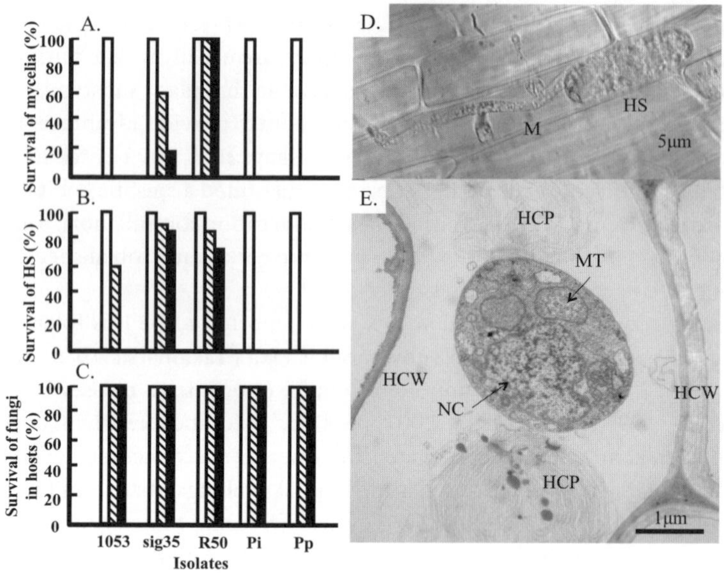

Figure 4.2 Freeze tolerance of cryophilic oomycetes *Pythium* spp. (revised from Tokura, 2009, and Murakami, 2012). (A) Mycelia. (B) Hyphal swellings. (C) Fungi in host plants. Open column: control (non-freezing), shaded column: one freeze–thaw treatment; closed column: two cycles of freeze–thaw. *Pythium* sp. OPU1053 from King George Is., sig35 from Signy Is., R50 from Adelaide Is., Pi: *P. iwayamai* and Pp: *P. padicum*. Mycelial discs (confirmed to be free of sexual and non-sexual organs) were frozen at −20°C in a programme freezer at a rate of 10°C/h in test tubes. They were the thawed at 2°C for 8h, transferred to fresh PDA plates, and incubated at 15°C. Hyphal swellings were prepared according to Kusunoki and Ichitani (1982). Fungi in the host: mycelial discs were inoculated to bentgrass, and incubated at 10°C for 1 month in the darkness. Ten infected bentgrass leaves were plated in sterilized test tubes and frozen at −20°C under the same condition as above. After freezing, infected moss shoots were plated on maize meal agar plates and incubated at 20°C for 3 days. (D) Light microscopy of hyphal swelling of *Pythium* sp. OPU1053 germinated in bentgrass leaf before freezing. HS, hyphal swelling, M, mycelia. (E) Electron microscopy of mycelium of *P. iwayamai* OPU1466 strain in infected bentgrass leaf after freezing. HCP, host chloroplast; HCW, host cell wall; MT, mitochondrion; NC, nucleus.

AFP from *Leucosporidium* sp. AY30 strain was crystrallized (Park *et al.*, 2011), and its crystal structure was analysed (Lee *et al.*, 2012). The fungal AFP had a common β-helical fold similar to that of canonical hyperactive antifreeze proteins, with a complex ice-binding site (Lee *et al.*, 2012).

AFPs were also cloned from an ice diatom (Janech *et al.*, 2005) and a bacterium in Antarctica (Raymond *et al.*, 2007), and these proteins showed high similarities to fungal AFPs. These findings suggest that fungal AFP homologues are widely distributed in several kingdoms, implying the possibility of horizontal gene transfer between eukaryotic microbes and prokaryotes (Raymond *et al.*, 2007; Sorhannus, 2011). Two major eukaryote lineages of AFP genes were proposed, i.e. from the diatom *Chaetoceros neogracile* to the copepod *Stephos longipes* and from a basidiomycete clade to a monophyletic group, consisting of the diatom species *Fragilariopsis* spp. Two other minor events included transfers from an ascomycete lineage to the *Proteobacterium Stigmatella aurantiaca* and from the *Proteobacterium*

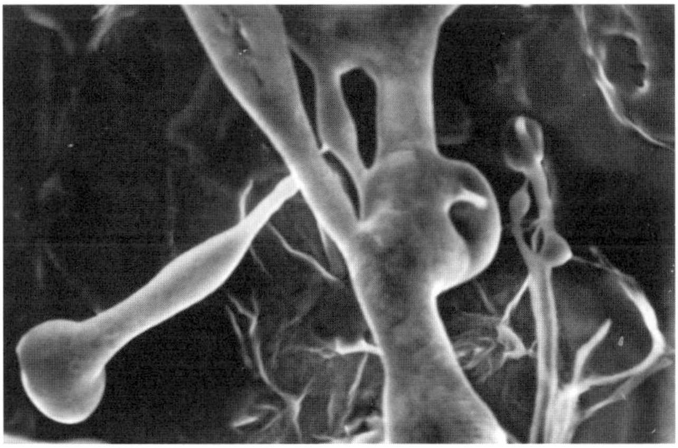

Figure 4.3 SEM image of extracellular cellulose covered mycelia of *Typhula ishikariensis* on a leaf of *Arabidopsis thaliana*. Arrow, normal mycelium; triangles, extracellular cellulose covered mycelia. Bar is 10μm.

Polaribacter irgensii to a group composed of four *Proteobacterium* species. The results imply complex evolutionary history of antifreeze protein genes (Sorhannus, 2011).

Specific strategy

The short growth period for fungi in polar regions favoured short life cycles (Kundsen, 2006). Lind (1934) noted that many rusts, smuts, and species of Dothideales have perennial mycelium in the host, enabling them to grow as soon as the season stars. Smut is a typical summer plant disease in low-altitude regions, and, interestingly, a smut fungus *Microbotryum bistortarum* was found adapted to Arctic climate and evolved to cause snow mould disease in Svalbard (Tojo and Nishitani, 2005). Many rusts have a simplified life cycle, producing only one kind of spores (micropuccinia) instead of a life cycle with three spore forms (eupuccinia) as commonly occurs in warm areas (Lind, 1934).

Sclerotinia borealis prevails where soil freezing is severe (Tomiyama, 1955; Röed, 1960; Nissinen, 1996). The fungus, however, does not produce extracellular AFPs (Snider *et al.*, 2000; Hoshino *et al.*, 2003a,b, 2009) and has another physiologic strategy to adapt to the frozen environment. Tomiyama (1955) cultured *S. borealis* and *T. incarnata* on both frozen and unfrozen PDA plates that were kept outside in Sapporo, Hokkaido, northern Japan; mycelial growth of *T. incarnata* was inhibited on frozen plates, but *S. borealis* grew faster on frozen plates than on unfrozen ones. His experiments were not carried out under controlled conditions and have not been reproduced by others. Under controlled conditions, we confirmed his results that *S. borealis* grew on frozen PDA (Hoshino *et al.*, 2009, 2010). *S. borealis* showed normal mycelial growth under the frozen condition, and mycelial growth rate on frozen plates at −1°C was faster than that on unfrozen PDA at the optimal growth temperature of 4–10°C. *S. borealis* can grow at low water potential on plates containing twice the concentration of medium ingredients (Tomiyama, 1955), sucrose, KCl (Bruehl and Cunfer, 1971), and D-mannitol (Namikawa *et al.*, 2004). An increase in intracellular osmosis enhanced mycelial growth and shifted the optimal mycelial growth temperature from 10–15°C to 4°C. However, mycelial growth of other snow moulds such as *S. nivalis* and

S. trifoliorum was inhibited at low water potential (Hoshino *et al.*, 2009, 2010). Therefore, osmophilism of *S. borealis* was probably a unique feature to adapt severe freezing.

AFP from Antarctic ascomycete *Antarctomyces psychrotrophicus* (AnpAFP) was glycosylated, and AnpAFP and Tis AFP were similar in that they both inhibit recrystallization (Xiao *et al.*, 2010b). Though AnpAFP was not as active as TisAFP, *A. psychrotrophicus* had higher freeze tolerance than *T. ishikariensis* and survived 15 freeze–thaw cycles (Xiao, 2012). Inhibition of recrystallization by AnpAFP may be more effective than that of TisAFP. Xiao (2012) cloned AnpAFP gene to find that it had low similarity (ca. 25% in amino acid sequence) to TisAFP gene. These two AFP genes from basidiomycete and ascomycete might have evolved independently from the same ancestor. Ascomycetes are largest fungal taxon in kingdom Fungi and probably have more variable mechanisms to adapt cold environment in the cryosphere than basidiomycetes.

Conclusion

We coined 'cryophilic fungi' to denote fungi well adapt to the cryosphere, and they are present on the snow and ice (snow fungi and cryoconite), under the snow (snow moulds, nivicolous species of myxomycetes) and around snowbanks (snowbank fungi). These fungi have general and specific ecophysiological strategies to adapt in cold environments. General strategy includes increase in cell size and natural cryoprotectants: the association of fungi within plant tissues remained in soil. Specific strategy is abbreviated life cycle and osmophilism. Antifreeze proteins evolved various molecules and are involved in both strategies.

Future trends

Cryophilic fungi developed cold adaptation mechanisms that differ at the phylum level. However, we do not know if this conclusion applies generally to all species in fungi and related kingdoms under snow: there must be exceptions. We should further investigate various fungi, especially chytridiomycetes, zygomycetes, myxomycetes and saprobes in other taxa adapted to cold environments.

Acknowledgements

We are very grateful to Dr Naoyuki Matsumoto, School of Agriculture, Hokkaido University (Sapporo, Japan) and Professor Dr Mototaki Tojo, Graduate School of Life and Environmental Sciences, Osaka Prefecture University (Sakai, Japan), for their constructive comments and warm encouragement. We thank Mr Yasunari Itsuki for permission to use the results of his Masters thesis. This research was financially supported in part by Grant-in Aid for Scientific Research (KAKENHI) (no. 23247012) from the Japanese Society for the Promotion of Science (JSPS).

Web resources

Bridge, P. (2010). List of non-lichenized fungi from the Antarctic region, version 2.3.3., http://www.antarctica.ac.uk/bas_research/data/access/fungi/index.html

CCFEE (Culture Collection of Fungi from Extreme Environments). http://www.sma.unitus.it/index.php/museo-nazionale-dellantartide-sezione-micologica.html

Index Fungorum. http://www.indexfungorum.org/Names/Names.asp

Mycobank (Fungal Databases Nomenclature and Species Banks Online Taxonomic Novelties Submission). http://www.mycobank.org/
Union of Japanese Societies for Systematic Biology (2003). Japanese Biota Species Number Survey, 1st Edition. http://research2.kahaku.go.jp/ujssb/
World Federation for Culture Collections. http://www.wfcc.info/

References

Aarnæs, J.O. (2002). Katalog over makro-og mikrosopp angitt for Norge og Svalbard (Synopsis fungorum 16) (Oslo, Norway: Fungiflora).
Årsvoll, K. (1975). Fungi causing winter damage on cultivated grasses in Norway. Meld. Norg. Landbrukshøgsk *54(9)*:1–49.
Barry, R., and Gan, T.Y. (2011). The Global Cryosphere Past, Present and Future (Cambridge, UK: Cambridge University Press).
Barry, R., Jania, J., and Birkenmajer, K. (2011). A.B. Dobrowolski – the first cryospheric scientist – and the subsequent development of cryospheric science. Hist. Geo. Space Sci. *2*, 75–79.
Baxter, R.M., and Gibbons, N.E. (1962). Observations on the physiology of psychrophilism in a yeast. Can. J. Microbiol. *8*, 511–517.
Beales, N. (2006). Adaptation of microorganisms to cold temperatures, weak acid preservatives, low pH, and osmotic stress: a review. Comp. Rev. Food Sci. Food. Safety *3*, 1–20.
Berry, J.A., and Magoon, C.A. (1934). Growth of microorganisms at below 0°C. Phytopathology *24*, 780–796.
Bommer, E., and Rousseau, M. (1900). Note preliminaire sur les champignons recueillis par l'Expedition Antarctique Belge. Bull. Cl. Sci. Acad. Roy. Belg. 640–646.
Bridge, P.D., Newsham, K.K., and Denton, G.J. (2008). Snow mould caused by a *Pythium* sp.: a potential vascular plant pathogen in the maritime Antarctic. Plant Pathol. *57*, 1066–1072.
Bruehl, G.W., and Cunfer, B.M. (1971). Physiologic and environmental factors that affect the severity of snow mold of wheat. Phytopathology *61*, 792–799.
Cairns, A.J., Howarth, C.J., and Pollock, C.J. (1995). Submerged batch culture of the psychrophile *Monographella nivalis* in a defined medium; growth, carbohydrate utilization and responses to temperature. New Phytol. *129*, 299–308.
Cavalier-Smith, T. (1998). A revised six-kingdom system of life. Biol. Rev. *73*, 203–266.
Cepák, V., and Lukavský, J. (2012). Cryoseston in the Sierra Nevada Mountains (Spain). Nova Hedw. *94*, 163–173.
Cha, J.Y., Lee, S.Y., Chun, K.W., Lee, S.Y., and Ohga, S. (2010). A new record of a snowbank fungus, *Mycena overholtsii*, from Japan. J. Fac. Agr. Kyushu Univ. *55*, 77–78.
Christner, B.C., Kvitko, B.H., and Reeve, J.N. (2003). Molecular identification of bacteria and eukarya inhabiting an Antarctic cryoconite hole. Extremophiles *7*, 177–183.
Cooke, W.B. (1944). Notes on the ecology of the fungi of Mount Shasta. Am. Midl. Nat. *31*, 237–249.
Cripps, C. (2009). Snowbank fungi revised. Fungi *2*, 47–53.
Deverall, B.J. (1968). Pscychrophiles. In The Fungi An Advanced Treatise. Vol. III The Fungal Population, G.C. Ainsworth, and A.S. Sussman eds. (New York, USA: Academic Press Inc.).
Dobrowolski, A.B. (1923). Historia naturalna lodu (Warsaw, Poland: Kasa Pomocy im. Dr J. Mianowskiego).
Duman, J.A., and Olsen, T.M. (1993). Thermal hysteresis protein activity in bacteria, fungi, and phylogenetically diverse plants. Cryobiology *30*, 322–328.
Eckblad, F.E. (1978). Soppøkologi (Oslo, Norway: Universitesforl).
Ekstrand, H. (1955). Höstsädens och vallgräsens överwintering. Statens Växtskyddsanstallt Meddelande *67*, 1–125.
Elvebakk, A., and Prestrud, P. (1996). A catalogue of Svalbard plants, fungi, algae and cyanobacteria (Skrifter nr. 198) (Oslo, Norway: Norsk Polarinstitutt).
Eriksson, J. (1879). En ny parasitsvamp å hvete, *Typhula graminum* Karst. Kongl. Landtbr.-Acad. Handl. o. Tidskr. *3*, 1–6.
Fenton, J.H.C. (1983). Concentric fungal rings in Antarctic moss communities. Tr. Br. Mycol. Soc. *80*, 413–420.
Fischer, B. (1888). Bakterienwachsthum bei 0°C. sowie über das Photographiren von Kulturen leuchtender Bakterien in ihren eigenen Lichte. Cent. Bakt. *4*, 89–92.
Friedmann, I. eds. (1993). Antarctic Microbiology (New York, USA: Wiley-Liss).

Frisvad, J.C. (2008). Fungi in cold ecosystems. In Psycrophiles: from Biodiversity to Biotechnology, R. Margesin, F. Schnner, J.-C. Marx, and C. Gerday eds. (Berlin, Germany: Springer), pp. 137–156.
Fukushima, H. (1963). Studies on Cryophyceae in Japan. J. Yokohama Municipal Univ. (C) *43*, 1–146.
Gamundi, I.J., and Spinedi, H.A. (1987). Sclerotinia antarctica sp. nov., The teleomorph of the first fungus described Antarctica. Mycotaxon *29*, 81–89.
Garric, R.K. (1965). The cryoflora of the Pacific Northwest. Am. J. Bot. *52*, 1–8.
Gerdel, R.W., and Drouet, F. (1960). The cryoconite of the Thule area, Greenland. Trans. Am. Microsc. Soc. *79*, 256–272.
Gounot, A.M. (1999). Microbial life in permanently cold soils. In Cold-adapted organisms: Ecology, physiology, enzymology and molecular biology, R. Margesin and F. Schinner, eds. (Berlin, Germany: Springer), pp.3–16.
Greenfield, L. (1983). Thermophilic fungi and actinomycetes from Mt. Erebus and fungus pathogenic to *Bryum antarcticum* at Cape Bird., NZ. Antarctic Record *4*, 10–11.
Griffin, D.H. (1981). Fungal Physiology (Wiley-Interscience Publication, New York: USA).
Hawksworth, D.L. (1973). *Thyronectria antarctica* (Speg.) Seeler var. *hyperantarctica* D. Hawksw. var. nov. Br. Antarc. Surv. Bull. *32*, 51–53.
Herbert, R.A. (1986). The ecology and physiology of psychrophilic microorganisms In: Microbes in Extreme Environments, R.A. Herbert, G.A. Codd, eds. (London, UK: Academic Press), pp. 1–24.
Hibbett, D.S., Binder, M., Bischoff, J.F., Blackwell, M., Cannon, P., Eriksson, O.E., Huhndorf, S., James, T., Kirk, P.M., Lücking, R., *et al.* (2007). A higher-level phylogenetic classification of the *Fungi*. Mycol. Res. *111*, 509–547.
Hirane, S. (1960). Studies on *Pythium* snow blight of wheat and barley, with special reference to the taxonomy of the pathogens (in Japanese with English abstract). Trans. Mycol. Soc. Jpn. *2*, 82–87.
Hoham, R.W. (1989). Snow microorganisms and their interaction with the environment. Proc. 57th Ann. Western Snow Conf. 31–35.
Hoham, R.W., and Blinn, D.W. (1979). Distribution of cryophilic algae in an arid region, the American Southwest. Phycologia *14*, 133–145.
Hoham, R.W., Laursen, A.E., Clive, S.O., and Duval, B. (1993). Snow algae and other microbes in several alpine areas in New England. Proc. Ann. 50th Eastern Snow Conf. 61st Western Snow Conf. 165–173.
Hornemann, J.W. (1836). Flora Danica fascicle 37 (Copenhagen, Denmark: Hauniae).
Hori, S. (1934). Old record of snow molds of winter wheat (in Japanese). J. Plant Protect. *21*, 165–166.
Hoshino, T., Tojo, M., Kanda, H., Okada, G., Ohgiya, S., and Ishizaki, K. (1999). A filamentous fungus, *Pythium ultimum* Trow var. *ultimum* isolated from moribund moss colonies from Svalbard, Northern islands of Norway. Polar Biosci. *12*, 68–75.
Hoshino, T., Tojo, M., and Tronsmo, A.M. (2000). *Pythium* blight of moss colonies (*Sanionia uncinata*) in Finnmark. Polarflokken *24*,161–164.
Hoshino, T., Tojo, M., Chen, B., and Kanda, H. (2001a). Ecological impact of phytopathogenic fungi in Antarctic terrestrial flora. Folia Fac. Sci. Nat. Univ. Masarykianae Brunensis Geogr. *25*, 95–102.
Hoshino, T., Tojo, M., Kanda, H., and Tronsmo, A.M. (2001b). Ecological role of fungal infections of moss carpet in Svalbard. Mem. Natl. Inst. Polar Res. Spec. Issue *54*, 507–513.
Hoshino, T., Tojo, M., Kanda, H., Herrero, M.L., Tronsmo, A.M., Kiriaki, M., Yokota, Y., and Yumoto, I. (2002). Chilling resistance of isolates of *Pythium ultimum* var. *ultimum* from the Arctic and Temperate Zones. Cryo-Lett. *23*, 151–156.
Hoshino, T., Kiriaki, M., and Nakajima, T. (2003a). Novel thermal hysteresis proteins from low temperature basidiomycete, *Coprinus psychromorbidus*. Cryo-Lett. *24*, 135–142.
Hoshino, T., Kiriaki, M., Ohgiya, S., Fujiwara, M., Kondo, H., Nihimiya, Y., Yumoto, I., and Tsuda, S. (2003b). Antifreeze proteins from snow molds. Can. J. Bot. *81*, 1175–1181.
Hoshino, T., Saito, I., and Tronsmo, A.M. (2003c). Two snow mold fungi from Svalbard. Lidia *6*, 30–32.
Hoshino, T., Saito, I., Yumoto, I., and Tronsmo, A.M. (2006a). New findings of snow mold fungi from Greenland. Medd. Grønland Biosci. *56*, 89–94.
Hoshino, T., Tojo, M., and Yumoto, I. (2006b). Blight of moss caused by *Pythium* sp. in Greenland. Medd. Grønland Biosci. *56*, 95–98.
Hoshino, T, Xiao, N., and Tkachenko, O.B. (2009). Cold adaptation in the phytopathogenic fungi causing snow molds. Mycoscience *50*, 26–38.
Hoshino, T., Terami, F., Tkachenko, O.B., Tojo, M., and Matsumoto, N. (2010). Mycelial growth of the snow mold fungus, *Sclerotinia borealis* improved at low water potentials: an adaptation to frozen environment. Mycoscience *51*, 98–102.

Hsiang, T., Matsumoto, N., and Millett, S. (1999). Biology and management of *Typhula* snow molds of turfgrass. Plant Dis. *83*, 788–798.

Humphrey, C.J., and Siggers, P.V. (1933). Temperature relations of wood-destroying fungi. J. Agric. Res. *47*, 997–1008.

Itsuki, Y. (1984). Studies on the snow mold disease of orchardgrass – with special reference to physiology and ecology of *Sclerotinia borealis*. Master thesis, Graduate School of Obihiro University of Agriculture and Veterinary Medicine.

Jamalainen, E.A. (1949). Overwintering of *Gramineae* plants and parasitic fungi. I. *Sclerotinia borealis* Bubák & Vleugel. J. Sci. Agric. Soc. Finl. *21*, 125–142.

Jamalainen, E.A. (1957). Overwintering of *Gramineae* plants and parasitic fungi. II. On the *Typhula* sp. fungi in Finland. J. Sci. Agric. Soc. Finl. *29*, 75–81.

Janech, M.G., Krell, A., Mock, T., Kang, J.-S., and Raymond, J.A. (2005). Icebinding proteins from sea ice diatoms (Bacillariophyceae). J. Phycol. *42*, 410–416.

Jennings, D.H., and Lysek, G. (1999). Fungal Biology: Understanding the Fungal Lifestyle, 2nd edn (Oxford, UK: Bios Scientific Publishers).

Karatygin, I.V., Nezdominogo, E.L., Novozhilov, Yu.K., and Zhurbenko, M.P. (1999). Russian Arctic fungi Check-list (in Russian) (St. Petersburg, Russia: V.L. Komarov Botanical Institute, RAS).

Kawahara, H., Takemura, T., and Obata, H. (2006). Function analysis and screening antifreeze material from fungi (in Japanse, English abstract). Cryobiol. Cryotechnol. *52*, 151–151.

Kirk, P.M., Cannon, P.F., Minter, D.W., and Stalpers, J.A., eds. (2008). Ainsworth & Bisby's Dictionary of Fungi, 10th edn (Wallingford, UK: CAB International).

Koike, M., Isobe, A., and Imai, R. (2003). A cotton jacket for winter survival: cellulose production of the snow mold fungus *Microdochium nivale*, is induced by the wheat defensing, TAD1, which accumulates during winter hardening. Functional Genomics and Breeding Strategies for Cold Tolerance in Plants (Sapporo, Japan: National Agricultural Research Center for Hokkaido Region), p. 46.

Kundsen, H. (2006). Mycology in Greenland – an introduction. Medd. Grønland Biosci. *56*, 7–16.

Kol, E. (1956). Sur la microvégetation de la neige des montagnes de la Bulgarie. I. Bull. Inst. Bot. Acad. Sci. Bulgarie *5*, 377–384.

Kol, E. (1966). Snow algae from the vally of the Morskie Oko lake in the high Tatra. Ann. Hist-Nat. Mus. Nat. Hung. *58*, 161–168.

Kol, E. (1968a). Kryobiologie. Biologie und Limnologie des Schnees und Elises. I. Kryovegetation. (Stuttgart, Germany: E. Schweizerbart'sche Verlagebuchhandlung).

Kol, E. (1968b). Cryobiological research in the Rocky Mountains. Arch. Hydrobiol. *60*, 278–285.

Kol, E. (1970). Vom roten Schnee der Tiroler Alpen. Ann. Hist-Nat. Mus. Nat. Hung. *62*, 130–136.

Kol, E. (1974). Trachiscia (Chlorophyta) red snow from Swedish Lapland. Ann. Hist-Nat. Mus. Nat. Hung. *66*, 59–63.

Kristinsson, H., and Guðleifsson, B.E. (1976). The activity of low temperature fungi under the snow cover in Iceland. Acta Bot. Isl. *4*, 44–57.

Kusunoki, M., and Ichitani, T. (1982). Preparation of mycrlium-free oospores of *Pythium bulteri* by a freezing method. Ann. Phytopath. Soc. Jpn. *48*, 695–698.

Lebeau, J.B., and Longston, C.E. (1958). Snow mold of forage crops in Alaska and Yukon. Phytopathology *48*, 148–150.

Lee, J.K., Park, K.S., Park, S., Park, H., Song, Y.H., Kang, S.-H., and Kim, H.J. (2010). An extracellular ice-binding glycoprotein from an Arctic psychrophilic yeast. Cryobiology *50*, 222–228.

Lee, J.H., Park, A.K., Do, H., Park, K.S., Moh, S.H., Chi, Y.M., and Kim, H.J. (2012). Structural basis for the antifreeze activity of an ice-binding protein from an Arctic yeast. J. Biol. Chem. doi: 10.1074/jbc.M111.331835.

Leya, T., Müller, T., Ling, H.U., and Fuhr, G. (2000). Taxonomy and biophysical properties of cryophilic microalgae and their environmental factors in Northwest Spitsbergen, Svalbard. Proc. 57th Ann. Eastern Snow Conf.

Levitt, J. (1980). Response of plants to environmental stress, vol 1.(New York, USA: Academic Press).

Light, J.J., and Belcher, J.H. (2007). A snow microflora in the Cairgorm Mountains, Scotland. Br. Phycol. Bull. 3, 471–473.

Lind, J. (1934). Studies on the geographical distribution of Arctic circumpolar Micromycetes. K. Dan. Vidersk. Selsk. Biol. Medd. *11*, 1–152.

Longton, R.E. (1973). The occurrence of radial infection patterns in colonies of polar bryophytes. Br. Antarc. Surv. Bull. *32*, 41–49.

Ludley, K.E., and Robinson, C.H. (2008). 'Decomposer' Basidiomycota in Arctic and Antarctic ecosystem. Soil Biol. Biochem. *40*, 11–1–29.

Lukavský, J. (2009). Algae, cyanobacteria and chytridiales of Černé lake in the Bohemian forest (Šumava, Czech Republic). Silva Gabreta *15*, 1–48.

Lukavský, J., and Cepák, V. (2010). Cryoseston in Stara Planina (Balkan) Mountains, Bulgaria. Acta Bot. Croat. *69*, 163–171.

Lukavský, J., Furnadzhieva, S., and Nedbalová, L. (2009). First record of cryoseston in the Vitosha Mountains (Bulgaria). Nova Hedw. *88*, 97–109.

Magan, N. (2007). Fungi in Extreme Environment. In Environmental and Microbial Relationships, The Mycota IV. C.P. Kubicek, and I.S. Druzhinina, eds. (Berlin, Germany: Springer-Verlag), pp. 85–103.

Mahewari, R. (2005). Fungal biology in the 21st century. Curr. Sci. *88*, 1406–1418.

Marchant, H.J. (1982). Snow algae from the Australian snowy mountains. Phycologia *21*, 178–184.

Margesin, R., Neuner, G., and Storey, K.B. (2007). Cold-loving microbes, plants, and animals-fundamental and applied aspects. Naturwissenschaften *94*, 77–99.

Matsumoto, N. (1989). Autecology of the pathogenic species of *Typhula*. Res. Bull. Hokkaido Natl. Agric. Exp. Stn. *152*, 91–162.

Matsumoto, N., and Hoshino, T. (2008). Fungi in snow environments: psychrophilic moulds. A group of pathogens affecting plants under snow. In: Fungi from different environments. vol 1. Progress in mycological research. J.K. Misra, and S.K. Deshmukh, eds. (Enfield, USA: Science Publisher), pp. 167–186.

Morita, R.Y. (1966). Marine psychrophilic bacteria. Oceanogr. Mar. Biol. Annu. Rev. *4*, 105–121.

Morita, R.Y. (1975). Psychrophilic bacteria. Bacteriol. Rev. *39*, 144–167.

Moser, M. (2004). Subalpine conifer forests in the Alps, the Altai, and the Rocky Mountains: a comparison of their fungal populations. In: Fungi in Forest Ecosystems: Systematics, Diversity and Ecology. C. Cripps ed. (Bronx, USA: New York Botanical Garden Press), pp. 151–158.

Mueller, D.R., Vincent, W.F., Pollard, W.H., and Fritsen, C.H. (2001). Glacial cryoconite ecosystems: a bipolar comparison of algal communities and habitats. Nova Hedw. *123*, 173–193.

Murakami, R. (2012). Studies on freezing resistance of *Pythium* species from Polar Regions (in Japanese). Master thesis, Graduate School of Life Sciencie, Hokkaido University.

Namikawa, Y., Watanabe, T., Saito, I., and Takasawa, T. (2004). Growth of the psychrophilic snow mold *Sclerotinia borealis* on the agar under xerophilic conditions (in Japanese with English abstract). Res. Bull. Obihiro Univ. *25*, 23–26.

Nedbalová, L., Kociánová, M., and Lukavský, J. (2008). Ecology of snow algae in the Giant Mts. Opera Corcontica *45*, 59–68.

Nissinen, O. (1996). Analyses of climatic factors affecting snow mould injury in first-year timothy (*Phleum pratense* L.) with special reference to *Sclerotinia borealis*. Acta Univ. Oul. A *289*, 1–115.

Onofri, S., Zucconi, L., and Tosi, S. (2007). Continental Antarctic Fungi (München, Germany: IHW-Verlag).

Ozerskaya, S., Kochkina,., Ivanushkina, N., and Gilichinsky, D.A. (2009). Fungi in Permafrost. In: Permafrost Soils, Soil Biology 16, R. Margesin, ed. (Berlin, Germany: Springer-Verlag), pp. 85–95.

Panasenko, V.T. (1944). The ecology of molds. Mikrobiologia *13*, 158–170.

Panasenko, V.T. (1967). Ecology of microfungi. Bot. Rev., *33*, 189–215.

Paul, Y.S., and Sharma, R.C. (2003). Mycoflora of Northwest Himalayas (Himachal Pradesh) (Dehra Dun, India: International Book Distributors).

Park, A.K., Park, K.S., Kim, H.J., Prak, H., Ahn, I.Y., Chi, Y.M., and Moon, J.H. (2011). Crystallization and preliminary X-ray crystallographic studies of the ice-binding protein from the Antarctic yeast *Leucosporidium* sp. AY30. Acta Cryst. *F67*, 800–802.

Petrov, V.F. (1983). Pathogenic microflora of root growth in perennial grasses in Khibiny (in Russian, with English abstract). Bull. Appl. Bot. Genet. Plant Breed. *82*, 38–45.

Raymond, J.A., and Janech, M.G. (2009). Ice-binding proteins from enoki and shiitake mushrooms. Cryobiology *58*, 151–156.

Raymond, J.A., Fritsen, C., and Shen, K. (2007). An ice-binding protein from an Antarctic sea ice bacterium. FEMS Microbiol. Ecol. *61*, 214–221.

Robinson, C. (2001). Cold adaptation in Arctic and Antarctic fungi. New Phytol. *151*, 341–353.

Röed, H. (1960). *Sclerotinia borealis* Bub. & Vleg., a cause of winter injuries to winter cereals and grasses in Norway. Acta Agric. Scand. *10*, 74–82.

Ronikier, A., and Ronikier, M. (2009). How 'alpine' are nivicolous myxomycetes? A worldwide assessment of altitudinal distribution. Mycologia *101*, 1–16.

Rose, A.H. (1968). Physiology of microorganisms at low temperatures. J. Appl. Bacteriol. *31*, 182–184.
Rostrup, E. (1888). Fungi Groenlandiae. Medd. om Grøonland *3*, 517–590.
Saito, I. (2001). Snow mold fungi in Sclerotiniaceae. In Low Temperature Plant Microbe Interaction Under Snow. N. Iriki, D.A. Gaudet, A.M. Tronsmo, N. Matsumoto, M. Yoshida, A. Nishimune eds. (Sapporo, Japan: Hokkaido National Agricultural Experimental Station), pp. 37–48.
Schmidt-Nielsen, S. (1902). Ueber einige psychrophile Microrganismen und ihr Vorkommen. Cent. Bakt., II, Abt. *9*, 145–147.
Schweiger-Hufnagel, U., Ono, T., Izumi, K., Hufnagel, P., Morita, N., Kaga, H., Morita, M., Hoshino, T., Yumoto, I., Matsumoto, N., Yoshida. M., Sawada, M.T., and Okuyama, H. (2000). Identifi cation of the extracellular polysaccharide produced by snow mold fungus *Microdochium nivalei*. Biotechnol. Lett. *22*, 183–187.
Singh, P., and Singh, S.M. (2011). Characterization of yeast and filamentous fungi isolated from cryoconite holes of Svalbard, Arctic. Polar Biol. doi: 10.1007/s00300–011–1103–1.
Smith, A.H. (1975). A Field Guide to Western Mushrooms. (Ann Arbor, USA: The University of Michigan Press).
Smith, J.D. (1986). Winter-hardiness and overwintering diseases of amenity turfgrass with special reference to the Canadian Prairies (Saskatoon, Canada: Research Branch Agriculture Canada).
Snider, C.S., Hsiang, T., Zhao, G.Y., and Griffith, M. (2000). Role of ice nucleation and antifreeze activities in pathogenesis and growth of snow molds. Phytopathology *90*, 354–361.
Sorhannus, U. (2011). Evolution of antifreeze protein genes in the diatom genus *Fragilaripsis*: Evidence for horizontal gene transfer, gene duplication and episodic diversity selection. Evol. Bioinformatics *7*, 279–289.
Stakhov, V.L., Gubin, S.V., Maksimovich, S.V., Rebrikov, D.V., Savilova, A.M., Kochkina, G.A., Ozerskaya, S.M., Ivanushkina, N.E., and Vorobyva, E.A. (2008). Microbial communities of ancient seeds derived from permanently frozen Pleistocene deposits. Mikrobiologia *77*, 348–355.
Stein, J.R., and Amundsen, C.C. (1967). Studies on snow algae and fungi from the front range of Colorado. Can. J. Bot. *45*, 2033–2045.
Stokes, J.L. (1963). General biology and nomenclature of psychrophilic micro-organisms. In Recent Progress in Microbiology. N.E. Gibbons ed. (Toronto, Canada: University of Toronto Press), pp. 187–192.
Takamatsu, S. (1989). Snow molds in winter wheat: studies on occurrence of *Pythium* snow rot (in Japanese with English summary). Spec. Bull. Fukui Agric. Exp. Stn. *9*, 1–135.
Takeuchi, N., Koshima, S., and Seko, K. (2001). Structure, formation, and darkening process of albedo-reducing material (cryconite) on a Himalayan glacier: A granular algal mat growing on the glacier. Arctic Anatrc. Alp. Res. *33*, 1150–122.
Tomiyama, K. (1955). Studies of the snow blight disease of winter cereals (in Japanese, English abstract). Hokkaido Natl. Agric. Exp. Stn. Rep. *47*, 1–234.
Tojo, M., and Newsham, K.K. (2012). Snow moulds in polar environments. Fungal Ecol., doi:10.1016/j.funeco.2012.01.003.
Tojo, M., and Nishitani, S. (2005). The effects of the smut fungus *Microbotryum bistotarum* on survival and growth of *Polygonum vivparum* in Svalbard. Can. J. Bot., *83*, 1513–1517.
Tokura, K. (2009). Species composition and ecology of *Pythium* spp. in a moss colony in the high Arctic region (in Japanese). Master thesis, Graduate School of Life and Environmental Sciences, Osaka Prefecture University.
Vidal-Leiria, M., Buckey, H., and van Uden, N. (1979). Distribution of the maximum temperature for growth among yeasts. Mycologia, *71*, 493–501.
Watson, K., Arthur, H., and Morton, H. (1978). Thermal adaptation in yeast: obligated psychrophiles are obligate aerobes, and obligate thermophiles are facultative anaerobes. J. Bacteriol., *136*, 815–187.
Wolf, F.A., and Wolf, F.T. (1947). The Fungi Volume II (New York: John Wiley & Sons, Inc.).
Xiao, N. (2012). Study on antifreeze proteins from cold adapted fungi and diatoms. PhD. thesis, Graduate School of Life Science, Hokkaido University.
Xaio, N., Inaba, S., Tojo, M., Degawa, Y., Fujiu, S., Hanada, Y., Kudoh, S., and Hoshino, T. (2010a). Antifreeze activities of various fungi and Stramenopila isolates from Antarctica. North Am. Fungi *5*, 215–220.
Xiao, N., Suzuki, K., Nishimiya, Y., Kondo, H., Miura, A., Tsuda, S., and Hoshino, T. (2010b). Comparison of functional properties of two fungal antifreeze proteins from *Antarctomyces psychrotrophicus* and *Typhula ishikariensis*. FEBS Lett. *277*, 394–403.
Yamagishi, T. (1972). A check list of the freshwater algae collected from the bogs and ponds of Mt. Poroshiri. Mem. Nat. Sci. Mus. *5*, 163–174.

Yamazaki, Y., Tojo, M., Hoshino, T., Kida, K., Sakamoto, T., Ihara, H., Yumoto, I., Tronsmo, A.M., and Kanda, H. (2011). Characterization of *Trichoderma polysporum* from Spitsbergen, Svalbard archipelago, Norway, on species identity, infectivity to moss, and polygalacturonase activity. Fungal Ecol. *4*, 15–21.

Part II

Physiology and Molecular Adaptation Mechanisms

5 Energy Metabolism at Low-temperature and Frozen Conditions in Cold-adapted Microorganisms

Pierre Amato

Abstract

Managing biochemical energy is a challenge that microorganisms have to face for maintaining activity at low temperatures, at which metabolic processes are altered by the decrease in reaction rates and by the rigidification of cellular structures, notably enzymes and membranes. Cold-adapted prokaryotes and eukaryotes, while exhibiting distinct trophic modes, share specificities for compensating these thermodynamic effects and keep the cellular machinery running: they elevate the concentration of adenylate compounds, the key molecules of the energy metabolism. This is achieved by tight adjustments acting at several levels of the metabolism in a global strategy of energy saving: strong orientation of the adenylate metabolism towards the production and regeneration of AMP and its phosphorylation, whereas its destruction is repressed; elevation of the respiration rate; intervention of specific enzymes allowing the rapid synthesis of ATP (polyphosphatases and interferases), shifts in the utilization of substrates; and rerouting of central metabolic pathways. These are presented here and illustrated by examples of metabolic regulations in cold-adapted microorganisms evidenced by recent transcriptomic and proteomic approaches.

Introduction

Microorganisms (i.e. unicellular organisms here) are at thermal equilibrium with their environment. Nevertheless, some can sustain metabolic activity at temperatures as low as −39°C (Panikov *et al.*, 2006) and they have colonized cold areas such as polar ice sheets (Priscu and Christner, 2004), sea ice (Junge *et al.*, 2002), permafrost (Gilichinsky *et al.*, 2005), glaciers and snow cover (Foght *et al.*, 2004; Amato *et al.*, 2007a), and clouds (Sattler *et al.*, 2001; Amato *et al.*, 2007b). Those microorganisms able to develop at low temperature (i.e. < 20°C) have been termed psychrophiles (Morita, 1975), which have later been subcategorized into eurypsychrophiles (psychrotolerant microorganisms) and stenopsychrophiles (psychrophilic microorganisms) whether they tolerate or not higher temperatures, respectively (Feller and Gerday, 2003; Cavicchioli, 2006). Psychrophiles have representatives in all kingdoms, from archaea to yeasts, bacteria, fungi and algae, and they exhibit a wide variety of trophic modes for generating biochemical energy: chemolithotrophs [methanogens in permafrost (Wagner *et al.*, 2003); nitrate reducers in glaciers (Foght *et al.*, 2004)], photolithotrophs [oxygenic phototrophic cyanobacteria on glaciers

(Stibal *et al.*, 2006); anoxygenic photosynthetic purple bacteria in Antarctic Lakes (Karr *et al.*, 2003)] or chemoorganotrophic aerobic bacteria like for example *Arthrobacter* species (Christner *et al.*, 2000; Foght *et al.*, 2004). Behind this apparent diversity, it was shown that cold-adapted microorganisms of different taxonomic affiliations can share more common proteins than they do with members of similar affiliation but exhibiting a mesophilic lifestyle, which attests of specific physiological needs (García-Descalzo *et al.*, 2011). Indeed, thriving at low temperatures at which enzymatic reaction rates are decreased and proteins, RNA, and membranes are rigidified is a real thermodynamic challenge. Most of our current knowledge about these needs comes from the study of a few cold-adapted species, the genome, proteome, phenotype and physiological properties of which have been deeply investigated. These revealed the unique particularities of psychrophiles, notably concerning the management of energy in the cold: besides modified enzymes and membranes (e.g. Feller and Gerday, 2003), psychrophiles also exhibit metabolic arrangements that likely reflect a global energy-saving strategy. Those fundamental adjustments of the energy and central metabolisms notably can have profound impacts in the functioning of psychrophiles compared with mesophiles.

This chapter is subdivided into three main sections, the first of which briefly recalls generalities that have to be kept in mind by readers about energy metabolism in microorganisms and about the impacts of low temperatures on cell structures and functioning. In the second part, the energy metabolism adaptations that allow cold-adapted microorganisms to maintain at low temperature are exposed. Finally, in a concluding fourth section, other aspects of metabolic adaptation to the cold are presented, such as similarities with the response to other factors which are encountered in cold environments, focusing on the oxidative stress, and questions related to evolution towards cold adaptation.

Energy metabolism in microorganisms and implications of low temperatures

Energy production in microorganisms

This section briefly recalls the most widespread systems deployed by microorganisms for producing and transporting the biochemical energy required by the cell machinery.

Metabolism is the sum of all the catabolic and anabolic processes occurring in a cell. Energy metabolism corresponds to the fraction of metabolic pathways that is directly intended to manage the energy. This always relies on redox reactions between electron donors and acceptors, and also on nucleotide compounds that mediate the storage and the transport of electrons and of chemical energy. The reducing power gained upon oxidation of organic compounds during catabolic processes is transferred to adenine based nucleotides, so called adenylates: nicotinamide adenine dinucleotide (phosphate) ($NAD(P)^+$) and flavine adenine dinucleotide (FAD). The molecular structures of NAD^+ and FAD are presented in Fig. 5.1. These are used as cofactors in most of the redox reactions of the metabolism and their oxidation can drives the respiratory chain responsible for the production of biochemical energy.

The biochemical energy itself is transported by other adenylates compounds: adenosine 5′-triphosphate (ATP, the molecular structure of which is shown in Fig. 5.1), adenosine 5′-diphosphate (ADP), adenosine-5′-monophosphate (AMP) and in a lesser extent also by

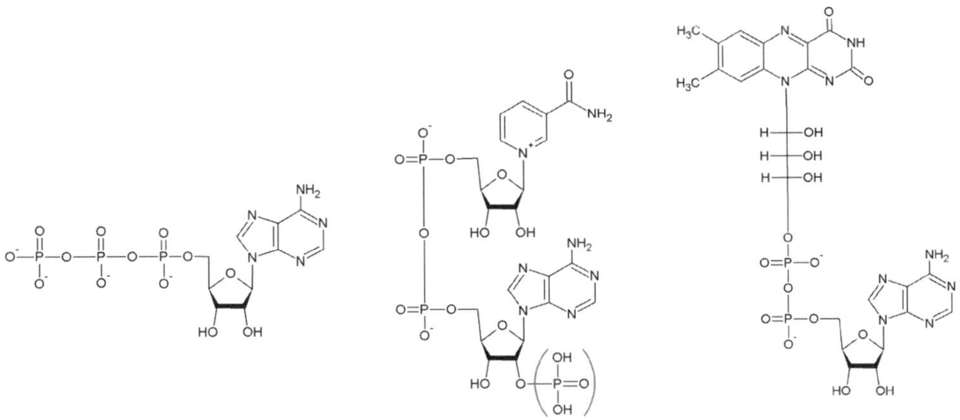

Adenosine 5' triphosphate (ATP) Nicotinamide adenine dinucleotide (phosphate) (NAD(P)⁺) Flavine adenine dinucleotide (FAD)

Figure 5.1 Molecular structures of adenylate compounds: ATP, $NAD(P)^+$ and FAD.

guanosine 5′-triphosphate (GTP). Most of the time, the term 'energy metabolism' refers to the pathways that directly contribute to the production of ATP. The chemical energy is stored in these molecules under the form of highly energetic phosphate bonds, the hydrolysis of which is highly exergonic (Gibbs free energy $\Delta G^{\circ\prime} = -32\,\text{kJ/mol}$ in ATP) and releases inorganic phosphate (Pi); this is thus coupled with most of the endergonic reactions taking place in the metabolism: 492 known enzymatic reactions involve ATP (Parry and Shain, 2011). The adenylate system comprised by ATP, ADP and AMP functions like an electrochemical storage cell common to all living organisms. In addition, the relative concentration of ADP to ATP participates to the regulation of many enzymatic processes such as those involving kinases (Atkinson, 1968).

The concentration of ATP in *Escherichia coli* was shown to be independent from the growth rate, with values about $\sim 2 \times 10^{-18}$ mol/cell, and ratios ATP–ADP–AMP of around 1:6:50 (Franzen and Binkley, 1961). The half-life of extracellular ATP in soil was estimated to less than 1 h (Conklin and MacGregor, 1972). Hence, measurements of the nucleotide content in environmental samples is commonly used as a mean for detecting and quantifying living cells in the environment (e.g. Karl, 1980; Karl *et al.*, 1999; Rakotonirainy *et al.*, 2003; Amato *et al.*, 2007c). For example, worldwide measurements of the ATP content in soil indicated a linear relationship with the amount of biomass [11 μmol ATP/(g biomass C)] (Contin *et al.*, 2001). In addition, the relative amount of potential chemical energy carried by adenylate compounds and normalized to their actual concentrations is often employed as a describer of the energetic state of cells, which was termed the adenylate energy charge {$\text{AEC} = ([\text{ATP}] + 0.5 \times [\text{ADP}])/([\text{ATP}] + [\text{ADP}] + [\text{AMP}])$; Atkinson, 1968; Karl, 1980}.

ATP is regenerated from ADP + P_i by two main metabolic processes in microorganisms: *substrate-level phosphorylation* and *chemiosmosis*.

- *Substrate-level phosphorylation* directly couples the oxidation of phosphorylated organic substrates to the synthesis of ATP, by enzymatic transfer of the phosphate moiety onto ADP, most frequently. For instance, the energy and Pi released upon hydrolysis of the phosphodiester bonds in molecules such as phosphoenolpyruvate ($\Delta G°' = -52$ kJ/mol) or acetyl-phosphate ($\Delta G°' = -45$ kJ/mol) are transferred to ADP for generating ATP. The enzymes involved (kinases or phosphatases) are widely shared within the living world and participate notably to the glycolysis and to the pentose phosphate pathways. Anaerobes use mainly fermentation pathways for regenerating their pool of ATP.

- *Chemiosmosis* couples the production of ATP with electrochemical gradients generated through the cytoplasmic membrane (or through the inner membrane of mitochondria or thylakoids in eukaryotes); this is more precisely termed oxidative phosphorylation (aerobic or anaerobic respirations) or photophosphorylation whether the initial source of energy used for fuelling the electrochemical gradient is a chemical compound or light, respectively (chemotrophic and phototrophic microorganisms, respectively). Phototrophs also employ the electron transport chain for regenerating their pool of reducing power (NADPH, H^+). The creation of membrane gradients implies a cascade of redox reactions involving an electron donor and specific electron transport systems embedded in the membrane, which are responsible for the translocation of energy-coupling ions (H^+ or Na^+) from the cytoplasm to the external space (Fig. 5.2). This gradient is termed proton motive force (PMF) in the case that the energy-coupling ions are H^+ and sodium motive force (SMF) in the case that the energy-coupling ions are Na^+. PMF or SMF exerts an inward force in most microorganisms and this is used for transporting specific solutes, rotating flagella, maintaining homeostasis and turgor, and for catalysing the phosphorylation of ADP into ATP using membrane bound ATP synthases (Konings *et al.*, 2002). In aerobic photosynthetic microorganisms, H_2O is used as the initial donor of electrons, while chemotrophs can utilize either organic or inorganic electron donors (NH_4^+, H_2, Fe^{2+} or H_2S), and they are thus subcategorized into organotrophic (all eukaryotes and some prokaryotes) or lithotrophic organisms (prokaryotes only), respectively. In aerobic respiration such as depicted in Fig. 5.2, the reducing power gained upon catabolic pathways, notably the tricarboxylic acid (TCA) cycle, is transferred to NADH, H^+, which in turn acts as the electron donor in the respiratory chain; O_2 is in this case the terminal acceptor of electrons and is reduced into H_2O. Anaerobic respirations rely on the reduction of compounds such as NO_3^- (denitrifiers), SO_4^{2-} (sulfate reducers), Fe^{3+} (iron reducers), CO_2 (methanogens) or others.

Decreased reaction rates

Molecular motion is dependent to temperature. Consequently the diffusion rates of substrates, metabolites and cellular components such as enzymes are slowed down in the cold, thereby altering metabolic processes. The rate of chemical reactions is also driven by temperature in accordance with the laws of thermodynamics, and the biochemical processes taking place in cells are no exception. The Arrhenius equation describes the variation of the rate of a chemical reaction (k) with temperature (T):

$$k = \kappa(k_B T/h)e^{-\Delta G*/RT}$$

where κ is the coefficient of transmission involving the transition state of the reaction, k_B is the Boltzmann's constant (1.381×10^{-23} J/K), h is Planck's constant (6.626×10^{-34} J/s), ΔG^* is the Gibbs free energy of activation of the reaction, and R is the universal gas constant

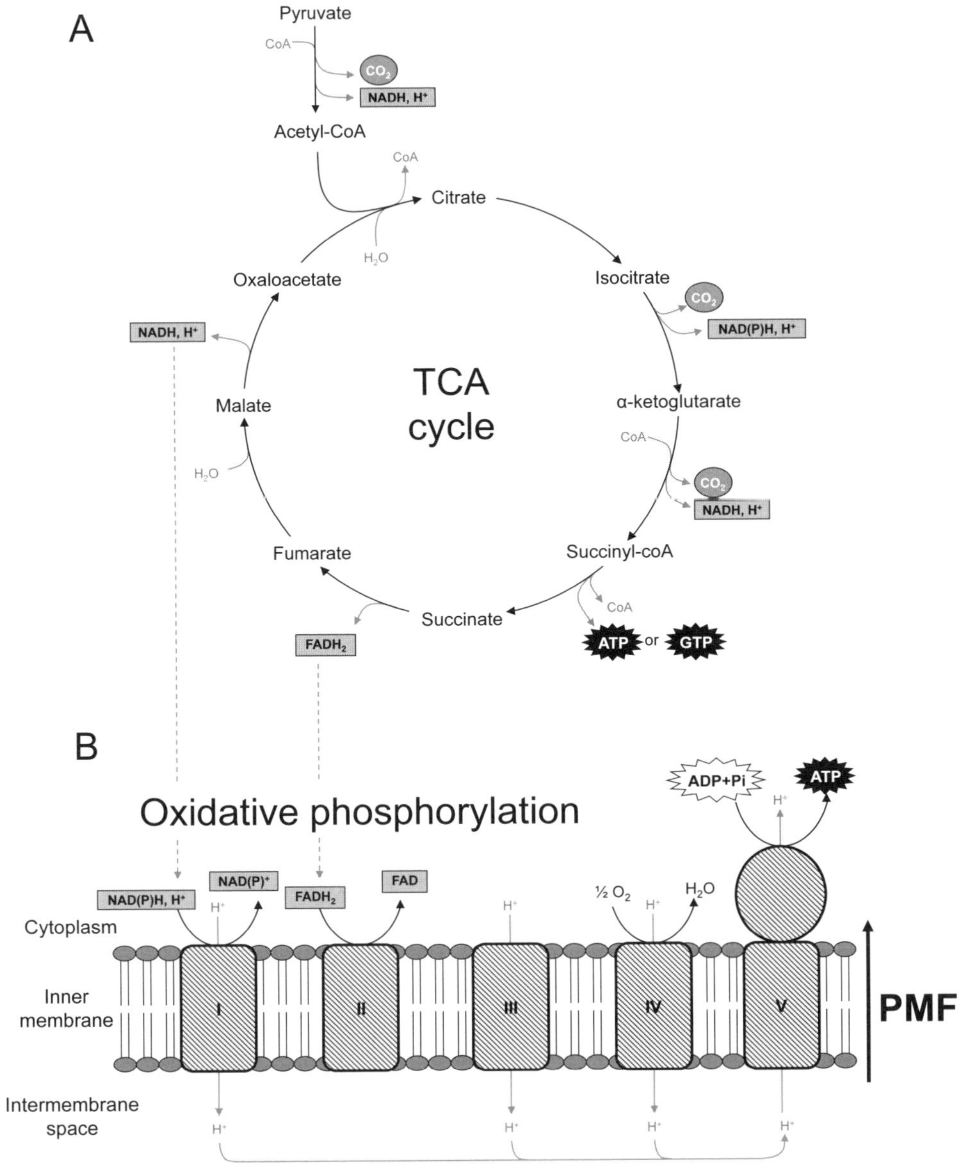

Figure 5.2 Schematic representations of (A) the TCA cycle, main provider of reducing power in most cellular systems, and (B) the respiration chain, here aerobic, which couple the oxidation of reduced compounds to a proton gradient (PMF) and to the phosphorylation of ADP into ATP. The electron chain is located in the inner membrane of bacteria or mitochondria and it is composed of several electron donor–acceptor complexes (Complex I, NADH dehydrogenase; Complex II, succinate dehydrogenase; Complex III, cytochrome *c* reductase; Complex IV, cytochrome *c* oxidase) and a membrane-bound ATP synthase (Complex V). The reaction of oxidation of succinate into fumarate in the TCA cycle actually takes place in the membrane in Complex II, which contains a bound cofactor FAD.

(8.314 J/K/mol). From this equation, it becomes obvious that the rate of a chemical reaction decreases with temperature. The unitless Q_{10} index has been defined as the variation of the rate of a reaction caused by a variation of temperature of 10°C; it generally reaches values between 2 and 3 in biological systems (Takacs and Priscu, 1998). In the absence of appropriate response from cells, the decrease of individual reaction rates implies a global diminution of the metabolic rate a low temperature, and this can have dramatic consequences for the production of energy.

Cold-adapted microorganisms can balance or counteract the deleterious effects of low temperatures by expressing enzymes with sequences modified from their mesophilic counterparts, so called cold-induced proteins (CIPs) (e.g. Hebraud *et al.*, 1994; Liu *et al.*, 2002; Homma *et al.*, 2003; Goodchild *et al.*, 2004). These are more precisely termed cold shock proteins (CSPs) or cold acclimation proteins (CAPs) depending on their kinetics of expression. CSPs are expressed directly upon cold exposure while the expression of CAPs can be delayed and it is maintained over time (Roberts and Inniss, 1992; Colucci and Inniss, 1996; Panoff *et al.*, 1997; Piette *et al.*, 2010). The *Gammaproteobacterium Colwellia psychrerythraea* 34H, which can grow at temperatures between −14°C and 19°C, has an optimum growth rate at 8°C and a maximum cell yield at −1°C; its psychrophilic lifestyle is likely due to a collection of multiple changes in its genome and in the amino acids composition of its proteins compared with its mesophilic counterparts (Methé *et al.*, 2005). The proteins of the cold-adapted *Archaea Methanogenium frigidum* and *Methanococcoides burtonii* and of the *Bacterium Psychrobacter arcticus* were found to contain a higher proportion of non-charged polar amino acids (Gln and Thr) and fewer hydrophobic residues (Leu, Pro) than their respective mesophilic counterparts (Saunders *et al.*, 2003; Ayala-del-Río *et al.*, 2010). Thanks to such sequence modifications, cold-adapted enzymes have more flexible structures. This elevates their thermolability and decreases the energy of activation needed for the reaction that they catalyse, resulting in increased catalytic efficiency (k_{cat}) and improved access to their substrate (decreased K_m) (e.g. Feller and Gerday, 2003). Cold-adapted proteins are treated elsewhere within this volume and will thus not be developed further here. A number of CIPs are not enzymes but chaperones intended to stabilize the three-dimensional structures of enzymes and RNAs and maintain them in a state compatible with activity. CIPs also include transport systems and regulation factors (e.g. Bakermans *et al.*, 2007), some of which directly contribute to central metabolism, substrate utilization and energy production.

Rigidification of macromolecules and membranes

Another consequence of low temperatures is the rigidification of cellular structures, notably proteins and membranes. Chaperones can help counteracting these effects on individual molecules, but the increased rigidity of membranes can have severe implications for the production of energy by chemiosmosis. Indeed, within the physiological range of temperatures of an organism, lipid membranes are in a semi-solid state and act as selective barriers between the cell and the extracellular compartment. This is essential for their function of control of the concentrations of intracellular molecules and for the maintenance of electrochemical gradients such as the PMF. At low temperature, the permeability of membranes is altered and they tend to be too rigid for ensuring proper functionality. A necessity for microorganisms is thus to adjust the viscosity of their membrane upon temperature changes by modifying the composition of lipids: rigidity has to be kept high enough for restricting the permeation rates (i.e. the passive diffusion) of ions and prevent the disruption of electrochemical gradients, but fluidity must remain compatible with the functions of transport and electron transfer. Since the permeation rate of sodium is much lower than that of protons, thermophilic microorganisms can save metabolic energy by converting PMF into SMF using H^+/Na^+ exchange transport systems, thereby avoiding 'leakages' of the energy-coupling ions at high temperatures. Actually, most thermophilic microorganisms have to cope permanently with temperatures too elevated for being efficiently balanced by membrane modifications and they use only SMF (Koning *et al.*, 2002). In contrast, psychrophilic microorganisms must prevent their membrane from being too rigid: the best-known achievement in cold adaptation consists of increasing the level of *cis*-unsaturation of fatty acids in lipid membranes, which was also revealed by genome analysis (Methé *et al.*, 2005; Ayala-del-Río *et al.*, 2010).

The phenomenon of rigidification of membranes at low temperature is the trigger indicating to the cell, through membrane-bound proteins, that appropriate adjustments have to be made. More than temperature itself, the fluidity of membranes was indeed demonstrated to initiate the cascade of inductions leading to modifications of their structure (Aguilar *et al.*, 2001; Browse and Xin, 2001; Suzuki *et al.*, 2001). In *Bacillus subtilis*, cold-induced genes include the gene *des*, coding for a lipid desaturase. The transduction regulators Desk and DesR act as a two-component signal transduction system: DesK is an autophosphorylatable histidine kinase located in the membrane, and is probably sensitive to changes in osmotic pressure; it constitutes the temperature sensor itself. DesR is a DNA-binding protein; it binds to DNA upstream the *des* genes and activates their expression. There is also a negative feedback loop played by the products of the lipid desaturase (unsaturated fatty acids) on the transcription of these genes (Aguilar *et al.*, 2001). In the *Cyanobacterium Synechocystis*, reduced turgor pressure of membranes such as that encountered at low temperature also induces desaturase enzymes (genes *desA*, *desB*, *desC* and *desD*). The process of induction is a cascade of phosphorylations involving two histidine kinases and a regulator (Browse and Xin, 2001 and references therein). First, the decrease of membrane fluidity leads to the autophosphorylation of the membrane-bound protein Hik33; the phosphate residue is then transferred to Hik19, a cytoplasmic kinase, and to the *Response regulator Rer1* responsible for the induction of the expression of *desB*. The study of a range of mutants demonstrated that several regulation systems actually acted simultaneously for regulating the expression of genes at low temperature, but Hik33 appeared to have a central role: it also participates to the response to other factors responsible for the rigidification of membranes like hyperosmotic shock (Suzuki *et al.*, 2001; Mikami *et al.*, 2002).

Regulation of the energy metabolism at low temperature in cold-adapted microorganisms

Energy metabolism response to low temperature

The relatively stable ATP concentration at changing environmental factors such as light and temperature in the *Cyanobacterium Anacystis nidulans* long ago suggested that very efficient regulation mechanisms exist in the energy metabolism (Bornefeld and Simonis, 1974). In the early 2000s, Napolitano and Shain (2004) investigated the adenylate profiles (ATP, ADP and AMP concentrations) in cold-adapted organisms and in their mesophilic counterparts subjected to temperature changes. The adenylate contents in the ice worm

Mesenchytraeus solifugus, the algae *Raphidonema* sp., the yeast *Cryptococcus* sp. and the bacterium *Pseudomonas* sp. were compared with those of *Lumbriculus variegatus*, *Chlamydomonas moewusii*, *Saccharomyces cerevisiae* and *Escherichia coli*, respectively. At the lower end of their physiological temperature range, all organisms exhibited slower growth rate than at optimal temperature, which was expectable by definition. However, contrarily to others, only cold-adapted organisms had adenylate concentrations increased upon cold exposure, by up to 12-fold, and their adenylate energy charge (AEC) was also enhanced. It was discovered that below their optimal growth temperature, cold-adapted microorganisms have adenylate levels invariably negatively correlated with their rate of growth. The fact that cold-adapted organisms from several distinct domains of life behave similarly to cold exposure suggests that this adaptation is crucial. It likely appeared several times in the evolution, probably through only one or a few evolutionary steps (Napolitano and Shain, 2004).

Accordingly with these, temperature has a great influence on the adenylate pool in the cold-adapted *Psychrobacter cryohalolentis* originating from saline water pockets, the cryopegs, in Siberian permafrost. ATP and ADP concentrations were found to increase with decreasing temperature from 22°C to −80°C, and the most significant change occurred at temperatures around freezing (Fig. 5.3A) (Amato and Christner, 2009). During the process of freezing, solutes are excluded from ice crystals and form interstitial microveins of concentrated saline liquid; microbial cells are also excluded into this network of veins (Junge *et al.*, 2006; Mader *et al.*, 2006; Amato *et al.*, 2009). Hence, cells are thought to experience hyperosmotic shock upon freezing. In order to simulate freezing, Amato and Christner (2009) exposed supercooled cell suspensions of *P. cryohalolentis* to hyperosmotic shock by addition of sodium chloride. Cells responded in a very similar way as they did upon freezing, i.e. by elevating the concentration of adenylate compounds, within minutes (Fig. 5.3B). The elevation of concentration of adenylate compounds upon temperature downshift around the freezing point is thereby also a response to hyperosmotic shock.

What are the processes via which this energy metabolism response to the cold is achieved? There are several means that are presented below: direct regulation of the adenylate metabolism, regulation of respiration, intervention of polyphosphatases and interferases enzymes, differential utilization of substrates and rerouting of metabolic pathways. These are complementary and probably act simultaneously in cells of cold-adapted microorganisms.

Regulation of the adenylate metabolism

In cold-adapted microorganisms, the gain of ATP observed at low temperature is probably dedicated to the fulfilment of specific needs. It implies a relative increase of catabolic processes versus anabolism involving energy production systems functioning in the cold. Potentially, many sources of ATP exist and could have been deployed in cold adaptation. In addition, the increased rigidity of membranes could virtually contribute to facilitate the gain of ATP by decreasing proton permeability and thus strengthen PMF (Napolitano and Shain, 2005b); however, the adenylate pool of cold-adapted microorganisms was found to be mainly driven by specific pivotal enzymes, which are largely responsible for this response. Indeed, observations of cold-adapted bacteria suggested that AMP is maintained in the pool of adenylates by selective inhibition of AMP degrading enzymes upon ATP drop (Napolitano and Shain, 2005a). This was confirmed by an extensive investigation of the *Gammaproteobacteria* genomes available in databases, which revealed that cold-adapted species were enriched with *de novo* AMP synthesis enzymes, mainly adenylosuccinate synthase (*purA*), whereas

A

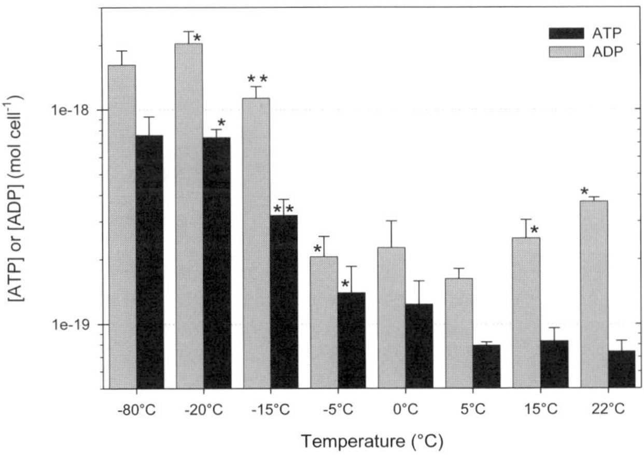

B

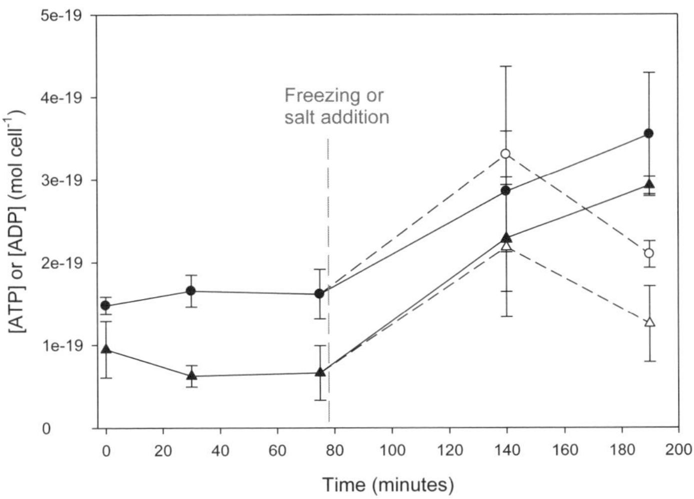

Figure 5.3 (A) Concentrations of ATP and ADP in cells of *Psychrobacter cryohalolentis* K5 after 88h of incubation at –80°C to 22°C. Samples at temperatures at or above –5°C remained liquid. Error bars are standard deviations from the mean of triplicates and asterisk (*) indicates significant difference (P <0.05) from the values measured at the next warmer or colder temperature. B- Concentrations of ATP (circles) and ADP (triangles) in *P. cryohalolentis* K5 cell cultured at 15°C, aliquoted and transferred to –20°C (t=0). After 78 min of incubation at –20°C, NaCl was added to half of the aliquots of the cell suspension (1 M final concentration; open symbols), while freezing was induced in the other half (closed symbols). Error bars are standard deviations from the mean of triplicates. From Amato and Christner (2009). Copyright © American Society for Microbiology, [Appl. Environ. Microbiol. *75*, 2009, 711–718, DOI: 10.1128/AEM.02193-08].

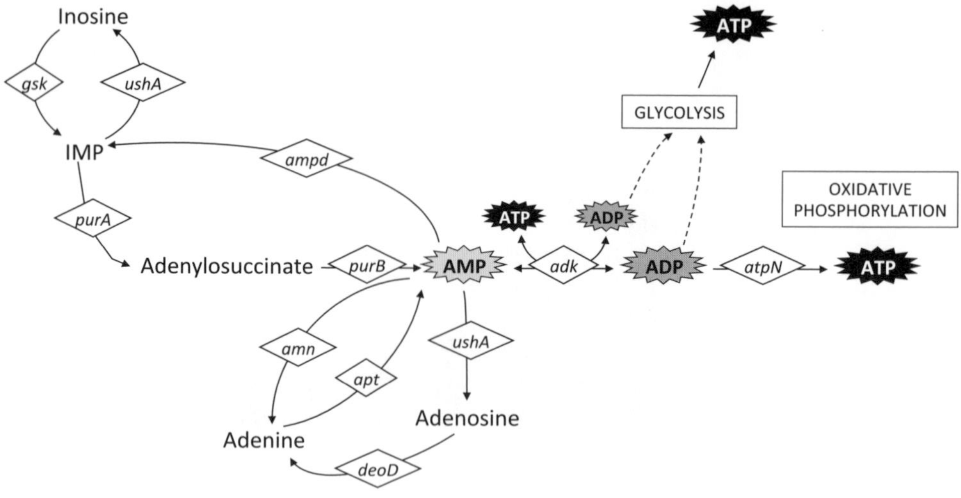

Figure 5.4 Schematic representation showing the main biochemical pathways and genes involved in the regulation of the adenylate pool in bacteria. IMP, inosine monophosphate; *adk*: adenylate kinase; *amn*, AMP nucleosidase; *ampd*, AMP deaminase; *apt*, adenine phosphoribosyltransferase; *atpN*, ATP synthase; *deoD*, purine nucleoside phosphorylase; *gsk*, inosine kinase; *purA*, adenylosuccinate synthase; *purB*, adenylosuccinate lyase; *ushA*, 5′ nucleosidase. After Parry and Shain (2011).

AMP degradative enzymes such as AMP nucleosidase (*amn*) were completely absent (Fig. 5.4) (Parry and Shain, 2011). It results that the balance between synthesis and degradation of AMP is entirely directed towards its production and can rapidly feed the regeneration of ATP. Animals like prairie dogs (*Cynomys leucurus*) also elevate ATP levels during hibernation by deactivating AMP deaminase enzyme and by keeping a high AMP kinase activity (English and Storey, 2000). In bacteria, the construction of mutants confirmed that this functioning was an adaptation to the cold. Mesophilic *Escherichia coli* mutant cells with knocked out AMP nucleosidase gene (*amn*) had steady-state ATP levels elevated by more than 30% compared with the wild type over their entire physiological temperature range, and by 2-fold at 0°C (Morrison and Shain, 2008). In addition, despite the fact that the mechanism through which it operates is not fully understood yet, the mutant was rendered more tolerant to the cold: it was significantly more viable than the wild type after exposure for several days at 0°C. Then, a gene responsible for AMP synthesis (*purA*; see Fig. 5.4) in the psychrotolerant *P. cryohalolentis* was integrated into this *E. coli* mutant (Parry and Shain, 2011). The resulting final mutant expressing AMP regulating pathways mimicking those of a psychrophile had increased growth rates at low temperatures (10°C and 15°C) and its survival to cold conditions (0°C for 8 days) was improved by a factor of 10. Hence, the concentration of ATP seems to be related to both tolerance to the cold and growth at low temperatures. The early evolutionary steps that led to this cold-adaptation in mesophilic microorganisms likely targeted *purA*, the main protein of AMP synthesis. This should have become more flexible for maintaining catalytic efficiency at low temperature. Another early adaptation was the loss of AMP nucleosidase genes, which was probably advantageous and favoured the multiplication of mutants in cold areas comparatively with original mesophilic cells.

Regulation of respiration

Respiration probably also plays a central role in energy metabolism at low temperature as it is often the main process leading to the regeneration of ATP. Some enzymes specific of the respiratory chain have indeed been found up-regulated at low temperature in *P. cryohalolentis* (H^+/Na^+-translocating ATP synthase; Bakermans *et al.*, 2007), *Bacillus subtilis* (NADH dehydrogenase analogue; Budde *et al.*, 2006) or again *Pseudoalteromonas haloplanktis* (electron transfer flavoprotein; Piette *et al.*, 2010) (Table 5.1). The phenomenon of respiratory homeostasis, which is the maintenance of a constant respiratory rate despite the temperature is changing, has also been directly observed. Increased respiration rates in response to low temperature have been reported in fish, plants, and yeasts notably. Cold-acclimated teleost fish had elevated numbers of mitochondria in the cold, thereby offsetting the lower diffusion rates and resulting in an increased production of ATP by aerobic respiration (Johnston and Dunn, 1987). Higher respiration rates were also observed in the leaves of *Arabidopsis thaliana* after acclimation at 5°C (Talts *et al.*, 2004) and in the roots of rice plants grown at 15°C versus 25°C, which were evidenced for being directly linked with the production of ATP (Kurimoto *et al.*, 2004). Elevation of ATP levels in the plants *Chenopodium* sp. and *Phaseolus vulgaris* were hypothesized to be also implicated in their vernalization and in the breaking of dormancy (Jones, 1970). Yeasts are other examples or organisms able to exhibit respiratory homeostasis (Römisch and Matheson, 2003). This is less clear in Bacteria, but this phenomenon likely also occurs: in the bacterium *P. cryohalolentis*, uncoupling the production of ATP from the respiratory chain by treating cells with an uncoupler like CCCP almost completely suppressed the increase of adenylate concentration following exposition to the cold (Amato and Christner, 2009).

Polyphosphatases and interferases as means of rapid production of ATP

Upon freezing, the increase of osmotic pressure can result in an increase of the extracellular concentrations of protons and of sodium, thus contributing to strengthen inward PMF and SMF. PMF can be directly used for producing ATP, and SMF can be converted into PMF via Na^+/H^+ antiport systems or be directly used for ATP production from ADP (Koning *et al.*, 2002). However, as both ATP and ADP concentrations increase upon exposition to low temperature in cold-adapted microorganisms, other mechanisms are also involved.

Polyphosphates (polyP) and pyrophosphates (PPi) represent sources of phosphate moieties that can serve for generating ATP within short time frames by processes other than respiration (Seufferheld *et al.*, 2008, and references therein). PolyP are chains of orthophosphate (Pi) residues linked together by high-energy phosphoanhydride bonds and stored in cells along with acidic calcium within membrane-bound organelles termed acidocalcisomes; these are shared by all types of organisms including bacteria and eukaryotes. PolyP and PPi can be used by polyphosphate kinase (PPK) and exopolyphosphatase (PPX) enzymes for regenerating ATP from ADP or GTP from GDP. This was suggested to participate to cold adaptation, notably in *Rhizobium* species, in which enzymes of the phosphate metabolism (polyphosphatases) are overexpressed at low temperature (Sardesai and Babu, 2000). They were also shown to be involved in the regulation of genes contributing to the survival of bacteria exposed to a range of stresses such as oxidative stress, thermal and acid shocks. Under oxidative stress, the presence of polyP notably induces the activation of RNA polymerase RpoS and the synthesis of guanosine nucleotides. In bacteria such as

Table 5.1 Microbial genes or proteins whose expression was found up-regulated at low temperatures

Gene or protein whose expression is up-regulated at low temperature	Gene or protein	Organism	Reference
3-Hydroxyacyl-CoA dehydrogenase analogue	*yusL*	*Bacillus subtilis*	Budde *et al.* (2006)
Acetaldehyde/alcohol dehydrogenase	AdhE	*Lactococcus piscium*	Garnier *et al.* (2010)
Acetate kinase	*ackA*	*Psychrobacter cryohalolentis* K5	Bakermans *et al.* (2007)
Acetyl-CoA carboxylase, carboxytransferase subunit alpha	*accA*	*Pseudoalteromonas haloplanktis*	Piette *et al.* (2010)
Acetylornithine transaminase	*dapC*	*Pseudoalteromonas haloplanktis*	Piette *et al.* (2010)
Acetyltransferase analogue	*ydaF*	*Bacillus subtilis*	Budde *et al.* (2006)
Acetyltransferase analogue	*ysnE*	*Bacillus subtilis*	Budde *et al.* (2006)
Aconitase	*acn*	*Propionibacterium freudenreichii*	Dalmasso *et al.* (2012)
Aconitate hydratase 2/2-methylisocitrate dehydratase	*acnB*	*Pseudoalteromonas haloplanktis*	Piette *et al.* (2010)
Alanine dehydrogenase	Ald/*ald*	*Propionibacterium freudenreichii*	Dalmasso *et al.* (2012)
Alcohol dehydrogenase analogue	*yhxC*	*Bacillus subtilis*	Budde *et al.* (2006)
Alcohol dehydrogenase analogue	*yhxD*	*Bacillus subtilis*	Budde *et al.* (2006)
Aspartate-ammonia lyase	Asp	*Propionibacterium freudenreichii*	Dalmasso *et al.* (2012)
Citrate synthase	*gltA*	*Pseudoalteromonas haloplanktis*	Piette *et al.* (2010)
Citrate synthase	*gltA*	*Propionibacterium freudenreichii*	Dalmasso *et al.* (2012)
Coenzyme A transferase	Cat	*Propionibacterium freudenreichii*	Dalmasso *et al.* (2012)
CTP synthetase	*pyrG*	*Pseudoalteromonas haloplanktis*	Piette *et al.* (2010)
Cytochrome M	*cytM*	*Synechocystis* sp. PCC6803	Mikami *et al.* (2002)
DAHP synthetase	*aroH*	*Pseudoalteromonas haloplanktis*	Piette *et al.* (2010
Dimethylamine corrinoid protein	DMA-CP	*Methanococcoides burtonii*	Goodchild *et al.* (2004
Electron transfer flavoprotein alpha-subunit	*etfA*	*Pseudoalteromonas haloplanktis*	Piette *et al.* (2010
Enolase	Eno/*eno*	*Propionibacterium freudenreichii*	Dalmasso *et al.* (2012
F_1/F_0 ATP synthase beta-subunit, H^+/Na^+ translocating	*atpF*	*Psychrobacter cryohalolentis* K5	Bakermans *et al.* (2007
F_{420} H_2 dehydrogenase	F_{420}-H_2 DH	*Methanococcoides burtonii*	Goodchild *et al.* (2004)
Formate dehydrogenase analogue	*yjgC*	*Bacillus subtilis*	Budde *et al.* (2006)
Fructose bisphosphate aldolase	Fba	*Lactococcus piscium*	Garnier *et al.* (2010)

Gene or protein whose expression is up-regulated at low temperature	Gene or protein	Organism	Reference
Fructose bisphosphate aldolase	*fba*	*Propionibacterium freudenreichii*	Dalmasso *et al.* (2012)
Fumarate hydratase	*fumB*	*Pseudoalteromonas haloplanktis*	Piette *et al.* (2010)
Fumarate hydratase	Fum C	*Propionibacterium freudenreichii*	Dalmasso *et al.* (2012)
Glucose-1-dehydrogenase analogue	*yhdF*	*Bacillus subtilis*	Budde *et al.* (2006
Glucose-1-dehydrogenase analogue	*ykuF*	*Bacillus subtilis*	Budde *et al.* (2006)
Glucose-1-dehydrogenase analogue	*yxbG*	*Bacillus subtilis*	Budde *et al.* (2006)
Glucose-1-dehydrogenase analogue	*yxnA*	*Bacillus subtilis*	Budde *et al.* (2006)
Glucose-1-phosphate adenyltransferase	*glgC*	*Propionibacterium freudenreichii*	Dalmasso *et al.* (2012)
Glucose-6-phosphate isomerase	PGI/*pgi*	*Propionibacterium freudenreichii*	Dalmasso *et al.* (2012)
Glutamate dehydrogenase (NADP specific)	GDH	*Methanococcoides burtonii*	Goodchild *et al.* (2004)
Glyceraldehyde-3-phosphate dehydrogenase	GAPDH	*Lactococcus piscium*	Garnier *et al.* (2010)
Glyceraldehyde-3-phosphate dehydrogenase	GAPDH	*Methanococcoides burtonii*	Goodchild *et al.* (2004)
Glycerol kinase	*glpK*	*Bacillus subtilis*	Budde *et al.* (2006)
Glycine dehydrogenase	*gcvP*	*Pseudoalteromonas haloplanktis*	Piette *et al.* (2010)
Glycolate oxidase subunit analogue	*ysfC*	*Bacillus subtilis*	Budde *et al.* (2006)
Glycosyltransferase	*glgA*	*Propionibacterium freudenreichii*	Dalmasso *et al.* (2012)
Histidine protein	HPr	*Lactococcus piscium*	Garnier *et al.* (2010)
Isocitrate lyase	*aceA*	*Psychrobacter cryohalolentis* K5	Bakermans *et al.* (2007)
Isocitrate lyase	*ICL*	*Colwellia psychrerythraea*	Watanabe *et al.* (2002)
Lactate dehydrogenase	LDH	*Rhizobium* spp.	Sardesai and Babu (2000)
Lactate dehydrogenase	*ldh*	*Propionibacterium freudenreichii*	Dalmasso *et al.* (2012)
Malate synthase G	*glcB*	*Pseudoalteromonas haloplanktis*	Piette *et al.* (2010
Malate/lactate dehydrogenase	*mdh*	*Psychrobacter cryohalolentis* K5	Bakermans *et al.* (2007
Methyl coenzyme M reductase	CH_3-CoM-R	*Methanococcoides burtonii*	Goodchild *et al.* (2004
Methylcobalamin:CoM methyltransferase	CH_3-CoM-MT	*Methanococcoides burtonii*	Goodchild *et al.* (2004
Methylmalonyl-CoA carboxytransferase	MmdA	*Propionibacterium freudenreichii*	Dalmasso *et al.* (2012
Methylmalonyl-CoA mutase	Mut	*Propionibacterium freudenreichii*	Dalmasso *et al.* (2012

Table 5.1 (Continued)

Gene or protein whose expression is up-regulated at low temperature	Gene or protein	Organism	Reference
Monomethylamine methyltransferase	MMA-MT	*Methanococcoides burtonii*	Goodchild *et al.* (2005
NADH dehydrogenase analogue	*yumB*	*Bacillus subtilis*	Budde *et al.* (2006
NADH dehydrogenase subunit 4	*ndhD2*	*Synechocystis* sp. PCC6803	Suzuki *et al.* (2001), Mikami *et al.* (2002)
NADH dehydrogenase subunit 4	*ndhD6*	*Synechocystis* sp. PCC6803	Suzuki *et al.* (2001)
NADH dehydrogenase subunit F	*ndhF*	*Synechocystis* sp. PCC6803	Mikami *et al.* (2002)
NH_3-dependent (glutamine-hydrolysing) NAD(+) synthase	*nadE*	*Psychrobacter cryohalolentis* K5	Bakermans *et al.* (2007)
NH_3-dependent (glutamine-hydrolysing) NAD(+) synthase	*nadE*	*Bacillus subtilis* (Firmicutes)	Budde *et al.* (2006)
Phosphoglucomutase	*pgm*	*Pseudoalteromonas haloplanktis*	Piette *et al.* (2010)
Phosphoglucomutase	Pgm/*pgm*	*Propionibacterium freudenreichii*	Dalmasso *et al.* (2012)
Phosphoglycerate kinase	PGK	*Lactococcus piscium*	Garnier *et al.* (2010)
Polynucleotide phosphorylase (PNPase)	*pnpA*	*Bacillus subtilis*	Budde *et al.* (2006)
Propionyl-CoA carboxylase	*pccB*	*Propionibacterium freudenreichii*	Dalmasso *et al.* (2012)
Propionyl-CoA:succinate-CoA transferase	*ygfH*	*Pseudoalteromonas haloplanktis*	Piette *et al.* (2010)
Pyruvate oxidase analogue	*ydaP*	*Bacillus subtilis*	Budde *et al.* (2006)
Pyruvate phosphate dikinase	*ppdK*	*Propionibacterium freudenreichii*	Dalmasso *et al.* (2012)
Ribulose-bisphosphate carboxylase analogue	*ykrW*	*Bacillus subtilis*	Budde *et al.* (2006)
Serine dehydratase	*sda*	*Propionibacterium freudenreichii*	Dalmasso *et al.* (2012)
Spermidine synthase	*speE*	*Pseudoalteromonas haloplanktis*	Piette *et al.* (2010)
Succinate dehydrogenase	*sdh*	*Propionibacterium freudenreichii*	Dalmasso *et al.* (2012)
Thiamine biosynthesis protein thiC	*thiC*	*Pseudoalteromonas haloplanktis*	Piette *et al.* (2010)
Transketolase 1	*tktA*	*Pseudoalteromonas haloplanktis*	Piette *et al.* (2010)
Trimethylamine corrinoid protein	TMA-CP	*Methanococcoides burtonii*	Goodchild *et al.* (2004)
Trimethylamine methyltransferase	TMA-MT	*Methanococcoides burtonii*	Goodchild *et al.* (2004)
UTP-glucose-1-phosphate uridylyltransferase	*gtaB*	*Bacillus subtilis*	Budde *et al.* (2006)

Agrobacterium tumefaciens and *Rhodospirillum rubrum*, a vacuolar pyrophosphatase (H^+-V-PPase) hydrolyses PPi to translocate protons, which generates a PMF across the membrane of acidocalcisomes.

Another means of producing ATP in the cold could involve RNA interferases (Amato and Christner, 2009). These are enzymes expressed in bacteria as salvage pathways under stress conditions. Their role is to cleave mRNA molecules, thereby altering protein synthesis to levels that can reach the cease of growth, to the entrance in a quasi-dormant state and eventually cell death (Inouye, 2006). Cells expressing RNA interferases are still capable of energy metabolism and biosynthesis of nucleic acids, and the degradation of RNA represents potentially a direct source of ATP. Despite the fact that this mechanism has never been clearly evidenced as a response to the exposition to low temperatures, it was strongly suggested by a reduction of ~50% of the RNA content in *E. coli* upon heat shock, simultaneous to a transient elevation of the concentration of adenylate compounds (Soini *et al.*, 2005).

Substrate utilization

As stressed by Feller and Gerday (2003) and Cavicchioli (2006), the cardinal temperatures of growth (T_{opt}, T_{max}) of an organism are not necessarily directly linked to its fitness for a cold habitat. Indeed, faster growth does not imply 'safer' growth as the general physiological state can be altered at T_{opt}, and additional factors like nutrient limitations can also constrain growth. This implies that the temperature considered to be optimal for the growth of an organism does not necessarily reflect its fitness for its habitat. As an illustration of this, the T_{opt} of most of the microorganisms recovered from cold environments and defined as 'cold-adapted' falls within the mesophilic range (Foght *et al.*, 2004; Groudieva *et al.*, 2004; Amato *et al.*, 2007b). However, even though the temperature is much lower than their T_{opt} in their natural habitat, they still dominate the microbial community. The bacterium *Sphingopyxis alaskensis* multiplies to high concentrations (10^5 cells/ml) in cold oceans, but it has a T_{opt} of 37°C under eutrophic conditions (Cavicchioli, 2006). This is indicative of the capacity of most of the cold-adapted microorganisms to thrive over a wide range of temperatures extending to the cold, which defines eurypsychrophile (psychrotolerant) microorganisms. This aspect is likely linked with the large temperature variations existing in cold environments and that microorganisms have to face to survive. Hence, rather than growth rate, a good describer of the optimum temperature for an organism would be the temperature at which the highest growth efficiency is reached (i.e. the relative proportion of carbon incorporated to the total amount of carbon ingested by the organism; Eiler *et al.*, 2003), as it corresponds to the temperature at which the lowest proportion of energy is dissipated (Del Giorgio *et al.*, 1997).

Microorganisms thriving in the cold have specific nutritive requirements related to a global strategy of energy-saving. The choice of the substrate for producing energy at low temperature obviously depends, at least in part, on the presence of transporters functioning at low temperature. Thanks to transport systems active in the cold, the uptake of NH_4^+ by the psychrophilic marine bacterium *Vibrio* sp. was observed to be faster at temperatures from 0 to 15°C than it was at higher temperature, while the apparent affinity for this substrate did not depend on temperature (Chou *et al.*, 1999). Another strategy for maintaining fluxes of substrates at lower diffusion rates consists of increasing the number of copies of the transport systems. A proteomic study of *P. cryohalolentis* revealed that several proteins involved in transport functions were overexpressed at low temperature in this psychrotolerant microorganism (Bakermans *et al.*, 2007). They comprised transporters of lipoproteins (LolD), dicarboxylates (DctP), iron (AfuA and FecA) and of other molecules carried by undetermined ABC transporters.

For most substrates, internalization is an energy demanding process. Hence, in a context of energy-saving strategy, it sounds relevant for microorganisms to adapt the substrates they use when exposed to temperature below optimal. Among 95 carbon sources tested, cold-adapted members of *Exiguobacterium* and *Psychrobacter* species isolated from Siberian permafrost and able to grow at temperature as low as −2.5°C, used six and seven of them, respectively, at 4°C but not at 24°C, while both temperatures were within their physiological range and allowed growth (Ponder *et al.*, 2005). The substrates preferentially used at the lowest temperature included lactic acid for *Psychrobacter* sp. and acetic acid for *Exiguobacterium* sp.. These compounds can rapidly be transformed into acetylcoenzyme A and/or pyruvate and enter central metabolic pathways such as fermentations, the glycolysis or the TCA cycle for producing energy; in *Corynebacterium glutamicum*, the uptake of these molecules involves active monocarboxylic acid–sodium symporters driven by the PMF (Jolkver *et al.*, 2009).

Rerouting of metabolic pathways

Below a certain temperature (termed critical), the slope of the growth rate function of an organism brutally changes (Bakermans and Nealson, 2004). This is probably the result of an increased need for energy aiming at synthesizing new proteins involved in some specific pathways of the central metabolism and in the response to stress (Goodchild *et al.*, 2004). Some microorganisms repress both anabolic and catabolic activities at low temperatures and, in turn, favour pathways that are involved in the general stress response; this was for example observed in the psychrotolerant but not fully cold-adapted *Bacillus subtilis* (Budde *et al.*, 2006). In contrast, true cold-adapted microorganisms can sustain metabolic activity at low temperature through deep modifications of their functioning by precise metabolic adjustments at the levels of transcription, translation and proteins (Lim *et al.*, 2000; Piette *et al.*, 2010, 2011). Some of the most important enzymes of the central metabolism that have been detected up-regulated in the cold by proteomic and transcriptomic methods are reported in Table 5.1.

In bacteria, the metabolic regulations existing in the cold are achieved in great part by chaperones like the heat shock protein HSP90, which targets and protects many enzymes of the central metabolism (García-Descalzo *et al.*, 2011). Other regulation mechanisms exist in *Synechocystis*, a *Cyanobacterium* widely studied as a model for plants. In this organism, most (65%) of the cold-inducible genes including the *des* genes involved in the desaturation of membrane lipids mentioned earlier (see 'Rigidification of macromolecules and membranes', above) and genes participating to the regulation of central metabolism pathways such as photosynthesis, the response to oxidative stress, and the production of energy are regulated by the temperature dependent histidine kinase Hik33 (Suzuki *et al.*, 2001; Mikami *et al.*, 2002). Osmotic stress inactivates the photosystems I and II of *Synechocystis* thereby causing important losses of the capacity to produce ATP; in response, Hik33 induces the NADH dehydrogenase *ndhD2* for maintaining the PMF and continue producing ATP. Hence, the role of Hik33 here is to keep the energy production machinery running by favouring oxidative phosphorylation pathways in this phototrophic organism while its main engine is shut down.

The metabolic adjustments that can be adopted in the cold are multiple and seem to depend on the microorganism. Fig. 5.5 combines different examples of metabolic pathways that are favoured in the cold in different microorganisms; these are in addition presented in details in the following paragraphs.

In the widely studied bacterium for its adaptation to the cold, *P. cryohalolentis*, the beta-subunit of the F_1/F_0 ATP synthase (AtpF), a NAD^+ synthase and the outer membrane efflux system protein TolC are up-regulated at −4°C compared with 4°C and 16°C, demonstrating the increased need of biochemical energy (Bakermans *et al.*, 2007) (Table 5.1). Simultaneously, the up-regulation of enzymes, notably an acetate kinase (AK), a malate dehydrogenase (MDH) and an isocitrate lyase (ICL) indicated that the energy was obtained through the glyoxylate cycle (Fig. 5.5). Cold-adapted *Rhizobium* spp. and *Colwellia psychrerythraea* also likely prefer this pathway in the cold (Sardesai and Babu, 2000; Watanabe *et al.*, 2002), as well as a psychrotolerant *Bacillus subtilis* (Budde *et al.*, 2006). Isocitrate lyase (ICL) catalyses the cleavage of isocitrate to succinate and glyoxylate, thus bypassing the two decarboxylation steps of the TCA cycle. In *C. psychrerythraea*, this enzyme is clearly cold adapted: it has a k_{cat}/K_m ratio optimum at 20°C, and its affinity for substrates is maximal at 15°C (Watanabe *et al.*, 2002).

Another cold-adapted bacterium, a strain of *Lactobacillus piscium* used for food biopreservation, favours glycolysis at low temperature (Garnier *et al.*, 2010) (Fig. 5.5). Glyceraldehyde-3-phosphate dehydrogenase (GAPDH), which catalyses the main temperature-limiting step of glycolysis, was found to be up-regulated along with other proteins directly involved in energy metabolism (Poolman *et al.*, 1987; Mair *et al.*, 2005). These included a fructose bisphosphate aldolase (FBA), a phosphoglycerate kinase (PGK), a histidine kinase (HPR) participating in the uptake of carbohydrates through a phosphotransferase system (PTS), and an aldehyde/alcohol dehydrogenase (ADH), which reflected the induction of acid fermentation pathways.

Some psychrotolerant *Rhizobium* species were shown to switch from aerobic to anaerobic respiration for generating energy in the cold (Sardesai and Babu, 2000): the activity of lactate dehydrogenase (LDH), which reversibly catalyses the oxidation of pyruvate into lactate using NADH as cofactor (Fig. 5.5), increased at low temperature, while simultaneously the activity of succinate dehydrogenase (SDH), an enzyme of the aerobic respiratory chain, decreased, and that of alcohol dehydrogenase remained slow. In parallel, the elevation of the membrane-bound ATPase activity suggested that active uptakes of solutes occurred. These microorganisms probably stop respiring oxygen in the cold but they can maintain metabolic activity by redirecting the regeneration of their pool of NAD^+ towards fermentative pathways.

In permafrost, methanogens play important roles in CH_4 production and participate in global warning. The functioning of the *Archaeon Methanococcoides burtonii* in the cold has been thoroughly characterized through proteomic and transcriptomic analyses. These revealed the presence of regulation systems involving a RNA helicase directed notably towards translation and biogenesis of ribosomes at low temperature (Lim *et al.*, 2000). PMF, which is easier to maintain at low temperature than SMF (Konings *et al.*, 2002), is preferred for generating ATP at 4°C compared with 23°C: proteins of the central metabolism such as trimethylamine methyltransferase (TMA-MT) involved in methanogenesis, glutamate dehydrogenase (GDH), glyceraldehyde-3-phosphate dehydrogenase (GAPDH) (Table 5.1 and Fig. 5.5) and proteins for PMF generation such as F_{420}-H_2 dehydrogenase (F_{420}-H_2

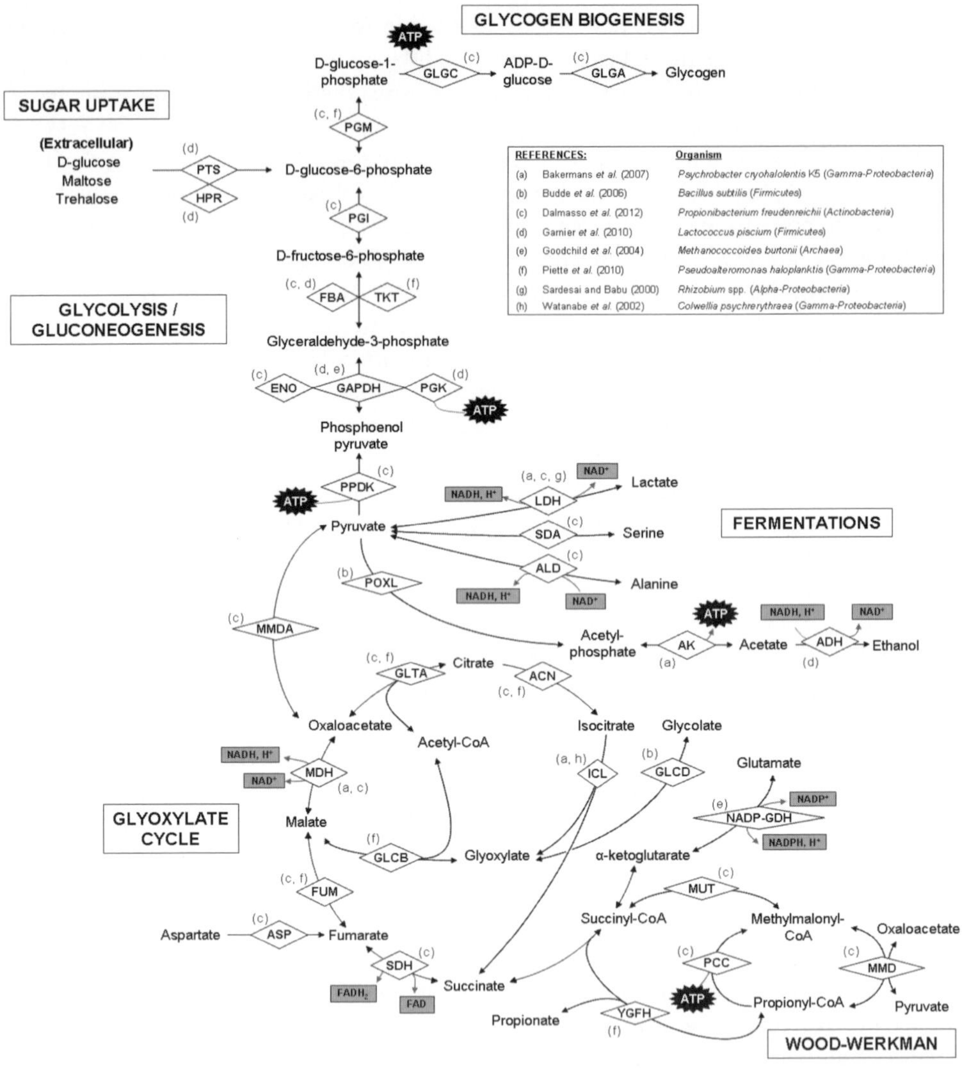

Figure 5.5 Schematic representation showing combined major pathways and enzymes of the central metabolism up-regulated at low temperature in different cold-adapted microorganisms. Up-regulated enzyme names are indicated within rhombuses. ACN, aconitase; ADH, acetaldehyde/alcohol dehydrogenase; AK, acetate kinase; ALD, alanine dehydrogenase; ASP, aspartate ammonia-lyase; ENO, enolase; FBA, fructose-bisphosphate aldolase; FUM, fumarate hydratase; GAPDH, glyceraldehyde-3-phosphate dehydrogenase; GLCB, malate synthase; GLCD, glycolate oxidase; GLGA, glucose-1-P-adenylyltransferase; GLGC, glycosyltransferase; GLTA, citrate synthase; HPR, histidine kinase; ICL, isocitrate lyase; LDH, lactate dehydrogenase; MDH, malate dehydrogenase; NAD(P)-GDH, NAD(P)-dependent glutamate dehydrogenase; MMD, methylmalonyl-CoA carboxytransferase; MUT, methylmalonyl-CoA mutase; PCC, propionyl-CoA carboxylase; PGI, glucose-6-phosphate isomerase; PGK, phosphoglycerate kinase; PGM, phosphoglucomutase; POXL, pyruvate oxidase; PPDK, pyruvate phosphate dikinase; PTS, phosphotransferase system; SDA, L-serine dehydratase; SDH: succinate dehydrogenase; TKT, transketolase; YGFH, propionyl-CoA:succinate-CoA transferase (see also Table 5.1).

DH), a membrane-bound proton pump, were found to be overexpressed at the lowest temperature (Table 5.1) (Goodchild *et al.*, 2004, 2005). Hence, catabolism was redirected towards methanogenesis and PMF generation, with transcriptional and translational regulation mechanisms responsible for an increase by 5.9-fold of the amount of GDH proteins at 4°C versus 23°C. This resulted in an elevation of the global activity of GDH by 1.8-fold. GAPDH was 2.4-fold more abundant and it was 1.8 times more active at 4°C than at 23°C in the forward (catabolic) reaction in this microorganism (Goodchild *et al.*, 2004). In contrast, pyruvate synthase, responsible for the conversion of acetyl-coenzyme A to pyruvate, was down-regulated in the cold and this had direct impacts on anabolic processes: acetyl-CoA was redirected towards the TCA cycle and amino acid metabolism, whereas oxoglutarate was left available for the anabolic functions of GDH.

The recent study of the *Actinobacterium Propionibacterium freudenreichii* by Dalmasso *et al.* (2012) provided other insights of the possible strategies that can be adopted in the cold. A total of 565 genes differentially expressed at 4°C and at 30°C were detected using proteomics, microarrays, RT-PCR and/or metabolomics, of which about 40% were up-regulated in the cold. The results revealed a functioning mainly directed towards biosynthetic pathways, rather than catabolism. Similarly to *Rhizobium* species (Sardesai and Babu, 2000), a lactate dehydrogenase (LDH) was induced in *P. freudenreichii* in the cold, and this was linked with an increased uptake of lactate. Other functions dedicated to the synthesis of pyruvate from lactate, alanine and serine were up-regulated and probably intended to feed gluconeogenesis and glycogen synthesis (Table 5.1 and Fig. 5.5). Catabolic processes were assured by up-regulation of enzymes of the TCA cycle [aconitase (ACN), fumarate hydratase (FUM), citrate synthase (GLTA), succinate dehydrogenase (SDH), malate dehydrogenase MDH)] and of the Wood–Werkman pathway [methylmalonyl-CoA mutase and carboxytransferase (MUT and MMD), propionyl-CoA carboxylase (PCC)] that ensures the interconversion between oxaloacetate and pyruvate. Enzymes of the energy metabolism were rather down-regulated (ATP synthases, oxidoreductases of the electron chain, metabolism of nucleotides, etc.). Hence, ATP synthesis was most probably obtained through phosphate metabolism pathways, as attested by the up-regulation of phosphate metabolism enzymes notably polyphosphatases, responsible for the phosphorylation of ADP into ATP (Seufferheld *et al.*, 2008) (see 'Polyphosphatases and interferases as means of rapid production of ATP', above).

Pseudoalteromonas haloplanktis from Antarctic sea water perhaps also promotes the synthesis of glycogen at 4°C, compared with 18°C, by up-regulating enzymes such as a transketolase (TKT) and a phosphoglucomutase (PGM) (Table 5.1 and Fig. 5.5) (Piette *et al.*, 2010). However, these enzymes could also be involved in their catabolic functions (glycolysis). In addition, energy is probably produced in this organism through the glyoxylate cycle and/or the TCA cycle, the enzymes of which were also detected up-regulated: citrate synthase (GLTA), aconitase (ACN), fumarate hydratase (FUM), glycolate oxidase (GLCD) (Piette *et al.*, 2010).

Concluding remarks

Similarities between cold-adaptation and response to oxidative stress

One can wonder whether it is legitimate to qualify cold environments of 'extreme' based on their low temperature only, since most of the planet is 'cold'. However, these are certainly extreme due to additional aspects that cold-adapted microorganisms have to face: high salinity, high UV levels, weak nutrient availability, etc. In many microorganisms, the protein profile expressed by cells at low temperature often suggests that they are also experiencing oxidative stress. For instance, the functions of transport of ferric iron are up-regulated at low temperature in *P. cryohalolentis* and this is probably intended too fulfil the increased demands for this transition metal as cofactor of the enzymatic reactions involved in the oxidative stress (superoxide dismutase, etc.) (Bakermans *et al.*, 2007). Furthermore, some of the cold-induced enzymes of the central metabolism are also frequently found up-regulated upon oxidative stress (Mikami *et al.*, 2002; Garnier *et al.*, 2010; Goodchild *et al.*, 2004; Bakermans *et al.*, 2007; Singh *et al.*, 2008; Piette *et al.*, 2010). Concerning more particularly the energy metabolism, in all organisms including humans an important response to oxidative stress consists of favouring the production of NADPH and prevent that of NADH (Singh *et al.*, 2008; Mailloux *et al.*, 2011 and references therein). Indeed, NADPH specifically participates to enzymatic reactions involved in the detoxification of toxic metabolic wastes (catalase, superoxide dismutase, glutathione peroxidases), whereas NADH promotes their production: in the aerobic respiratory chain, initiated by the oxidation of NADH, 0.2–2% of the oxygen consumed forms reactive oxygen species (ROS), notably superoxide radicals ($O_2\cdot^-$) that can severely damage cell structures and proteins and cause death; in addition, enzymes stimulated by the presence of NADH, such as alpha-ketoglutarate dehydrogenase (α-KGDH), also generate ROS. Hence, the response to oxidative stress consists of converting the pool of NAD into NADP by up-regulating NAD kinases (NADK) and down-regulating NADP phosphatases (NADPase) (Fig. 5.6). Central metabolism functions involving enzymes of the TCA cycle, glycolysis and gluconeogenesis are also engaged. These funnel metabolic pathways towards the production of NADP rather than NAD, by the implication of NADP-dependent enzymes. NADPH is produced, along with ROS scavengers such as alphaketoglutarate [isocitrate dehydrogenase (NADP-ICDH) and glutamate dehydrogenase (NADP-GDH), which both produce alphaketoglutarate], and pyruvate [pyruvate carboxylase (PC), pyruvate kinase (PK), malic enzyme (ME), malate dehydrogenase (MDH)]. In addition, this is coupled to the down-regulation of phosphoenolpyruvate carboxykinase (PEPCK) and pyruvate dehydrogenase (PDH) for ensuring that the substrates converge towards this metabolic pathway (Fig. 5.6).

The response to oxidative stress deployed by cells varies with the carbon source utilized. In *Pseudomonas fluorescens* fed with citrate as the sole source of carbon, NADP-dependent isocitrate dehydrogenase (NADP-ICDH) is favoured, while malic enzyme (MDH) and NADP-dependent glutamate dehydrogenase (NADP-GDH) are preferred if malate is the source of carbon. In the presence of glucose, in *Saccharomyces cerevisiae,* the activities of glucose-6-phosphate dehydrogenase and aldehyde dehydrogenase (glycolysis) are favoured for fighting against oxidants, or NADP-dependent isocitrate lyase (NADP-ICL) (glyoxylate cycle) in the presence of acetate (Mailloux *et al.*, 2011, and references therein). The responses are multiple and depend on the environmental conditions, but they are all intended to adjust the ratio NAD^+ to $NADP^+$.

Evolution towards cold adaptation

The absence of major rearrangement and of horizontal transfer in the genome of cold-adapted bacteria such as *Psychrobacter arcticus* confirmed the strong selection pressure exerted by cold environments (Ayala-del-Río *et al.*, 2010). Acclimation to the cold seems

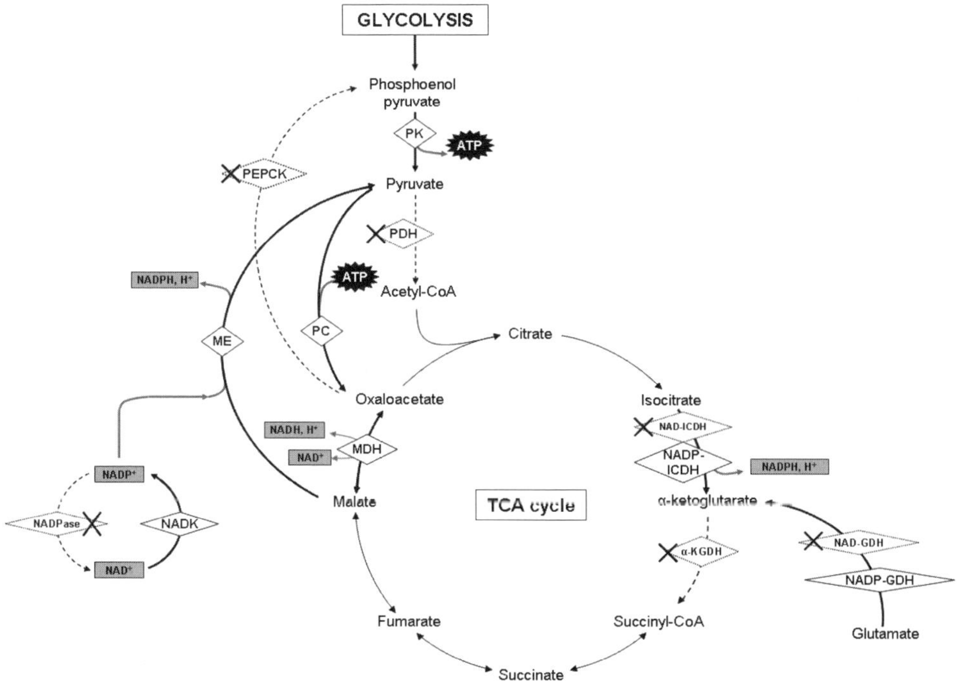

Figure 5.6 Schematic representation showing major pathways and enzymes of metabolic adjustments made in response to oxidative stress in microorganisms and intended to favour the production of NADPH. Arrows in bold indicate up-regulated pathways, those in dashed lines down-regulated pathways, and a '×' on enzyme name indicates its down-regulation. α-KGDH, alpha-ketoglutarate dehydrogenase; MDH, malate dehydrogenase; ME, malic enzyme; NADK, NAD kinase; NADPase, NADP phosphatase; NAD(P)-GDH, NAD(P)-dependent glutamate dehydrogenase; NAD(P)-ICDH, NAD(P)-dependent isocitrate dehydrogenase; PC, pyruvate carboxylase; PDH, pyruvate dehydrogenase; PK: pyruvate kinase; PEPCK, phosphoenolpyruvate carboxykinase. (After Singh *et al.*, 2008 and Mailloux *et al.*, 2011.)

to derive from the response to cold shock of mesophilic microorganisms. Indeed cold-acclimation proteins (CAP) and cold shock proteins (CSP) often refer to the same molecules, suggesting common origin in the evolution towards cold adaptation (Roberts and Inniss, 1992; Colucci and Inniss, 1996; Panoff *et al.*, 1997; Piette *et al.*, 2010). As an illustration, a protein, although categorized as a CSP (*cspA*) in mesophiles, was found to be constitutive in the psychrophilic *Gamma-Proteobacterium Psychrobacter arcticus*: its presence is required for protein synthesis independently from temperature (Ayala-del-Río *et al.*, 2010). Another example comes from the *Gamma-Proteobacterium Pseudoalteromonas haloplanktis*, which was isolated from sea water in the Antarctic. This up-regulates the chaperone protein called trigger factor (Tig) when developing in the cold. Tig was found to be essential for survival at low temperature, and not only upon cold shock. It rescues the chaperone functions at temperatures at which other chaperones cannot work properly. Tig also promotes the folding of all newly synthesized proteins on the ribosome, and it exhibits a *cis–trans* isomerase activity, which is usually the limiting factor for correct protein folding and thus for growth under cold conditions (Piette *et al.*, 2010).

Finally, although this is not clearly evidenced yet, the atmosphere and particularly clouds and precipitations may have played important roles in the evolution of mesophilic microorganisms towards cold adaptation. Microorganisms can be aerosolized from surfaces like plants or soils, transported upwards by turbulent fluxes, and reach distant environments that they eventually colonize if they remain alive. Clouds and precipitations, which are largely responsible for the deposition onto the surface of airborne particles of the size range of microbial cells (e.g. Renoux and Boulaud, 1998), may have exerted a selection pressure on those microorganisms disseminating by air means. Indeed, in clouds, cells are subject to conditions very similar to those encountered in extreme cold environments (low temperature, high UV levels, freeze–thaw cycles), and to numerous shocks due to rapid variations of environmental conditions. The dispersion of phenotypes resistant to multiple stress and capable of maintaining at low temperatures, would have been favoured, thus proliferating over the planet. In addition, the high levels of UV at high altitude could have boosted evolution processes. This hypothesis is supported notably by the large fraction of the ~10^4 cells per milliliter of cloud that intriguingly resemble those commonly retrieved in polar areas (Amato *et al.*, 2007b; Vaitilingom *et al.*, 2012).

References

Aguilar, P.S., Hernandez-Arriaga, A.M., Cybulski, L.E., Erazo, A.C., and de Mendoza D. (2001). Molecular basis of thermosensing: a two component signal transduction thermometer in *Bacillus subtilis*. EMBO J. *20*, 1681–1691.

Amato, P., Hennebelle, R., Magand, O., Sancelme, M., Delort, A.-M., Barbante, C., Boutron, C., and Ferrari, C. (2007a). Bacterial characterization of the snow cover at Spitzberg, Svalbard. FEMS Microbiol. Ecol. *59*, 255–264.

Amato, P., Parazols, M., Sancelme, M., Laj, P., Mailhot, G., and Delort, A.-M. (2007b). Microorganisms isolated from the water phase of tropospheric clouds at the Puy de Dôme: major groups and growth abilities at low temperature. FEMS Microbiol. Ecol. *59*, 242–254.

Amato, P., Parazols, M., Sancelme, M., Mailhot, G., Laj, P., and Delort, A.-M. (2007c). An important oceanic source of micro-organisms for cloud water at the puy de Dôme (France). Atmos. Environ. *41*, 8253–8263.

Amato, P., and Christner, B.C. (2009). Energy metabolism response to low-temperature and frozen conditions in *Psychrobacter cryohalolentis*. Appl. Environ. Microbiol. *75*, 711–718.

Amato P., Doyle, S., and Christner, B.C. (2009). Macromolecular synthesis by yeast under frozen conditions. Environ. Microbiol. *11*, 589–596.

Atkinson, D.E. (1968). The adenylate charge of the adenylate pool as a regulatory parameter: interaction with feedback modifiers. Biochemistry *7*, 4030–4034.

Ayala-del-Río, H., Chain, P.S., Grzymski, J.J., Ponder, M.A., Ivanova, N., Bergholz, P.W., Di Bartolo, G., Hauser, L., Land, M., Bakermans, C., Rodrigues, D., Klappenbach, J., Zarka, D., Larimer, F., Richardson, P., Murray, A., Thomashow, M., and Tiedje, J.M. (2010). The genome sequence of *Psychrobacter arcticus* 273–4, a psychroactive Siberian permafrost bacterium, reveals mechanisms for adaptation to low-temperature growth. Appl. Environ. Microbiol. *76*, 2304–2312.

Bakermans, C., and Nealson, K.H. (2004). Relationship of critical temperature to macromolecular synthesis and growth yield in *Psychrobacter cryopegella*. J. Bacteriol. *186*, 2340–2345.

Bakermans, C., Tollaksen, S.L., Giometti, C.S., Wilkerson, C., Tiedje, J.M., and Thomashow, M.F. (2007). Proteomic analysis of *Psychrobacter cryohalolentis* k5 during growth at subzero temperatures. Extremophiles *11*, 343–354.

Bornefeld, T., and Simonis, W. (1974). Effects of light, temperature, pH, and inhibitors on the ATP level on the blue-green alga *Anacystis nidulans*. Planta *115*, 309–318.

Browse, J., and Xin, Z. (2001). Temperature sensing and cold acclimation. Curr. Opin. Plant Biol. *4*, 241–246.

Budde, I., Steil, L., Scharf, C., Völker, U., and Bremer, E. (2006). Adaptation of *Bacillus subtilis* to growth at low temperature: a combined transcriptomic and proteomic appraisal. Microbiology *152*, 831–853.

Cavicchioli, R. (2006). Cold-adapted Archaea. Nat. Rev. Microbiol. *4*, 331–343.

Chou, M., Matsunaga, T., Takada, Y., and Fukunaga, N. (1999). NH_4^+ transport system of a psychrophilic marine bacterium, *Vibrio* sp. strain ABE-1. Extremophiles *3*, 89–95.
Christner, B.C., Mosley-Thompson, E., Thompson, L.G., Zagorodnov, V., Sandman, K., and Reeve, J.R. (2000). Recovery and identification of viable bacteria immured in glacial ice. Icarus *144*, 479–485.
Colucci, M.S., and Inniss, W.E. (1996). Ethylene glycol utilization, cold and ethylene glycol shock and acclimation proteins in a psychrotophic bacterium. Curr. Microbiol. *32*, 179–182.
Conklin, A.R., and MacGregor, A.N. (1972). Soil adenosine triphosphate: extraction, recovery and half-life. Bull. Environ. Contam. Toxicol. *72*, 296–300.
Contin, M., Todd, A., and Brookes, P.C. (2001). The ATP concentration in the soil microbial biomass. Soil Biol. Biochem. *33*, 701–704.
Dalmasso, M., Aubert, J., Briard-Bion, V., Chuat, V., Deutsch, S.-M., Even, S., Falentin, H., Jan, G., Jardin, J., Maillard, M.-B., Parayre, S., Tanskanen, J., and Thierry, A. (2012). A temporal –omic study of *Propionibacterium freudenreichii* CIRM-BIA1T adaptation strategies in conditions mimicking cheese ripening in the cold. PLoS One *7*, e29083.
Eiler, A., Langenheder, S., Bertilsson, S., and Tranvik, L.J. (2003). Heterotrophic bacterial growth efficiency and community structure at different natural and organic carbon concentrations. Appl. Environ. Microbiol. *69*, 3701–3709.
Del Giorgio, P.A., Cole, J.J., and Cimbleris, A. (1997). Respiration rates in bacteria exceed phytoplankton production in unproductive aquatic systems. Nature *385*, 148–151.
English, T.E., and Storey, K.B. (2000). Enzymes of adenylate metabolism and their role in hibernation of the white tailed prairie dog, *Cynomys leucurus*. Arch. Biochem. Biophys. *376*, 91–100.
Feller, G., and Gerday, C. (2003). Psychrophilic enzymes: hot topics in cold adaptation. Nature Reviews Microbiology *1*, 200–208.
Foght, J., Aislabie, J., Turner, S., Brown, C.E., Ryburn, J., Saul, D.J., and Lawson, W. (2004). Culturable bacteria in subglacial sediments and ice from two Southern hemisphere glaciers. Microb. Ecol. *47*, 329–340.
Franzen, J.S., and Binkley, S.B. (1961). Comparaison of the acid-soluble nucleotides in *Escherichia coli* at different growth rates. J. Biol. Chem. *236*, 515–519.
García-Descalzo, L., Alcazar, A., Baquero, F., and Cid, C. (2011). Identification of in vivo HSP90-interacting proteins reveals modularity of HSP90 complexes is dependent on the environment in psychrophilic bacteria. Cell Stress and Chaperones *16*, 203–218.
Garnier, M., Matamoros, S., Chevret, D., Pilet, M.-F., Leroi, F., and Tresse, O. (2010). Adaptation to cold and proteomic responses of the psychrotrophic biopreservative *Lactococcus piscium* strain CNCM I-4031. Appl. Environ. Microbiol. *76*, 8011–8018.
Gilichinsky, D., Rivkina, E., Bakermans, C., Shcherbakova, V., Petrovskaya, L., Ozerskaya, S., Ivanushkina, N., Kochkina, G., Laurinavichuis, K., Pecheritsina, S., Fattakhova, R., and Tiedje, J.M. (2005). Biodiversity of cryopegs in permafrost. FEMS Microbiol. Ecol. *53*, 117–128.
Goodchild, A., Saunders, N.F.W., Ertan, H., Raftery, M., Guilhaus, M., Curmi, P.M.G., and Cavicchioli, R. (2004). A proteomic determination of cold adaptation in the Antarctic archaeon, *Methanococcoides burtonii*. Mol. Biol. *53*, 309–321.
Goodchild, A., Raftery, M., Saunders, N.F.W., Guilhaus, M., and Cavicchioli, R. (2005). Cold adaptation of the Antarctic Archaeon, *Methanococcoides burtonii* assessed by proteomics using ICAT. J. Prot. Res. *4*, 473–480.
Groudieva, T., Kambourova, M., Yusef, H., Royter, M., Grote, R., Trinks, H., and Antranikian, G. (2004). Diversity and cold-active hydrolytic enzymes of culturable bacteria associated with Arctic sea ice, Spitzbergen. Extremophiles *8*, 475–488.
Hebraud, M., Dubois, E., Potier, P., and Labadie, J. (1994). Effect of growth temperatures on the protein levels in a psychrotrophic bacterium, *Pseudomonas fragi*. J. Bacteriol. *176*, 4017–4024.
Homma, T., Iwahashi, H., and Komatsu, Y. (2003). Yeast gene expression during growth at low temperature. Cryobiology *46*, 230–237.
Jolkver, E., Emer, D., Ballan, S., Krämer, R., Eikmanns, B.J., and Marin, K. (2009). Identification and characterization of a bacterial transport system for the uptake of pyruvate, propionate, and acetate in *Corynebacterium glutamicum*, J. Bacteriol. *191*, 940–948.
Inouye, M. (2006). The discovery of mRNA interferases: implication in bacterial physiology and application to biotechnology. J. Cell. Physiol. *209*, 670–676.
Jones, P.C.T. (1970). The effects of light, temperature, and anaesthetics on ATP levels in the leaves of *Chenopodium rubrum* and *Phaseolus vulgaris*. J. Exp. Bot. *21*, 58–63.

Johnston, I.A., and Dunn, J. (1987). Temperature acclimation and metabolism in ectotherms with particular reference to teleost fish. Symposia of the Society for Experimental Biology *41*, 67–93.

Junge, K., Imhoff, F., Staley, T., and Deming, J.W. (2002). Phylogenetic diversity of numerically important Arctic sea-ice bacteria cultured at subzero temperature. Microb. Ecol. *43*, 315–321.

Junge, K., Eicken, H., Swanson, B.D., and Deming, J.W. (2006). Bacterial incorporation of leucine into protein down to −20°C with evidence for potential activity in sub-eutectic saline ice formations. Cryobiology *52*, 417–429.

Karl, D.M. (1980). Cellular nucleotide measurements and applications in microbial ecology. Microbiol. Rev. *44*, 739–796.

Karl, D.M., Bird, D.F., Björkman, K., Houlihan, T., Shackelford, R., and Tupas, L. (1999). Microorganisms in the accreted ice of Lake Vostok, Antarctica. Science *286*, 2144–2147.

Karr, E.A., Sattley, W.M., Jung, D.O., Madigan, M.T., and Achenbach, L.A. (2003). Remarkable diversity of phototrophic purple bacteria in a permanently frozen Antarctic Lake. Appl. Environ. Microbiol. *69*, 4910–4914.

Konings, W.N., Albers, S.-V., Koning, S., and Driessen, A.J.M. (2002). The cell membrane plays a crucial role in survival of bacteria and archaea in extreme environments. Ant. Van Leeuwenhoek *81*, 61–72.

Kurimoto, K., Harvey Millar, A., Lambers, H., Day, D.A., and Noguchi, K. (2004). Maintenance of growth rate at low termpertaure in rice and wheat cultivars with a high degree of respiratory homeostasis is associated with a high efficiency of respiratory ATP production. Plant Cell Physiol. *45*, 1015–1022.

Lim, J., Thomas, T., and Cavicchioli, R. (2000). Low temperature regulated DEAD-box RNA helicase from the Archaeon, *Methanococcoides burtonii*. J. Mol. Biol. *297*, 553–567.

Liu, S., Graham, J.E., Bigelow, L., Morse II, P.D., and Wilkinson, B.J. (2002). Identification of *Listeria monocytogenes* genes expressed in response to growth at low temperature. Appl. Environ. Microbiol. *68*, 1697–1705.

Mader, H.M., Pettitt, M.E., Wadham, J.L., Wolff, E.W., and Parkes, R.J. (2006). Subsurface ice as a microbial habitat. Geology *34*, 169–172.

Mair, T., Warnke, C., Tsuji, K., and Müller, S.C. (2005). Control of glycolytic oscillations by temperature. Biophys. J. *88*, 639–646.

Mailloux, R.J., Lemire, J., and Appanna, V.D. (2011). Metabolic networks to combat oxidative stress in *Pseudomonas fluorescens*. Ant. Van Leeuwenhoek *99*, 433–442.

Methé B.A., Nelson, K.E., Deming, J.E., Momen, B., Melamud, E., Zhang, X., Moult, J., Madupu, R., Nelson, W.C., Dodson, R.J., Brinkac, L.M., Daugherty, S.C., Durkin, A.S., DeBoy, R.T., Kolonay, J.F., Sullivan, S.A., Zhou, L., davidsen, T.M., Wu, M., Huston, A.L., Lewis, M., Weaver, B., Weidman, J.F., Khouri, H., Utterback, T.R., Feldblyum, T.V., and Fraser, C.M. (2005). The psychrophilic lifestyle as revealed by the genome sequence of *Colwellia psychrerythraea* 34H through genomic and proteomic analyses. Proc. Nat. Acad. Sci. USA *102*, 10913–10918.

Mikami, K., Kanesaki, Y., Suzuki, I., and Murata, N. (2002). The histidine kinase Hik33 perceives osmotic stress and cold stress in *Synechocystis* sp. PCC 6803. Mol. Biol. *46*, 905–915.

Morita, R.Y. (1975). Psychrophilic bacteria. Bacteriol. Rev. *39*, 144–167.

Morrison, B.A., and Shain, D.H. (2008). An AMP nucleosidase gene knockout in *Escherichia coli* elevates intracellular ATP levels and increases cold tolerance. Biol. Lett. *4*, 53–56.

Napolitano, M.J., and Shain, D.H. (2004). Four kingdoms on glacier ice: convergent energetic processes boost energy levels as temperatures fall. Proc. R. Soc. Lond. Ser. B *271* (Suppl.), S273–S276.

Napolitano, M.J., and Shain, D.H. (2005a). Distinctions in adenylate metabolism among organisms inhabiting temperature extremes. Extremophiles *9*, 93–98.

Napolitano, M.J., and Shain, D.H. (2005b). Quantitating adenylate nucleotides in diverse organisms. J. Biochem. Biophys. Methods *63*, 69–77.

Panikov, N.S., Flanagan, P.W., Oechel., W.C., Mastepanov, M.A., and Christensen, T.R. (2006). Microbial activity in soils frozen to below −39°C. Soil Biol. Biochem. *38*, 785–794.

Panoff, J.-M., Corroler, D., Thammavongs, B., and Boutibonnes, P. (1997). Differentiation between cold shock proteins and cold acclimation proteins in a mesophilic Gram-positive bacterium *Enterococcus faecalis* JH2–2. J. Bacteriol. *179*, 4451–4454.

Parry, B.R., and Shain, D.H. (2011). Manipulations of AMP metabolic genes increase growth rate and cold tolerance in *Escherichia coli*: implications for psychrophilic evolution. Mol. Biol. Evol. *28*, 2139–2145.

Piette, F., D'Amico, S., Struvay, C., Mazzucchelli, G., Renaut, J., Tutino, M.L., Danchin, A., Leprince, P., and Feller, G. (2010). Proteomics of life at low temperatures: trigger factor is the primary chaperone in the Antarctic bacterium *Pseudoalteromonas haloplanktis* TAC125. Mol. Microbiol. *76*, 120–132.

Piette, F., D'Amico, S., Mazzucchelli, G., Danchin, A., Leprince, P., and Feller, G. (2011). Life in the cold: a proteomic study of cold-repressed proteins in the Antarctic bacterium *Pseudoalteromonas haloplanktis* TAC125. Appl. Environ. Microbiol. 77, 3881–3883.

Ponder, M.A., Gilmour, S.J., Bergholz, P.W., Mindcock, C.A., Hollingsworth, R., Thomashow, M.F., and Tiedje, J.M. (2005). Characterization of potential stress responses in ancient Siberian permafrost psychroactive bacteria. FEMS Microbiol. Ecol. *53*, 103–115.

Poolman, B., Bosman, B., Kiers, J., and Konings, W.N. (1987). Control of glycolysis by glyceraldehydes-3-phosphate dehydrogenase in *Streptococcus cremoris* and *Streptococcus lactis*. J. Bacteriol. *169*, 5887–5890.

Priscu, J.C., and Christner, B.C. (2004). Earth's icy biosphere. In Microbial diversity and bioprospecting, A.T. Bull, eds. (Washington, DC, USA: ASM Press), pp. 130–145.

Rakotonirainy, M.S., Héraud, C., and Lavédrine, B. (2003). Detection of viable fungal spores contaminant on documents and rapid control of the effectiveness of an ethylene oxide disinfection using ATP assay. Luminescence *18*, 113–121.

Renoux, A., and Boulaud, D. (1998). Les aérosols – Physique et Métrologie. Lavoisier Tech & Doc Editions, Paris.

Roberts, M.E., and Inniss, W.E. (1992). The synthesis of cold shock proteins and cold acclimation proteins in the psychrophilic bacterium *Aquaspirillum arcticum*. Curr. Microbiol. *25*, 275–278.

Römisch, K., and Matheson, T. (2003). Cell biology in the Antarctic: studying life in the freezer. Nat. Cell Biol. *4*, 33–36.

Sardesai, N., and Babu, C.R. (2000). Cold stress induces switchover of respiratory pathway to lactate glycolysis in psychrotrophic *Rhizobium* strains. Folia Microbiol. *45*, 177–182.

Sattler, B., Puxbaum, H., and Psenner, R. (2001). Bacterial growth in supercooled cloud droplets. Geophys. Res. Lett. *28*, 239–242.

Saunders, N.F.W., Thomas, T., Curmi, P.M.G., Mattick, J.S., Kuczek, E., Slade, R., Davis, J., Franzmann, P.D., Boone, D., Rusterholtz, K., Feldman, R., Gates, C., Bench, S., Sowers, K., Kadner, K., Aerts, A., Dehal, P., Detter, C., Glavina, T., Lucas, S., Richardson, P., Larimer, F., Hauser, L., Land, M., and Cavicchioli, R. (2003). Mechanisms of thermal adaptation revealed from the genomes of the Antarctic Archaea *Methanogenium frigidum* and *Methanococcoides burtonii*. Genome Res. *13*, 1580–1588.

Seufferheld, M.J., Alvarez, H.M., and Farias, M.E. (2008). Role of polyphosphates in microbial adaptation to extreme environments. Appl. Environ. Microbiol. *74*, 5867–5874.

Singh, R., Lemire, J., Mailloux, R.J., and Appanna, V.D. (2008). A novel strategy involved anti-oxidative defense: the conversion of NADH into NADPH by a metabolic network. PLoS One 3, e2682.

Soini, J., Falschlehner, C., Mayer, C., Böhm, D., Weinel, S., Panula, J., Vasala, A., and Neubauer, P. (2005). Transient increase of ATP as a response to temperature up-shift in *Escherichia coli*. Microb. Cell Fact. *4*, 9. doi:10.1186/1475-2859-4-9.

Stibal, M., Šabacká, M., and Kaštovská, K. (2006). Microbial communities on glacier surfaces in Svalbard: Impact of physical and chemical properties on abundance and structure of Cyanobacteria and algae. Microb. Ecol. *52*, 644–654.

Suzuki, I., Kanesaki, Y., Mikami, K., Kanehisa, M., and Murata, N. (2001). Cold-regulated genes under control of the cold sensor Hik33 in *Synechocystis*. Mol. Biol. *40*, 235–244.

Takacs, C.D., and Priscu, J.C. (1998). Bacterioplankton dynamics in the McMurdo Dry Valley Lakes, Antarctica: production and biomass loss over four seasons. Microb. Ecol. *36*, 239–250.

Talts, P., Pärnik, T., Gardeström, P., and Keerberg, O. (2004). Respiratory acclimation in *Arabidopsis thaliana* leaves at low temperature. J. Plant Physiol. *161*, 573–579.

Vaïtilingom, M., Attard, E., Gaiani, N., Sancelme, M., Deguillaume, L., Flossmann, A.I., Amato, P., and Delort, A.-M. (2012). Long-term features of cloud microbiology at the puy de Dôme (France). Atmos. Environ. *56*, 88–100.

Wagner, D., Kobabe, S., Pfeiffer, E.-M., and Hubberten, H.-W. (2003). Microbial controls on methane fluxes from a polygonal tundra of the Lena delta, Siberia. Permafrost and periglacial processes *14*, 173–185.

Watanabe, S., Yamaoka, N., Fukunaga, N., and Takada, Y. (2002). Purification and characterization of a cold-adapted isocitrate lyase and expression analysis of the cold-inducible isocitrate lyase gene from the psychrophilic bacterium *Colwellia psychrerythraea*. Extremophiles *6*, 397–405.

CSPs in psychrophiles

Many studies have been carried out to identify CSPs in order to gain insight into the cold adaptation mechanisms of psychrophilic microorganisms. To date, the genomic sequences

Proteins Involved in Cold Adaptation

6

Kazuaki Yoshimune, Jun Kawamoto and Tatsuo Kurihara

Abstract

This chapter primarily describes cold shock proteins (CSPs), which are induced in response to temperature downshift in both psychrophiles and mesophiles. These proteins are important for various cellular processes, including transcription, translation, protein synthesis and folding, and membrane functions, to maintain their viability under cold conditions. Here, the CSPs in psychrophilic and mesophilic microorganisms are represented on the basis of the results of proteome analyses. In particular, CSPs that were isolated from the Antarctic psychrophile *Shewanella livingstonensis* Ac10 are described in detail. A number of microorganisms induce molecular chaperones in response to the cold, although the majority of chaperones are induced as a result of heat shock. Chaperones with peptidyl prolyl *cis–trans* isomerase activity are often induced by cold to accelerate protein folding by interconverting the *cis* and *trans* isomers of proline imidic peptide bonds, and these cold shock chaperones are also described here. Cold-adapted proteins often have higher flexibility as compared to mesophilic proteins. The features of cold-adapted proteins are briefly described. Finally, the production of recombinant proteins in psychrophiles is shown to suggest a future application of psychrophiles.

Cold shock proteins

Normal cellular function decreases at low temperatures, which leads to reduced growth of microorganisms. Psychrophiles and mesophiles also face these problems, but they induce cold shock proteins (CSPs) to partly compensate for these disadvantages. Although the temperatures at which cold shock occurs in psychrophiles are lower than those in mesophiles, their induced CSPs seem to share similar functions. Fig. 6.1 illustrates the functions of psychrophilic and mesophilic CSPs such as nucleic acid binding, energy metabolism, catabolism, and protein translation and folding (Kim *et al.*, 2005; Stubs *et al.*, 2005; Wouters *et al.*, 1998; Graumann *et al.*, 1997). In this section, we review proteomic studies of both psychrophiles and mesophiles, which will provide important clues regarding the mechanisms by which they adapt to extreme-cold environments. In this chapter, psychrophilic and psychrotolerant microorganisms are not distinguished; instead, the term 'psychrophilic microorganism' or 'psychrophile' is used throughout the text.

CSPs in psychrophiles

Many studies have been carried out to identify CSPs in order to gain insight into the cold-adaptation mechanisms of psychrophilic microorganisms. To date, the genomic sequences

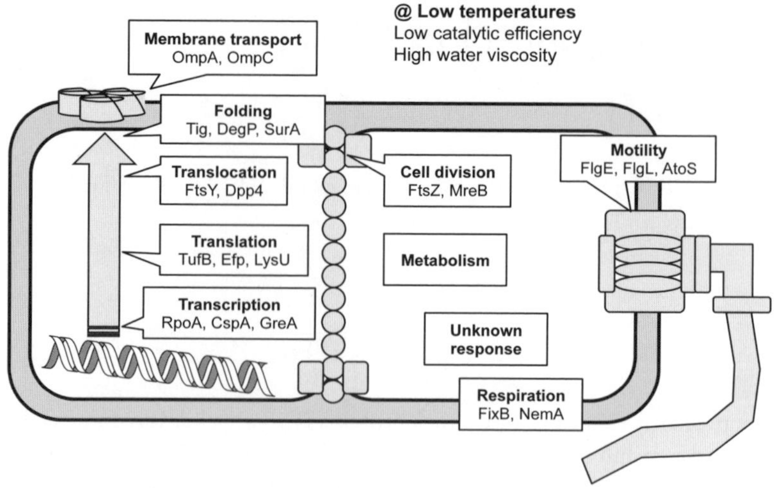

Figure 6.1 Illustrated functions of CSPs. *Shewanella livingstonensis* Ac10 inducibly produces proteins involved in transcription, translation, protein translocation, and protein folding. A RpoA homologue, one of the low-temperature specific proteins, which is probably involved in the expression of the genes required for the growth at low temperatures. CspA most likely acts as an RNA chaperone to maintain the conformation of RNA structure at low temperatures. Tig (TF) is supposed to act as the protein chaperones by its PIPase activity. TF probably triggers the translocation of Pro-OmpA, which is the precursor of the major-outer-membrane porin, OmpA. OmpA and OmpC may modulate the outer membrane permeability for hydrophilic compounds at low temperatures. The induction of OmpA and OmpC production at low temperatures might be a counteracting against the increase in water viscosity and the decrease in diffusion efficiency of solutes at low temperatures. Proteins involved in cell division, mobility, respiration, and metabolisms are also inducibly produced at low temperatures, which may support the cold adaptation of *S. livingstonensis* Ac10. Moreover, functionally unknown proteins identified as the cold-inducible proteins imply the possibility of the presence of the cold-adaptation mechanism of which remains unclear so far.

of about 30 psychrophilic organisms have been fully determined, and at the least, the study of additional 12 strains is under way (Casanueva *et al.*, 2010). In addition, a considerable number of genomic fragment data have also been obtained from metagenomic studies in various cold environments (Berlemont *et al.*, 2011). These sequence data provide a framework for fundamental and application studies of psychrophilic microorganisms. With the aid of these data, it is possible to make a global identification of the proteins produced under particular growth conditions. Recent proteomic studies identified a set of cold-inducible proteins, which illustrate how the cellular processes of psychrophilic microorganisms, namely *Sphingopyxis alaskensis* (Ting *et al.*, 2010) and *Pseudoalteromonas haloplanktis* TAC125 (Wilmes *et al.*, 2011), function at low temperatures. Proteomic studies, which were conducted on the proteins obtained from other psychrophilic microorganisms with available whole genomic DNA sequences, will clarify more distinctive features of the proteomes of psychrophilic microorganisms. Future proteomic studies will undoubtedly provide us more clues that will improve our understanding of the mechanisms employed by psychrophilic microorganisms to adapt to cold environments. This information will also be useful for engineering these microorganisms for biotechnology applications.

CSPs in *Shewanella livingstonensis* Ac10

S. livingstonensis Ac10, a psychrophilic bacterium isolated from Antarctic seawater, can grow at temperatures ranging from 4°C to 25°C, with an optimum growth temperature of 18°C. *Shewanella* species are widely distributed in various environments on earth, and complete genomic DNA sequences of various *Shewanella* species have been determined. These include mesophilic *S. oneidensis* MR-1 (Heidelberg *et al.*, 2002) and piezophilic *Shewanella violacea* DSS12 (Aono *et al.*, 2010). Comparative studies of these phylogenetically related strains that inhabit different environments would increase the understanding of the mechanisms they employ to adapt to various environments. In this respect, *S. livingstonensis* Ac10 is a fascinating model microorganism for studying microbial cold adaptation mechanisms.

Soluble and insoluble proteins of *S. livingstonensis* Ac10, which were grown at 4°C and 18°C, were subjected to two-dimensional gel electrophoresis (2DE), and the proteins whose relative abundance increased at 4°C by more than two-fold were identified using peptide mass fingerprinting (PMF; Kawamoto *et al.*, 2007). Twenty-six soluble proteins and two outer membrane proteins were inducibly produced at 4°C (Table 6.1). Two transcription-related proteins, RpoA and GreA, were up-regulated at 4°C; RpoA is the α-subunit of RNA polymerase that plays an important role in RNA polymerase-mediated protein assembly and promoter recognition, and GreA induces cleavage and removal of the 3′-proximal dinucleotide from nascent RNA to control transcriptional fidelity. Since the secondary structure of mRNA is greatly influenced by temperature, different sets of proteins in different amounts are probably required to ensure efficient and accurate transcription at various temperatures. The RNA chaperone CspA is also inducibly synthesized at 4°C, suggesting that this protein is essential to render the secondary structure of mRNA suitable for translation. Cold-inducible Tig probably contributes to efficient translation at low temperatures. Tig is a ribosome-associated chaperone with peptidyl prolyl *cis–trans* isomerase (PPIase; EC 5.2.1.8) activity that facilitates proper folding of newly synthesized proteins. It is composed of three domains: an N-terminal domain that mediates its association with the large ribosomal subunit, a central substrate-binding domain with homology to FK506-binding protein (FKBP) that has the PPIase activity, and a C-terminal domain of unknown function. PPIase, FKBP22 is also inducibly produced at low temperatures in a strain related to *S. livingstonensis* Ac10, namely *Shewanella* sp. SIB1, and the psychrophilic *Pseudoalteromonas haloplanktis* TAC125. Although Tig of *S. livingstonensis* Ac10 is similar to FKBP22 in that it is reported to have PPIase activity, Tig has two additional N-terminal and C-terminal domains that are not found in FKBP22.

Two putative outer membrane porin homologues, namely OmpA and OmpC, were inducibly produced at 4°C. Homologues of these proteins from other bacteria can form channels for hydrophilic solutes that play important roles in the uptake of nutrients. Up-regulation of these proteins probably counteracts the low diffusion rate of solutes at low temperatures and enables efficient uptake of nutrients.

The amounts of the hook-related proteins of flagella, namely FlgE and FlgL, were increased at low temperatures, suggesting that *S. livingstonensis* Ac10 modulates its motility by altering the expression of these proteins. Consistent with this speculation, we found that the cells were more motile at 4°C than at 18°C. The physiological significance of higher motility of this strain at low temperatures remains to be clarified.

S. livingstonensis Ac10 is a Gram-negative bacterium with a cell surface structure composed of three morphologically distinct layers: an inner membrane bordering the cytoplasm,

Table 6.1 Cold-inducible proteins of *Shewanella livingstonensis* Ac10

Gene	Protein[a]	Accession No.	Localization[b,c]
Group 1. RNA synthesis and folding			
rpoA	DNA-directed RNA polymerase α subunit	AB284093	CY
greA	Transcription elongation factor	AB284102	CY
cspA	Cold shock protein	AB284087	CY
Group 2. Protein synthesis and folding			
tufB	GTPase – translation elongation factor	AB284113	CY
tig	FKBP type peptidyl-prolyl *cis–trans* isomerase (trigger factor)	AB284101	CY
Efp	Translation elongation factor P/Ttranslation initiation factor	AB284108	CY
lysU	Lysyl -tRNA synthase class II	AB284089	CY
pspA	Phage shock protein A	AB646434	IM
degP	Protease Do	AB646446	IM
dpp4	Peptidase S9 prolyl oligopeptidase	AB646447	IM
surA	PpiC-type peptidyl-prolyl *cis–trans* isomerase	AB648914	IM
Group 3. Membrane transport			
ompC	Porin	AB284090	OM
ompA	Outer membrane protein A	AB284075	OM
sliv_c417088	Porin	AB646440	IM
Group 4. Motility			
flgE	Flagellar basal body and hook protein	AB284085	CY
flgL	Flagellin and related hook-associated protein	AB284088	CY
atoS	Methyl-accepting chemotaxis sensory transducer, PAS domain	AB646472	IM
Group 5. Cell division			
mreB	Cell shape-determining protein	AB646452	IM
ftsY	Signal recognition particle-docking protein	AB646454	IM
ftsZ	Cell division GTPase	AB284074	CY
Group 6. Metabolism			
ppx1	Inorganic pyrophosphatase/exopolyphosphatase	AB284081	CY
purD	Phosphoribosylamine–glycine ligase	AB284115	CY
deoC	Deoxyribose phosphate aldolase	AB284104	CY
fixB	Electron transfer flavoprotein α-subunit	AB284109	CY
nemA	NADPH: flavin oxidoreductase	AB284084	CY
pdxJ	Pyridoxal phosphate biosynthesis protein	AB284100	CY
ald	Alanine dehydrogenase	AB646461	IM
sdhA	Succinate dehydrogenase/fumarate reductase	AB646463	IM
pykF	Pyruvate kinase	AB646458	IM
leuB	Isocitrate dehydrogenase	AB646462	IM
sucB	Dihydrolipoamide acyltransferase	AB646465	IM

Gene	Protein[a]	Accession No.	Localization[b,c]
ribP_PPkin	Ribose phosphate pyrophosphokinase	AB646468	IM
Group 7. Other function			
ac10nuc	Predicted extracellular nuclease	AB284091	CY
ac10cp	Predicted carboxypeptidase	AB284086	CY
ac10C417081	6Fe-6S prismane cluster-containing protein	AB284110	CY
ac10omp	Predicted outer membrane protein	AB284079	CY
tsx	Nucleotide-binding outer membrane protein	AB284082	CY
ac10tn	Transposase	AB284094	CY
Group 8. Unknown function			
ac10C130109	–	AB284083	CY
ac10C097028	–	AB284078	CY
ac10C346010	–	AB284107	CY
ac10C084124	–	AB284077	CY

[a]Proteins in the database showing the highest sequence similarity to the proteins from *S. livingstonensis* Ac10.
[b]Localization of proteins identified by proteomic analysis are characterized as follows: OM, outer membrane; IM, inner membrane; CY, cytoplasm.
[c]Inner membrane proteins identified by proteomic analysis of the inner membrane fraction (Park *et al*., 2012).

a periplasm containing the peptidoglycan layer external to the inner membrane, and an outer membrane on the external surface of the cell. The inner membrane proteins are involved in many key cellular processes such as energy generation and conversion in the respiratory chain, cell division, signal transduction, and transport processes. Because of the presence of large amount of outer membrane proteins in the crude membrane extracts, proteome analysis of inner membrane proteins is often difficult. Therefore, it is desirable to separate the inner and outer membranes in order to facilitate identification of inner membrane proteins. Recently, we established a method for separating the inner and outer membranes, and successfully identified the inner membrane proteins that are inducibly produced at low temperatures, thus providing a better understanding of the cold adaptation mechanism of this strain (Park *et al.*, 2012).

Inner membrane proteins inducibly produced at low temperatures that might support the adaptation of *S. livingstonensis* Ac10 were identified by proteomic analysis of the phospholipid-enriched fraction obtained from cells grown at 4°C and 18°C. Using PMF analysis, 52 protein spots were identified, 19 of which were predicted inner membrane proteins; six were predicted outer membrane proteins; 22 were predicted cytoplasmic proteins; and 5 were predicted periplasmic proteins. The following proteins were more numerous (>2-fold) at 4°C than at 18°C (Table 6.1): AtoS, PspA, MreB, FtsY, Ald, SdhA, DegP, SurA, Dpp4, PykF, LeuB, SucB, RibP_PPkin, and Porin Sliv_c417088. These proteins might contribute to protein folding, morphogenesis, environmental sensing, and other cellular processes at low temperatures, as discussed in more detail below.

AtoS, a sensor kinase of the AtoSC two-component system, was inducibly produced at 4°C, and is a member of the family of Per-ARNT-Sim (PAS) domain-containing proteins that monitor changes in light, redox potential, oxygen, small ligands, and overall energy levels in a cell. AtoSC is reported to participate in flagella synthesis and chemotaxis in *Escherichia coli.* As mentioned above, *S. livingstonensis* Ac10 inducibly produces changes in the flagella-related proteins FlgE and FlgL, which lead to changes in its motility at 4°C. AtoS may be involved in the inducible production of these flagella-related proteins, thereby contributing to the increased motility of *S. livingstonensis* Ac10 at low temperatures. This might be a method of counteracting the lower diffusion rates of nutrient compounds in this strain at low temperatures. Under such unfavourable environments, *S. livingstonensis* Ac10 might show increased motility to facilitate movement towards more favourable conditions, i.e. more favourable rates of nutrient uptake.

The periplasmic chaperones DegP and SurA are expressed at low temperatures. These proteins are located adjacent to membrane phospholipids and/or the inner membrane secretion system such as the Sec apparatus, and are involved in the targeting pathway of outer membrane integral β-barrel proteins. In this strain, two putative outer membrane porins, namely, OmpA and OmpC, are produced at high levels at low temperatures (Table 6.1), which may contribute to the efficient uptake of nutrients at low temperatures. Cold-inducible DegP and SurA are thought to be important for the increased production of OmpA and OmpC at low temperatures, because low thermal energy and high aqueous viscosity at low temperatures may decrease the efficiency of protein folding.

An actin-like protein, MreB, was inducibly produced at low temperatures. In bacteria, the process of cell elongation is modulated by MreB, which is believed to localize the components of the peptidoglycan synthetic machinery along the lateral cell wall, thereby governing the geometry of cell wall growth. The cell of *S. livingstonensis* Ac10 becomes shorter at 4°C than it was at 18°C. MreB may play a role in regulating the morphology of this strain in response to its growth temperatures.

CSPs in mesophiles

Mesophilic *E. coli* grows at warm temperatures up to 49°C (Toshimune *et al.*, 2005b); this temperature is approximately 20°C higher than the maximum growth temperature of *S. livingstonensis* Ac10 (Kulakova *et al.*, 1999). Despite this difference in maximum growth temperatures, mesophiles seem to face problems similar to those faced by the psychrophiles at lower temperatures. Folding rates of some proteins decrease under conditions of low temperature, and chaperones that facilitate the folding of these proteins are induced. These cold shock chaperones are explained in the following section. RNA tends to form secondary structures that cause transcription termination and protein stabilization at low temperatures. Cold-inducible CspA binds to RNA to avoid these negative outcomes (Bae *et al.*, 2000; Phadtare, 2011; Uh *et al.*, 2010). In *E. coli*, CspA is induced in more than 10% of the total cellular protein synthesis (Goldstein *et al.*, 1990). The cold shock *E. coli* protein CsdA unwinds double-stranded RNA in the absence of ATP (Jones *et al.*, 1996).

Low temperatures lead to rigidified cell membrane structures that decrease their physiological functions. Increase in the amounts of unsaturated and branched-chain lipids in the lipid bilayer, which introduces steric hindrance, is found to preserve the fluidity in cell membranes (Russell, 2007). At cold temperatures, when the solubility of gases increases, the microorganisms are exposed to higher concentrations of oxygen that can lead to increased production of reactive oxygen species. To eliminate the excess toxic active oxygen species, *Acidithiobacillus ferrooxidans* induces the antioxidant proteins AhpC-Tsa (Mykytczuk *et al.*, 2011).

Many CSPs are also induced by low concentrations of a specific group of antibiotics that include chloramphenicol, tetracycline, erythromycin, spiramycin, and fusidic acid, which inhibit ribosome formation, thus reducing the rate of translation (Van Bogelen and Neidhardt, 1990). Because these antibiotics inhibit the induction of heat shock proteins upon increase in temperature, ribosomes are thought to be thermosensors of low temperatures (Van Bogelen and Neidhardt, 1990).

Chaperones in cold adaptation

The rate of protein folding decreases at low temperatures, resulting in slower metabolic rates. To overcome these disadvantages, microorganisms induce some chaperones at these low temperatures. Chaperones that are CSPs are listed in Table 6.2. Trigger factor (TF) associates with nascent polypeptides on ribosomes to facilitate their folding, and is suggested to be a rate-limiting step in the growth of the psychrophilic bacterium *P. haloplanktis* TAC125 at cold temperatures (Piette *et al.*, 2010). In addition to the chaperone function, TF has PPIase activity. As shown in Fig. 6.2, PPIase catalyses the interconversion of *cis–trans* peptide bonds that precede proline residues in the protein to facilitate its folding. PPIase is also found in the periplasm of both psychrotrophic and mesophilic *A. ferrooxidans* strains (Lazar *et al.*, 1998), and these proteins, designated as SurA, are induced at cold temperatures (Mykytczuk *et al.*, 2011). The activity of PPIase is significantly reduced at low temperatures (Budiman *et al.*, 2011). The induction of TF should compensate for the decreased function at cold temperatures. In *E. coli*, TF is a CSP, and its cellular levels increase at temperatures below 37°C (Kandror and Goldberg, 1997). This could be explained by the fact that overproduction of TF protects against cold and reduces viability at 50°C (Kandror and Goldberg, 1997).

Extensive mutational analyses of chaperones in *E. coli* have been performed to reveal their functions. Although TF is not essential for growth at all temperatures, double-defective mutants of the genes encoding TF and DnaK lead to the formation of lethal protein aggregates (Teter *et al.*, 1999, Deuerling *et al.*, 1999) at temperatures above 30°C (Vorderwulbecke *et al.*, 2004). DnaK is a member of the 70-kDa heat shock protein (Hsp70) family, and Hsp70 has ATPase activity that is associated with conformational changes to capture and release the substrate peptide. In *E. coli*, DnaK is necessary for growth at temperatures higher than 43°C and lower than 16°C (Itikawa and Ryu, 1979; Bukau and Walker, 1990). However, the expression of DnaK is suppressed at cold temperatures (Kim *et al.*, 2005). This fact can be explained by assuming that the function of DnaK is to suppress the activity of heat shock expression factor σ^{32} at low temperatures (Piette *et al.*, 2011). Cold sensitivity of the *dnaK*-defective mutant is suppressed by a mutation in the *rpoH* gene that encodes σ^{32} (Bukau and Walker, 1990). In contrast, the σ^{32}-defective mutant cannot grow at temperatures above 20°C (Zhou *et al.*, 1988). This is because heat shock chaperones that are induced by the function of σ^{32} are necessary at higher temperatures (Mogk *et al.*, 1999).

S. livingstonensis Ac10 induces DnaK (SheDnaK) at 24°C, which is approximately 20°C lower than the temperature causing heat shock in mesophilic *E. coli* (Yoshimune *et al.*,

Table 6.2 Cold shock chaperones in psychrophiles and mesophiles

Protein	Function	Class	Reference
Chaperones with PPIase activity			
FKBP22	PPIase	P	Budiman *et al.* (2011)
PpiB	PPIase	M	Kim *et al.* (2005)
SurA	Periplasmic PPIase	M, P	Mykytczuk *et al.* (2011)
Trigger factor	Folding of newly synthesized	M, P	Mykytczuk *et al.* (2011)
	proteins with PPIase activity	M	Kandror and Goldberg (1997)
		P	Kawamoto *et al.* (2007)
		P	Qiu *et al.* (2006)
Mbur_1485	PPIase	P	Williams *et al.* (2010)
Mbur_2256	PPIase	P	Williams *et al.* (2010)
YP_616522	PPIase	P	Ting *et al.* (2010)
Other chaperones			
Clp	ATP-dependent protease chaperone	M	Porankiewicz *et al.* (1998)
		P	Ting *et al.* (2010)
DnaK	Folding of newly synthesized proteins, Hsp70 family	P	Qiu *et al.* (2006)
DnaJ	Co-chaperone of DnaK	P	Ting *et al.* (2010)
GrpE	Co-chaperone of DnaK	P	Ting *et al.* (2010)
GroEL	Folding of partially folded proteins, Hsp60 family	P	Mykytczuk *et al.* (2011)
		P	Zheng *et al.* (2007)
		P	Ting *et al.* (2010)
GroES	Co-chaperone of GroEL	M	Stubs *et al.* (2005)
		P	Seo *et al.* (2004)
		M, P	Mykytczuk *et al.* (2011)
Hsc66	Iron–sulfur cluster assembly	M	Lelivert and Kawula (1995)
			Reyda *et al.* (2009)
SecB	Protein translocation	P	Mykytczuk *et al.* (2011)

Psychrophiles and mesophiles are represented by ‘P’ and ‘M’, respectively.

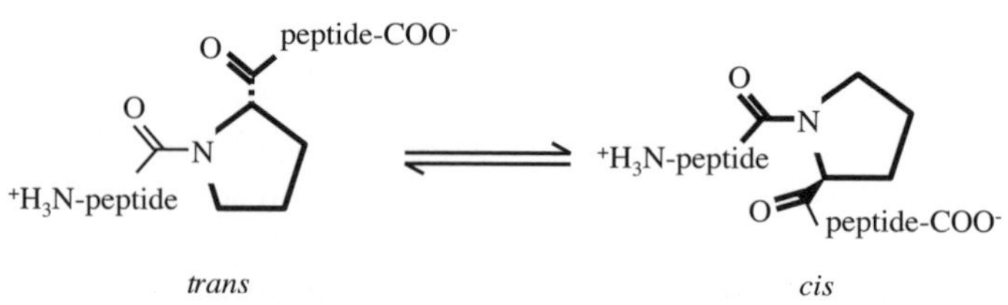

Figure 6.2 Reactions catalysed by PPIase. PPIase accelerates the conversion between *cis* and *trans* conformations. In denatured proteins, all peptide bonds are assumed to be in the energetically stable *trans* conformation, while approximately 20% of peptides are assumed to be in the *cis* conformation. The *trans*-to-*cis* conversion is a rate-limiting step in protein folding (Budiman, 2011).

2005). DnaK from psychrotolerant bacteria exhibits higher ATPase activities under cold temperatures. The optimum temperatures for the ATPase activities of SheDnaK and DnaK from *E. coli*, and DnaK from psychrotolerant *Vibrio proteolyticus* – which cannot grow at temperatures above 40°C (Yoshimune *et al.*, 2005a) are 35°C, 60°C, and 45°C, respectively. It is worth noting that *E. coli* has three Hsp70 homologues, namely, DnaK, Hsc66, and Hsc62 (Yoshimune *et al.*, 2002) with the optimum temperature for ATPase activity at 60°C, 45°C, and 37°C, respectively (Yoshimune *et al.*, 1998). Hsc66 is a CSP (Lelivelt and Kawula, 1995) that forms an operon with the genes of proteins required for proper functioning of iron–sulfur cluster proteins; it is also suggested to function in the assembly, stabilization, and transfer of Fe/S clusters (Silberg *et al.*, 2001). HscC (Hsc62) is required for cellular resistance to Cd^{2+} ions and UV irradiation (Kluck *et al.*, 2002).

In *E. coli*, protein folding is mainly facilitated by TF, DnaK, and GroEL. GroEL functions as a chaperone with its co-chaperone GroES. These chaperones are also necessary for growth at cold temperatures. Overproduction of GroEL and GroES restored growth of a TF and DnaK double mutant (Vorderwulbecke *et al.*, 2004), and GroEL and GroES are essential for the growth of *E. coli* at all temperatures (Fayet *et al.*, 1989).

Features of cold-adapted proteins

Some metabolic enzymes are induced during cold shock, suggesting that their activities are insufficient for growth at low temperatures. Proteins in psychrophiles should adapt to low temperature to function under these growth conditions. The primary strategies for cold adaptation are increased flexibility at low temperatures (D'Auria *et al.*, 2009), compared to mesophilic and thermophilic proteins. Protein flexibility is obtained by a decrease in the strength of weak interactions such as salt bridges and hydrogen bonds. This adaptation mechanism seems to be the opposite of the mechanisms adopted by thermophilic proteins that mainly undergo increase in rigidity. Thermostability is increased when hydrophobicity is increased in the cores of these proteins (Ohkuri and Yamagishi, 2007), by the introduction of an inter-subunit disulfide bond (Kabashima *et al.*, 2001), or by substitution of the rigid amino acid proline (Watanabe *et al.*, 1996). Comparisons between the amino acid sequences of the psychrophilic and mesophilic proteins revealed that psychrophilic proteins have higher compositions of amino acid residues with smaller side chains, such as Ser, Asp, Thr and Ala, instead of Glu and Leu residues, which have larger side chains in their coil regions (Metpally and Reddy, 2009).

Enhanced flexibilities of these cold-adapted proteins often lead to weak thermostabilities (Watanabe *et al.*, 2002; Huston *et al.*, 2008). However, some psychrophilic enzymes have higher thermostability compared with their mesophilic counterparts (Kazuoka *et al.*, 2003; Yamanaka *et al.*, 2002). The structures of these enzymes suggest that enhanced flexibility of a selected area, but not of the overall protein structure, is adequate for its activity under cold conditions. Some random mutations increase the activity of thermophilic enzymes at low temperatures (Theriot *et al.*, 2010). Low-temperature activity of 3-isopropylmalate dehydrogenase from *Thermus thermophilus* is improved by the modulation of coenzyme-binding pockets by using site-specific mutation (Hayashi *et al.*, 2011). These facts also support the hypothesis that increased flexibility in a selected area, but not in the overall protein structure, is adequate for adaptation to the cold conditions.

Production of recombinant proteins in cold conditions

Heterogeneous proteins are often produced from mesophiles such as *E. coli*, yeast, and Chinese hamster ovary cells. However, these mesophilic systems have disadvantages regarding the production of thermolabile proteins, which tend to result in degradation by the host proteases and formation of insoluble inclusion bodies (Speed *et al.*, 1996). Production of recombinant proteins in psychrophiles at low temperatures partly solves these problems. Transformation procedures for *S. livingstonensis* were established, and a heterologous protein expression system was developed (Miyake *et al.*, 2007). Target gene disruption methods were also established, which may be useful to genetically modify this strain to improve the productivity of recombinant proteins (Kawamoto *et al.*, 2009). The expression system of the psychrotolerant *S. livingstonensis* Ac10 produces higher amounts of recombinant proteins from the psychrophilic *Desulfotalea psychrophila*, compared to those produced by *E. coli* at 18°C (Miyake *et al.*, 2007). The secretion system of the Antarctic bacterium *P. haloplanktis* is useful for the production of recombinant proteins (Cusano *et al.*, 2006). Protein expression in psychrophile is an efficient alternative method for producing recombinant proteins (Tamura *et al.*, 2011, Inouye *et al.*, 2008, Hayashi and Kojima, 2008).

Conclusions

Recent proteomic studies have revealed increasing numbers of CSPs in the psychrophiles. These studies showed that CSPs are associated with various cellular processes such as gene expression, protein synthesis and folding, membrane transport, motility, and cell division that support psychrophile growth at low temperatures (Fig. 6.1). Mesophiles also induce CSPs that have functions similar to those of the CSPs in psychrophiles. Both psychrophiles and mesophiles induce some chaperones as a result of cold shock, although the majority of chaperones are induced by heat shock. Chaperones with PPIase activity are often detected as CSPs, suggesting that this activity is important at cold temperatures to facilitate protein folding. Cold-adapted proteins show higher flexibilities than their mesophilic counterparts, and these high flexibilities often lead to low thermostabilities. However, some enzymes in psychrophiles show higher thermostabilities than their mesophilic counterparts do. This fact suggests that increased flexibility in specific parts of a protein is adequate for cold adaptation. Cold-adapted proteins with low thermostabilities can be produced in psychrophiles at low temperatures. Production of recombinant proteins in psychrophiles can provide a useful alternative to that in mesophiles.

Future trends

Recent proteomic studies have identified a set of cold-inducible proteins, and they illustrate the mechanisms by which the cellular processes of the psychrophilic microorganisms *S. alaskensis* and *P. haloplanktis* TAC125 survive low temperatures. Proteomic studies in many other psychrophilic microorganisms whose whole genomic DNA sequences are available will allow more distinctive features of their proteomes to be clarified. Future proteomic studies will undoubtedly provide us with more clues to help understand how psychrophilic microorganisms adapt to cold environments; this information will also be useful for engineering these microorganisms for biotechnology applications. After targeting the genes that are important for their viability at cold temperatures, gene disruption will provide further information regarding their cold adaptation mechanisms.

Cold-adapted enzymes have high flexibilities, and thermotolerant enzymes have rigid structures. Some proteins in psychrophiles have high thermostabilities. These proteins seem to adapt to cold conditions by increasing the flexibility and rigidity of certain areas in their structures. It is possible that enzymes with high flexibilities in the catalytic part and their surrounding rigid structures show higher activities under cold conditions and higher stabilities at high temperatures. Such enzymes could be useful for industrial applications that require higher activities at low temperature and higher stabilities. Future research will elucidate the cold-adaptation mechanisms of these enzymes, which will allow development of methods to improve enzyme activity at cold temperatures and enzyme thermostability at high temperatures.

Psychrophiles can potentially be used for the efficient production of recombinant proteins under cold conditions. The transformation procedures and heterologous protein expression system in the psychrophile *S. livingstonensis* Ac10 provide an efficient way of analysing the function of expressed genes. Furthermore, these methods are useful for the production of thermolabile proteins.

References

Aono, E., Baba, T., Ara, T., Nishi, T., Nakamichi, T., Inamoto, E., Toyonaga, H., Hasegawa, M., Takai, K., Okumura, Y., *et al.* (2010). Complete genome sequence and comparative analysis of *Shewanella violacea*, a psychrophilic and piezophilic bacterium from deep sea floor sediments. Mol. Biosyst. *6*, 1216–1226.

Bae, W., Xia, B., Inouye, M., and Severinov, K. (2000). *Escherichia coli* CspA-family RNA chaperones are transcription antiterminators. Proc. Natl. Acad. Sci. U.S.A. *97*, 7784–7789.

Berlemont, R., Pipers, D., Delsaute, M., Angiono, F., Feller, G., Galleni, M., and Power, P. (2011). Exploring the Antarctic soil metagenome as a source of novel cold-adapted enzymes and genetic mobile elements. Rev. Argent. Microbiol. *43*, 94–103.

Budiman, C., Koga, Y., Takano, K., and Kanaya, S. (2011). FK506-binding protein 22 from a psychrophilic bacterium, a cold shock-inducible peptidyl prolyl isomerase with the ability to assist in protein folding. Int. J. Mol. Sci. *12*, 5261–5284.

Bukau, B., and Walker, G.C. (1990). Mutations altering heat shock specific subunit of RNA polymerase suppress major cellular defects of *E. coli* mutants lacking the DnaK chaperone. EMBO J. *9*, 4027–4036.

Casanueva, A., Tuffin, M., Cary, C., and Cowan, D.A. (2010). Molecular adaptations to psychrophily: the impact of 'omic' technologies. Trends Microbiol. *18*, 374–381.

Cusano, A.M., Parrilli, E., Marino, G., and Tutino, M.L. (2006). A novel genetic system for recombinant protein secretion in the Antarctic *Pseudoalteromonas haloplanktis* TAC125. Microb. Cell Fact. *5*, 40.

D'Auria, S., Aurilla, V., Marabotti, A., Gonnelli, M., and Strambini, G. (2009). Structure and dynamics of cold-adapted enzymes as investigated by phosphorescence spectroscopy and molecular dynamics studies. 2. The case of an esterase from *Pseudoalteromonas haloplanktis*. J. Phys. Chem. B *113*, 13171–13178.

Deuerling, E., Schulze-Specking, A., Tomoyasu, T., Mogk, A., and Bukau, B. (1999). Trigger factor and DnaK cooperate in folding of newly synthesized proteins. Nature *400*, 693–696.

Fayet, O., Ziegelhoffer, T., and Georgopoulos, C. (1989). The groES and groEL heat shock gene products of *Escherichia coli* are essential for bacterial growth at all temperatures. J. Bacteriol. *171*, 1379–1385.

Goldstein, J., Pollitt, N.S., and Inouye, M. (1990). Major cold shock protein of *Escherichia coli*. Proc. Natl. Acad. Sci. U.S.A. *87*, 283–287.

Graumann, P., Wendrich, T.M., Weber, M.H., Schroder, K., and Marahiel, M.A. (1997). A family of cold shock proteins in *Bacillus subtilis* is essential for cellular growth and for efficient protein synthesis at optimal and low temperatures. Mol. Microbiol. *25*, 741–756.

Hayashi, S., Akanuma, S., Onuki, W., Tokunaga, C., and Yamagishi, A. (2011). Substitutions of coenzyme-binding, nonpolar residues improve the low-temperature activity of thermophilic dehydrogenases. Biochemistry *50*, 8583–8593.

Hayashi, K., and Kojima, C. (2008). pCold-GST vector: a novel cold shock vector containing GST tag for soluble protein production. Protein Expr. Purif. *62*, 120–127.

Heidelberg, J.F., Paulsen, I.T., Nelson, K.E., Gaidos, E.J., Nelson, W.C., Read, T.D., Eisen, J.A., Seshadri, R., Ward, N., Methe, B., et al. (2002). Genome sequence of the dissimilatory metal ion-reducing bacterium *Shewanella oneidensis*. Nat. Biotechnol. *20*, 1118–1123.

Huston, A.L., Haeggstrom, J.Z., and Feller, G. (2008). Cold adaptation of enzymes: structural, kinetic and microcalorimetric characterizations of an aminopeptidase from the Arctic psychrophile *Colwellia psychrerythraea* and of human leukotriene. Biochim. Biophys. Acta. *1784*, 1865–1872.

Inouye, S., and Sahara, Y. (2008). Soluble protein expression in *E. coli* cells using IgG-binding domain of protein A as a solubilizing partner in the cold induced system. Biochem. Biophy. Res. Commun. *376*, 448–453.

Itikawa, H., and Ryu, J. (1979). Isolation and characterization of a temperature-sensitive *dnaK* mutant of *Escherichia coli* B. J. Bacteriol. *138*, 339–344.

Jones, P.G., Mitta, M., Kim, Y., Jiang, W., and Inouye, M. (1996). Cold shock induces a major ribosomal-associated protein that unwinds double-stranded RNA in *Escherichia coli*. Proc. Natl. Acad. Sci. U.S.A. *93*, 76–80.

Kabashima, T., Li, Y., Kanada, N., Ito, K., and Yoshimoto, T. (2001). Enhancement of the thermal stability of pyroglutamyl peptidase I by introduction of an intersubunit disulfide bond. Biochim. Biophys. Acta *1547*, 214–220.

Kandror, O., and Goldberg, A.L. (1997). Trigger factor is induced upon cold shock and enhances viability of *Escherichia coli* at low temperatures. Proc. Natl. Acad. Sci. U.S.A. *94*, 4978–4981.

Kawamoto, J., Kurihara, T., Kitagawa, M., Kato, I., and Esaki, N. (2007). Proteomic studies of an Antarctic cold-adapted bacterium, *Shewanella livingstonensis* Ac10, for global identification of cold-inducible proteins. Extremophiles *11*, 819–826.

Kawamoto, J., Kurihara, T., Yamamoto, K., Nagayasu, M., Tani, Y., Mihara, H., Hosokawa, M., Baba, T., Sato, S.B., and Esaki, N. (2009). Eicosapentaenoic acid plays a beneficial role in membrane organization and cell division of a cold-adapted bacterium, *Shewanella livingstonensis* Ac 10. J. Bacteriol. *191*, 632–640.

Kazuoka, T., Masuda, Y., Oikawa, T., and Soda, K. (2003). Thermostable aspartase from a marine psychrophile, *Cytophaga* sp. KUC-1: molecular characterization and primary structure. J. Biochem. *133*, 51–58.

Kim, Y.H., Han, K.Y., Lee, K., and Lee, J. (2005). Proteome response of *Escherichia coli* fed-batch culture to temperature downshift. Appl. Microbiol. Biotechnol. *68*, 786–793.

Kluck, C.J., Patzelt, H., Genevaux, P., Brehmer, D., Rist, W., Schneider-Mergener, J., Bukau, B., and Mayer, M.P. (2002). Structure–function analysis of HscC, the *Escherichia coli* member of a novel subfamily of specialized Hsp70 chaperones. J. Biol. Chem. *277*, 41060–41090.

Kulakova, L., Galkin, A., Kurihara, T., Yoshimura, T., and Esaki, N. (1999). Cold-active serine alkaline protease from the psychrotrophic bacterium *Shewanella* strain Ac10: gene cloning and enzyme purification and characterization. App. Environ. Microbiol. *65*, 611–617.

Lazar, S.W., Almiron, M., Tormo, A., and Kolter, R. (1998). Role of the *Escherichia coli* SurA protein in stationary-phase survival. J. Bacteriol. *180*, 5704–5711.

Lelivelt, M.J., and Kawula, T.H. (1995). Hsc66, an Hsp70 homolog in *Escherichia coli*, is induced by cold shock but not by heat shock. J. Bacteriol. *177*, 4900–4907.

Metpally, R.P., and Reddy, B.V. (2009). Comparative proteome analysis of psychrophilic versus mesophilic bacterial species: Insights into the molecular basis of cold adaptation of proteins. BMC Genomics *10*, 11.

Miyake, R., Kawamoto, J., Wei, Y.L., Kitagawa, M., Kato, I., Kurihara, T., and Esaki, N. (2007). Construction of a low-temperature protein expression system using a cold-adapted bacterium, *Shewanella* sp. strain Ac10, as the host. Appl. Environ. Microbiol. 73, 4849–4856.

Mogk, A., Tomotasu, T., Goloubinoff, P., Rudiger, S., Roder, D., Langen, H., and Bukau, B. (1999). Identification of thermolabile *Escherichia coli* proteins: prevention and reversion of aggregation by DnaK and ClpB. EMBO J. *18*, 6934–6949.

Mykytczuk, N.C., Trevors, J.T., Foote, S.J., Leduc, L.G., Ferroni, G.D., and Twine, S.M. (2011). Proteomic insights into cold adaptation of psychrotrophic and mesophilic *Acidithiobacillus ferrooxidans* strains. Antonie Van Leeuwenhoek *100*, 259–277.

Ohkuri, T., and Yamagishi, A. (2007). The effects of mutations at position 253 on the thermostability of the *Bacillus subtilis* 3-isopropylmalate dehydrogenase subunit interface. J. Biochem. *141*, 791–797.

Park, J., Kawamoto, J., Esaki, N., and Kurihara, T. (2012). Identification of cold-inducible inner membrane proteins of the psychrotrophic bacterium, *Shewanella livingstonensis* Ac10, by proteomic analysis. Extremophiles *16*, 227–236.

Phadtare, S. (2011). Unwinding activity of cold shock proteins and RNA metabolism. RNA Biol. *8*, 394–397.

Piette, F., D'Amico, S., Mazzucchelli, G., Danchin, A., Leprince, P., and Feller, G. (2011). Life in the cold: a proteomic study of cold-repressed proteins in the Antarctic bacterium *Pseudoalteromonas haloplanktis* TAC125. Appl. Environ. Microbiol. *77*, 3881–3883.

Piette, F., D'Amico, S., Struvay, C., Mazzucchelli, G., Renaut, J., Tutino, M.L., Danchin, A., Leprince, P., and Feller, G. (2010). Proteomics of life at low temperatures: trigger factor is the primary chaperone in the Antarctic bacterium *Pseudoalteromonas haloplanktis* TAC125. Mol. Microbiol. *76*, 120–132.

Porankiewicz, J., Schelin, J., and Clarke, A.K. (1998). The ATP-dependent Clp protease is essential for acclimation to UV-B and low temperature in the cyanobacterium *Synechococcus*. Mol. Microbiol. *29*, 275–283.

Qiu, Y., Kathariou, S., and Lubman, D.M. (2006). Proteomic analysis of cold adaptation in a Siberian permafrost bacterium--*Exiguobacterium sibiricum* 255–15 by two-dimensional liquid separation coupled with mass spectrometry. Proteomics *6*, 5221–5233.

Reyda, M.R., Fugate, C.J., and Jarrett, J.T. (2009). A complex between biotin synthase and the iron–sulfur cluster assembly chaperone HscA that enhances in vivo cluster assembly. Biochemistry *48*, 10782–10792.

Russell, N.J. (2007). Psychrophiles: membrane adaptations. In Physiology and Biochemistry of Extremophiles. Gerday, C., and Glansdorff, N. (eds). Washington, D.C.: ASM Press, pp. 155–164.

Seo, J.B., Kim, H.S., Jung, G.Y., Nam, M.H., Chung, J.H., Kim, J.Y., Yoo, J.S., Kim, C.W., and Kwon, O. (2004). Psychrophilicity of *Bacillus psychrosaccharolyticus*: a proteomic study. Proteomics *4*, 3654–3659.

Silberg, J.J., Hoff, K.G., Tapley, T.L., and Vickery, L.E. (2001). The Fe/S assembly protein IscU behaves as a substrate for the molecular chaperone Hsc66 from *Escherichia coli*. J. Biol. Chem. *276*, 1696–1700.

Speed, M.A., Wang, D.I., and King, J. (1996). Specific aggregation of partially folded polypeptide chains: the molecular basis of inclusion body composition. Nat. Biotechnol. *14*, 1283–1287.

Stubs, D., Fuchs, T.M., Schneider, B., Bosserhoff, A., and Gross, R. (2005). Identification and regulation of cold-inducible factors of *Bordetella bronchiseptica*. Microbiology *151*, 1895–1909.

Tamura, M., Iyo, K., Kunihiro, S., Tamasaki, C., and Haragauchi, M. (2011). Production of human β-actin and a mutant using a bacterial expression system with a cold shock vector. Protein Expr. Purif. *78*, 1–5.

Teter, S.A., Houry, W.A., Ang, D., Tradier, T., Rockabrand, D., Fischer, G., Blum, P., Georgopoulos, C., and Hartl, F.U. (1999). Polypeptide flux through bacterial Hsp70: DnaK cooperates with trigger factor in chaperoning nascent chains. Cell *97*, 755–765.

Theriot, C.M., Du, X., Tove, S.R., and Grunden, A.M. (2010). Improving the catalytic activity of hyperthermophilic *Pyrococcus* prolidases for detoxification of organophosphorus nerve agents over a broad range of temperatures. Appl. Microbiol. Biotechnol. *87*, 1715–1726.

Ting, L., Williams, T.J., Cowley, M.J., Lauro, F.M., Guilhaus, M., Raftery, M.J., and Cavicchioli, R. (2010). Cold adaptation in the marine bacterium, *Sphingopyxis alaskensis*, assessed using quantitative proteomics. Environ. Microbiol. *12*, 2658–2676.

Uh, J.H., Jung, Y.H., Lee, Y.K., Lee, H.K., and Im, H. (2010). Rescue of cold-sensitive mutant at low temperatures by cold shock proteins from *Polaribacter irgensii* KOPRI 22228. J. Microbiol. *48*, 798–802.

Van Bogelen, R.A., and Neidhardt, F.C. (1990). Ribosomes as sensors of heat and cold shock in *Escherichia coli*. Proc. Natl. Acad. Sci. U.S.A. *87*, 5589–5593.

Voderwulbecke, S., Kramer, G., Merz, F., Kurz, T.A., Rauch, T., Zachmann-Brand, B., Bukau, B., and Deuerling, E. (2004). Low temperature or GroEL/ES overproduction permits growth of *Escherichia coli* cells lacking trigger factor and DnaK. FEBS Lett. *559*, 181–187.

Watanabe, K., Kitamura, K., and Suzuki, Y. (1996). Analysis of the critical sites for protein thermostabilization by proline substitution in oligo-1,6-glucosidase from *Bacillus coagulans* ATCC 7050 and 1the evolutionary consideration of proline residues. Appl. Environ. Microbiol. *62*, 2066–2073.

Watanabe, S., Yamaoka, N., Fukunaga, N., and Takada, Y. (2002). Purification and characterization of a cold-adapted isocitrate lyase and expression analysis of the cold-inducible isocitrate lyase gene from the psychrophilic bacterium *Colwellia psychrerythraea*. Exremophiles *6*, 397–405.

Williams, T.J., Burg, D.W., Raftery, M.J., Poljak, A., Guilhaus, M., Pilak, O., and Cavicchioli, R. (2010). Global proteomic analysis of the insoluble, soluble, and supernatant fractions of the psychrophilic archaeon *Methanococcoides burtonii*. Part I: the effect of growth temperature. J. Proteome Res. *9*, 640–652.

Wilmes, B., Kock, H., Glagla, S., Albrecht, D., Voigt, B., Markert, S., Gardebrecht, A., Bode, R., Danchin, A., Feller, G., Hecker, M., and Schweder, T. (2011). Cytoplasmic and periplasmic proteomic signatures of exponentially growing cells of the psychrophilic bacterium *Pseudoalteromonas haloplanktis* TAC125. Appl. Environ. Microbiol. 77, 1276–1283.

Wouters, J.A., Sanders, J.W., Kok, J., de Vos, M.M., Kuipers, O.P., and Abee, T. (1998). Clustered organization and transcriptional analysis of a family of five *csp* genes of *Lactococcus lactis* MG1363. Microbiology *144*, 2885–2893.

Yamanaka, Y., Kazuoka, T., Yoshida, M., Yamanaka, K., Oikawa, T., and Soda, K. (2002). Thermostable aldehyde dehydrogenase from psychrophile, *Cytophaga* sp. KUC-1: enzymological characteristics and functional properties. Biochem. Biophys. Res. Commun. *298*, 632–637.

Yoshimune, K., Galkin, A., Kulakova, L., Yoshimura, T., and Esaki, N. (2005a). Cold-active DnaK of an Antarctic psychrotroph *Shewanella* sp. Ac10 supporting the growth of *dnaK*-null mutant of *Escherichia coli* at cold temperatures. Extremophiles 9, 145–150.

Yoshimune, K., Galkin, A., Kulakova, L., Yoshimura, T., and Esaki, N. (2005b). DnaK from *Vibrio proteolyticus*: complementation of a *dnaK*-null mutant of *Escherichia coli* and the role of its ATPase domain. J. Biosci. Bioeng. *99*, 136–142.

Yoshimune, K., Yoshimura, T., and Esaki, N. (1998). Hsc62, a new DnaK homologue of *Escherichia coli*. Biochem. Biophys. Res. Commun. *250*, 115–118.

Yoshimune, K., Yoshimura, T., Nakayama, T., Nishino, T., and Esaki, N. (2002). Hsc62, Hsc56, and GrpE, the third Hsp70 chaperone system of *Escherichia coli*. Biochem. Biophys. Res. Commun. *293*, 1389–1395.

Zheng, S., Ponder, M.A., Shih, J.Y., Tiedje, J.M., Thomashow, M.F., and Lubman, D.M. (2007). A proteomic analysis of *Psychrobacter articus* 273–4 adaptation to low temperature and salinity using a 2-D liquid mapping approach. Electrophoresis *28*, 467–488.

Zhou, Y.N., Kusukawa, N., Erickson, J.W., Gross, C.A., and Yura, T. (1988). Isolation and characterization of *Escherichia coli* mutants that lack the heat shock sigma factor sigma 32. J. Bacteriol. *170*, 3640–3649.

Heat Shock Response in Psychrophilic Microorganisms

7

Seiji Yamauchi, Shinsuke Fukata and Hidenori Hayashi

Abstract

Psychrophilic microorganisms can optimally grow at temperatures below 15°C. In these microorganisms, heat stress occurs at relatively low temperatures in comparison with that in mesophilic microorganisms. The majority of psychrophilic microorganisms possess genes encoding a complete set of heat shock proteins (Hsps). Therefore, psychrophilic microorganisms respond to heat stress by producing Hsps like other microorganisms; however, they need a specific system to enable the expression and function of Hsps at relatively low temperatures. In this chapter, we summarize the heat shock response of psychrophilic microorganisms, focusing on how this response starts working at relatively low temperatures and what the features of psychrophilic Hsps are.

Introduction

Temperature is one of the major factors affecting the growth and survival of microorganisms. Although a variety of microorganisms can be found growing under temperatures ranging from subzero to over 100°C, individual microorganisms have an optimum growth temperature. Microorganisms can be classified into five groups on the basis of their optimum growth temperature: hyperthermophiles (optimum growth temperature is above 80°C), thermophiles (optimum growth temperature is approximately 60°C), mesophiles (optimum growth temperature is approximately 37°C), psychrotrophs (psychrotolerant microorganisms) (optimum growth temperature is in the mesophilic range, but they are capable of growing at low temperatures), and psychrophiles (optimum growth temperature is below 15°C). If the environmental temperature deviates from the optimum temperature, it is great obstacle for microorganism growth. Temperature stresses are classified into two types, namely heat stress and cold stress. In the case of heat stress, a temperature increase of only a few degrees causes fatal damage to the proliferation of microorganisms. To cope with the effects of these temperature stresses, all living microorganisms have developed sophisticated mechanisms called the heat shock response and the cold shock response.

One of the types of fatal damage from heat stress is protein inactivation by denaturation and aggregation. In the heat shock response, the synthesis of a set of highly conserved proteins called heat shock proteins (Hsps) is transiently increased in the cells. These events in gene regulation are commonly found in eukaryotes, prokaryotes, and archaea, suggesting that the heat shock response is a universal mechanism in all living organisms (Richter *et al.*, 2010). Major Hsps work as molecular chaperones, which inhibit protein aggregation and/

or repair denatured proteins, and as proteases, which degrade denatured proteins (Walter and Buchner, 2002; Hengge and Bukau, 2003). On the other hand, under cold stress conditions, a specific set of proteins called cold shock proteins (Csps) are transiently induced (Phadtare, 2004). The Csps function as RNA chaperones to facilitate translation at low temperatures by preventing the formation of secondary structures of mRNA (Jiang *et al.*, 1997). Thus, the response systems for heat stress and cold stress are completely different from one another.

Psychrophiles are defined as organisms having an optimum growth temperature below 15°C, and a maximal temperature for growth below 20°C (Morita, 1975). Since the temperature falls below 15°C on almost 80% of the earth, the study of psychrophilic microorganisms cannot be disregarded within the field of microbiology. Psychrophilic microorganisms possess cold-adapted enzymes that show highly specific activities at low temperatures in order to grow effectively under cold environmental conditions. It has been suggested that general properties of cold-adapted enzymes enhance the flexibility of their structure, which enables them to accomplish binding of substrates and conformational changes for their activities at low temperatures (D'Amico *et al.*, 2006; Feller and Gerday, 1997; Fields and Somero, 1998; Georlette *et al.*, 2003). However, these properties cause pronounced thermolability of cold-adapted proteins (D'Amico *et al.*, 2003; Collins *et al.*, 2003). For example, isocitrate lyase of a psychrophilic bacterium, *Colwellia psychrerythraea* NRC 1004, showed maximum activity at 25°C and was completely inactivated by incubation even at 30°C for 2 min (Watanabe *et al.*, 2002a). Cold-adapted and thermolabile enzymes were also reported for citrate synthase from an Antarctic bacterial strain DS2–3R (Gerike *et al.*, 1997, 2001), and for malate dehydrogenase from the psychrophilic bacterium *Flavobacterium frigidimaris* KUC-1 (Oikawa *et al.*, 2005). Our group has revealed that the *E. coli* cells expressing thermolabile enzymes from psychrophilic bacteria exhibit the heat shock response around 30°C, suggesting that the expressed psychrophilic enzymes denatured at physiological temperatures (Y. Ueda, unpublished). These findings clearly indicate that several proteins in psychrophilic microorganisms exhibit pronounced thermolability. Thus, in psychrophilic microorganisms, heat stress occurs at relatively low temperatures in comparison with that in mesophilic microorganisms; in other words, we can consider that they possess the most sensitive response system against heat stress.

Although many reports have described the mechanisms of heat shock response of mesophilic and thermophilic microorganisms, little is known about the response to heat stress in psychrophilic microorganisms. Genome sequences of several psychrophilic microorganisms have now been determined (Mathé *et al.*, 2005; Médigue *et al.*, 2005; Aono *et al.*, 2010; Allen *et al.*, 2009). From this information, it has been revealed that psychrophilic microorganisms have genes encoding a complete set of Hsps. This means that Hsps play important roles in the lives of psychrophilic microorganisms under cold environmental conditions.

In this chapter, we summarize the heat shock response of psychrophilic microorganisms, focusing on the following points: (i) the regulatory mechanism of Hsp genes in psychrophilic microorganisms; and (ii) the function and character of psychrophilic Hsps.

Effects of heat stress and Hsp functions in general

Temperature is a major factor that determines the natural distribution of organisms, and individual organisms survive best at their own optimum growth temperature. The environmental temperature is not constant throughout the year; for example, the atmospheric temperature in summer is often markedly different from that in winter. Thus, even in regions with temperature conditions suitable for a particular organism to grow, the organism will encounter temperature stress, that is, yearly, monthly and daily fluctuations in temperature.

Heat stress, a rapid increase in the environmental temperature, results in severe damage to cellular proteins and membranes (Richter *et al.*, 2010). Because proteins need to be conformationally flexible to perform their functions in the cell, they should be evolutionarily optimized to be only marginally stable at individual growth temperatures. Even a small increase in temperature can cause denaturation and aggregation of some proteins, and eventually results in functional inactivation. It has been shown that many phenotypic effects of heat stress are attributable to an imbalance of protein homeostasis. In addition to the protein inactivation, heat stress causes severe types of damage of functional states of cellular membranes; for example, changes in membrane morphology cause higher fluidity of the membranes and a functional defect of membrane proteins (Kruuv *et al.*, 1983). Furthermore, the resulting enhanced membrane permeability leads to a drop in cytosolic pH and changes in ion homeostasis (Coote *et al.*, 1991). Accumulation of these physical and chemical changes in proteins and membranes leads to arrest of the cell cycle, and eventually results in cell death. To avoid fatal damage caused by heat stress, microorganisms possess a specific mechanism called heat shock response to protect and repair their damaged proteins and membranes. In this mechanism, the damage in cells, such as accumulation of denatured proteins and aberrance of cellular membranes, is recognized by specific sensing systems, and they then transiently induce the synthesis of Hsps (Morimoto, 1998; Narberhaus *et al.*, 2006; Vigh *et al.*, 2007).

Under heat stress conditions, organisms rapidly respond to temperature changes and induce Hsp synthesis to protect cells from damage. The predominant class of Hsps is the molecular chaperones, which comprise five major families, namely, Hsp104, Hsp90, Hsp70, Hsp60, and small heat shock protein (sHsp) (Fig. 7.1). The most characterized molecular chaperones are the Hsp70 and Hsp60 families, including extensively studied DnaK and GroEL of *E. coli*, respectively (Frydman and Hartl, 1994; Genevaux *et al.*, 2007; Horwich

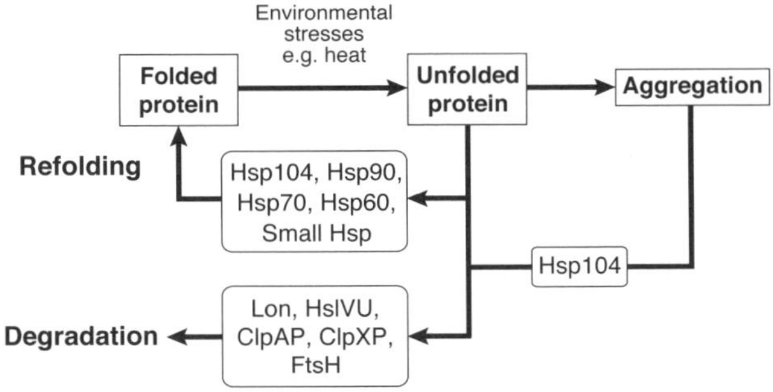

Figure 7.1 Protein quality control by Hsps. Cellular proteins are unfolded by various environmental stresses, for example heat stress. Hsps work as molecular chaperones, which inhibit protein aggregation and/or repair unfolded proteins, and as proteases, which degrade unfolded proteins. In cooperation with other molecular chaperones, Hsp104 proteins can repair the denatured proteins that form aggregates.

et al., 2006). These molecular chaperones play an important role in the protection of cellular proteins and/or in the repair of unfolded proteins. Recently, considerable cooperation between Hsp families has also been reported. For example, in cooperation with the Hsp70 family, Hsp104 proteins (called ClpB in *E. coli*) can re-solubilize denatured proteins, even in aggregated forms (Goloubinoff *et al.*, 1999). The sHsp family also cooperatively functions with the Hsp70 and Hsp104 families, and promotes efficiencies of protein folding (Ratajczak *et al.*, 2009). Furthermore, these Hsps appear to contribute to the folding process of newly translated proteins under physiological conditions (Houry, 2001).

While DnaK and GroEL function as molecular chaperones, other kinds of Hsps, such as Lon, HslVU, ClpAP, ClpXP, and FtsH in microorganisms, function as proteases that degrade the denatured proteins under heat stress conditions in an ATP-dependent manner (Fig. 7.1) (Hengge and Bukau, 2003), and additionally exhibit major regulatory functions of gene expression by controlling the level of specific transcriptional factors (Kanemori *et al.*, 1997).

Heat shock response in psychrophilic microorganisms

In psychrophilic microorganisms, the heat shock response should occur at relatively low temperatures, such as 20°C, at which the cold shock response is generally induced in most mesophilic microorganisms (Kim *et al.*, 2005; Sakamoto and Bryant, 1997; Aguilar *et al.*, 2001). Thus, psychrophilic microorganisms need a specific regulatory system to enable the induction of Hsp expression at relatively low temperatures. In this section, we summarize the transcriptional patterns and regulatory mechanisms of Hsp genes of psychrophilic microorganisms.

Transcriptional pattern of Hsp genes in *Colwellia maris* ABE-1

Our group has investigated the heat shock response of a psychrophilic bacterium, *Colwellia maris* ABE-1, isolated from the Okhotsk Sea in Hokkaido, Japan. *C. maris* grows optimally at 10–15°C, and its maximum temperature for growth is about 25°C (Yumoto *et al.*, 1998). Naturally, treatment at 30°C causes lethal damage to *C. maris* cells. However, when the cells were pre-treated at 20°C for 1 h, a larger number of cells survived after 30°C treatment. This suggests that the heat shock response occurs at approximately 20°C in *C. maris*.

Fig. 7.2 shows the transcriptional pattern of *groESL*, *dnaK*, and *dnaJ* of *C. maris*. Transcription of three Hsp genes was induced by exposing the *C. maris* cells grown at 10°C to 20°C for 60 min, suggesting that *C. maris* responds to heat stress by producing Hsps like other bacteria (Yamauchi *et al.*, 2003, 2004). It was determined that *groES* is co-transcribed with *groEL* in *C. maris*, as found in other mesophilic and thermophilic bacteria (Segal and Ron, 1996). In contrast, transcriptional patterns of the *dnaK* and *dnaJ* in *C. maris* were somewhat different from those of other bacteria. Although *dnaK* is co-transcribed with *dnaJ* in most mesophilic and thermophilic bacteria (Segal and Ron, 1996), *dnaK* and *dnaJ* are transcribed separately as monocistronic mRNA in *C. maris*. The level of *dnaK* mRNA increased about 5-fold by a temperature increase from 10°C to 20°C for 60 min, while the induction of *dnaJ* was much lower than that of *dnaK*. The level of *dnaJ* mRNA reached a maximum of only 1.5-fold after the same temperature treatment. These results indicate that the expression of *dnaK* and *dnaJ* is regulated separately in *C. maris*.

In *E. coli*, Hsp genes are positively regulated by the σ^{32} subunit of RNA polymerase

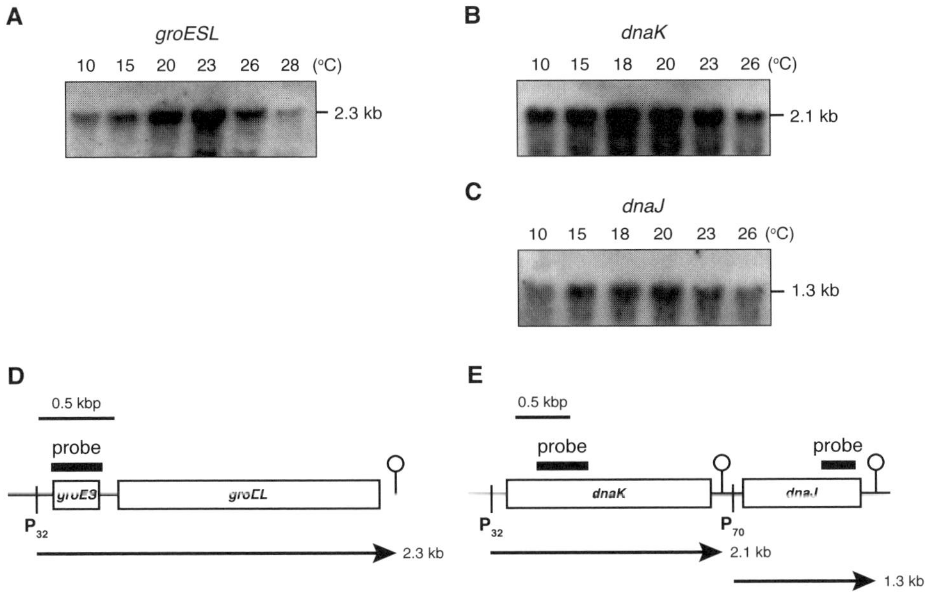

Figure 7.2 Transcriptional patterns of Hsp genes in *C. maris*. *C. maris* cells grown at 10°C were exposed to the indicated temperatures for 60 min. Transcriptions of *groESL* (A), *dnaK* (B), and *dnaJ* (C) were examined by Northern blot analysis. The sizes of each transcript are indicated on the right of each panel. (D and E) Schematic representations of the *groES* and *groEL* (D), and the *dnaK* and *dnaJ* (E). The DNA fragments used for the Northern blot analyses are shown by black boxes. The regions of detected mRNAs are indicated by arrows and their sizes are also shown on the right. P_{32} and P_{70} indicate a heat-shock promoter and a housekeeping promoter, respectively. Stem–loop structures located downstream of the *groEL*, *dnaK*, and *dnaJ*, which may act as the transcriptional terminator, are also shown.

specific to the heat shock promoter (Arsène *et al.*, 2000). We found nucleotide sequences similar to the heat shock promoter in upstream regions of both *groES* and *dnaK* of *C. maris* (Table 7.1). This result indicates that the Hsp genes of *C. maris* are regulated by the σ^{32} system like those of *E. coli*, although the temperature for their induction is over 20°C lower than is the case in *E. coli*. On the other hand, the housekeeping σ^{70}-type promoter was found in the upstream region of *C. maris dnaJ* (Table 7.1). These results indicate that the regulatory mechanism of the *dnaJ* differs from that of the *dnaK*, and the different induction by heat stress between the *dnaK* and *dnaJ* might be due to the different types of promoter.

Mechanism for regulating the expression of Hsp genes in *C. maris*

The transcription of Hsp genes in *E. coli* was induced at 42°C by the σ^{32} system, whereas their transcription in *C. maris* was induced at a much lower temperature by the same system. How does the σ^{32} system of *C. maris* enable induction of the expression of Hsp at relatively low temperatures, such as 20°C? We discuss the mechanism that induces the transcription of Hsp genes at relatively low temperatures from the following three perspectives: (i) mechanisms regulating the level of *rpoH* mRNA that encodes σ^{32}; (ii) translational regulation of σ^{32} by a secondary structure of *rpoH* mRNA; and (iii) G + C contents of the 5′-untranslated region of Hsp genes.

Table 7.1 Comparison of the putative promoters of *groESL* and *dnaK* from *C. maris*, *C. psychrerythraea*, *P. haloplanktis* and *E. coli*

	Promoter	–35 region	Spacing (nucleotides)	–10 region
C. maris	*groESL*	TAGCACTTCTGAA	— 12 —	CCCCACTTC
	dnaK	TTTATCTATTGAA	— 14 —	CCCCATTTC
C. psychrerythraea	*groESL*	TAGGGCTTGTGAA	— 12 —	CCCCATTTC
	dnaK1	TCATATTTATGAC	— 15 —	CCTTGAGTA
	dnaK2	TTTATCTATTGAA	— 14 —	CCCCACTTA
P. haloplanktis	*groESL*	TTTTACGCTTGGA	— 14 —	CCCCATTAT
E. coli	*groESL*	TTTCCCCCTTGAA	— 13 —	CCCCATTTC
	dnaK	TCTCCCCCTTGAT	— 14 —	CCCCATTTC
σ^{32} consensus[1]		TNNCNCCCTTGAA	— 13–15 —	CCCCATNTA
C. maris	*dnaJ*	TTGATT	— 20 —	AATCAT
σ^{70} consensus[1]		TTGACA	— 16–19 —	TATAAT

[1]Consensus sequences for σ^{32} and σ^{70} promoter region as reported in Hawley and McClure (1993).

The first mechanism is regulation of the level of *rpoH* mRNA. In most mesophilic bacteria, *rpoH* transcription is strictly regulated by the use of several promoters (Reisenauer *et al.*, 1996; Manzanera *et al.*, 2001; Ramírez-Santos *et al.*, 2001). For example, transcription of *E. coli rpoH* is regulated by at least four promoters: three are σ^{70}-type and one is an σ^{E}-type promoter (Fig. 7.3A). Because the σ^{E}-type promoter is known to be strongly activated at high temperatures, transcription of *rpoH* is induced by heat stress in *E. coli* (Erickson *et al.*, 1987). In contrast, only one housekeeping σ^{70}-type promoter was found in *C. maris rpoH* (Fig. 7.3A), suggesting that the transcription of *C. maris rpoH* is constant. However, the level of *rpoH* mRNA increased 3-fold after a temperature increase from 10°C to 20°C. Further experiments using an inhibitor of transcription, rifampicin, explained this discrepancy; namely, the rate of degradation of *rpoH* mRNA was much lower at 20°C than at 10°C (Yamauchi *et al.*, 2006). These results indicate that the promoter constitution of *C. maris rpoH* differs from those of mesophilic bacterial species, and the level of its mRNA is mainly regulated by the rate of degradation.

The second mechanism is translational regulation of σ^{32}. In *E. coli*, the level of σ^{32} is rapidly increased by a temperature increase from 30 to 42°C, and its level is mainly regulated at the translational stage. The *rpoH* of *E. coli* contains a sequence region called a 'downstream box', which is positioned immediately downstream of the *rpoH* translational initiation codon. This region is complementary to the 3′-region of 16S rRNA and acts as a positive regulator of σ^{32} translation (Fig. 7.3B) (Nagai *et al.*, 1991). In *C. maris rpoH*, two regions of putative downstream box were found (downstream box1 and downstream box2) (Fig. 7.3B). The first downstream box was found to be located one nucleotide downstream from the initiation codon, and about 74% of the nucleotides were complementary to the 3′ sequence of 16S rRNA in *C. maris*. The second downstream box was found six nucleotides downstream from the first downstream box, and about 88% of the nucleotides were complementary to the 16S rRNA in *C maris*. The additional downstream box region may ensure the translation of *C. maris* σ^{32} at relatively low temperature.

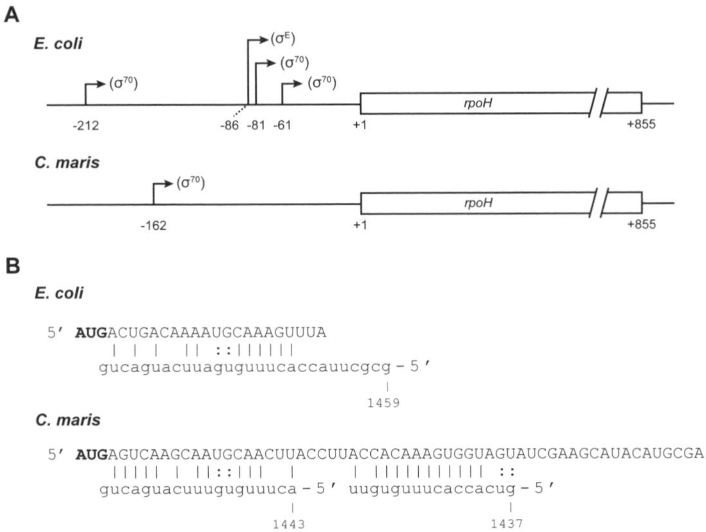

Figure 7.3 (A) Promoter constitution of the *rpoH* of *E. coli* and *C. maris*. The transcriptional start sites are shown by arrows, and the type of promoter regulating each transcript is indicated in parentheses. Numbers below indicate the positions of nucleotides from the translational start codon of *rpoH*. (B) Putative downstream boxes of the *rpoH* of *E. coli* and *C. maris*. Nucleotide sequence of downstream boxes (upper sequence) is shown to be complementary to part of the 3′-end of 16S rRNA (lower sequence). Lines indicate canonical base pairing, and two dots denote allowed U–G pairing. Numbers below the 16S rRNA sequence indicate the positions of nucleotides.

It has been demonstrated that the 5′-region of *rpoH* mRNA of *E. coli* forms a secondary structure, and blocking the downstream box by base pairings is the most important factor regulating the translation efficiency of σ^{32} (Morita *et al.*, 1999; Terao-Morita *et al.*, 1999). When the predicted secondary structure of the *rpoH* mRNA of *C. maris* was compared with that of *E. coli*, an interesting feature was found in the downstream box regions (Fig. 7.4A) (Yamauchi *et al.*, 2006). In the case of *E. coli*, the downstream box was perfectly sequestered by base pairing (Fig. 7.4B) (Morita *et al.*, 1999). On the other hand, the stems formed in downstream box1 and 2 of *C. maris rpoH* were interrupted by one or more short unpaired regions (Fig. 7.4A and C). Thus, in *C. maris*, it seems that the base pairing of two downstream boxes should be less stable than that in *E. coli*. It is easy to imagine that imperfect base pairing is necessary to guarantee the melting of mRNA secondary structure within the appropriate temperature range. Additional unpaired regions in the downstream boxes of *C. maris rpoH* might serve as 'wedges', conferring more instability on the stem structure compared with those of mesophilic bacteria. These features of the secondary structure of *rpoH* mRNA play critical roles for the translation of σ^{32} followed by the expression of Hsp at relatively low temperature in *C. maris* (Fig. 7.5A).

An interesting feature was also found in the nucleotide sequences of the 5′-untranslated region of Hsp genes (Table 7.2). In *C. maris*, the G+C contents of the 5′-untranslated region of *groES* and *dnaK* were 23.6 mol% and 29.6 mol%, respectively. These values were about 20 mol% lower than those of the 5′-untranslated regions in *E. coli* (43.8 mol% for *groES* and 48.1 mol% for *dnaK*, respectively) and in *Salmonella typhimurium* (51.9 mol% for

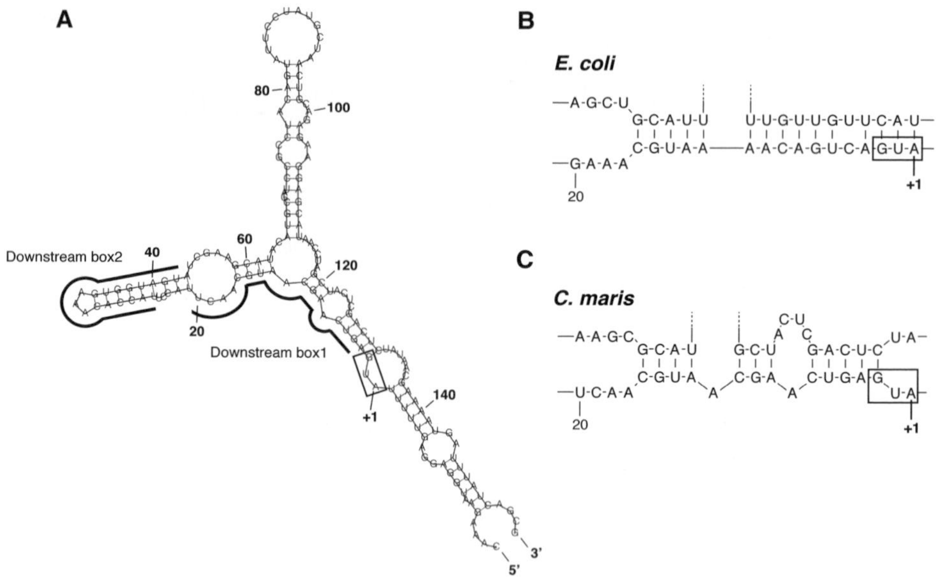

Figure 7.4 (A) Possible secondary structure of the 5′-region of *C. maris rpoH* mRNA. The 5′-region of *rpoH* mRNA of *C. maris* (nt –20 to +144) was analysed by using the Vienna RNA secondary structure prediction program on the Internet (http://rna.tbi.univie.ac.at/cgi-bin/RNAfold.cgi/). The initiation codon is boxed, and the two putative downstream boxes are underlined. (B), (C) Enlarged view of the secondary structure of *rpoH* mRNA of *E. coli* (B) and *C. maris* (C). Base pairings of the region including the initiation codon and the downstream box (the downstream box1 for *C. maris*) are shown.

groES and 43.3 mol% for *dnaK*, respectively) (Yamauchi *et al.*, 2003, 2004). In contrast, the G + C contents of the coding regions of *groES, groEL, dnaK*, and *dnaJ* of *C. maris* are 41.8, 40.2, 40.9, and 41.6 mol%, respectively, which are only about 10 mol% lower than those of mesophilic bacteria. The G + C contents of the 5′-untranslated region of genes encoding isocitrate lyase (*icl*) and isocitrate dehydrogenase (*idhI* and *idhII*) of *C. maris*, which are not heat shock genes, were found to be 35.7 mol%, 37.1 mol%, and 38.5 mol%, respectively (Watanabe *et al.*, 2002b; Ishii *et al.*, 1993), suggesting that the particularly low G + C content in the 5′-untranslated region may be specific to Hsp genes in *C. maris*. The low G + C content results in a decrease in the melting point of the region (Wada and Suyama, 1986). Since a low melting point in the 5′-untranslated region facilitates the unfolding of the DNA double strands, it might allow the expression of these Hsp genes to be induced at relatively low temperatures (Fig. 7.5B). It will be of interest to examine the correlation between low G + C content and the expression of Hsp genes at low temperature by further comparison of nucleotide sequence as well as gene manipulation.

Hsp expression in other psychrophilic bacteria

Transcriptional patterns of the *groESL* and *dnaK* in *Colwellia psychrerythraea* 34H and of *groESL* in *Pseudoalteromonas haloplanktis* TAC125 have been investigated (Tosco *et al.*, 2003; Yamauchi *et al.*, 2012).

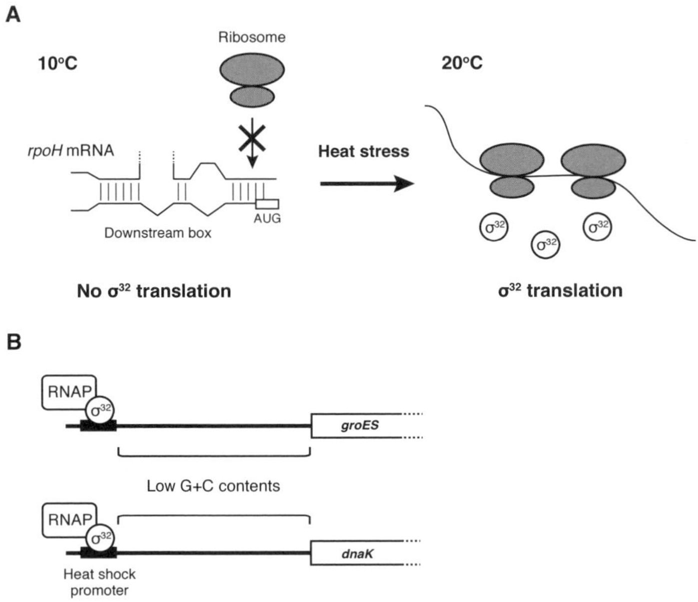

Figure 7.5 Mechanisms for regulating Hsp expression in *C. maris*. (A) Translational regulation of the σ^{32} of *C. maris* by the secondary structure of *rpoH* mRNA. In *C. maris*, the translation of σ^{32} is induced at 20°C, since the secondary structure of *rpoH* mRNA is less stable than that of mesophilic bacteria. (B) Features of the 5′-untranslated regions of the *groES* and *dnaK* of *C. maris*. Specific low G+C contents in the 5′-untranslated regions of Hsp genes should also contribute to Hsp expression in *C. maris* at relatively low temperature. See text for more detailed information.

Table 7.2 Comparison of the G+C content (%) of *groES*, *groEL*, *dnaK*, and *dnaJ*, and the 5′-untranslated region of the *groES* and *dnaK* from *C. maris*, *E. coli*, and *S. typhimurium*

	Coding region				5′-Untranslated region	
	groES (%)	*groEL* (%)	*dnaK* (%)	*dnaJ* (%)	*groES* (%)	*dnaK* (%)
C. maris	41.8	40.2	40.9	41.6	23.6	29.6
E. coli	52.8	48.6	51.1	55.2	48.1	43.8
S. typhimurium	54.4	46.9	52.6	54.0	51.9	43.3

C. psychrerythraea 34H was isolated from Arctic marine sediments, and its optimum and lethal temperatures are 8°C and 19°C, respectively (Huston *et al.*, 2000). This bacterium possesses one *groESL* and two *dnaK* (designated as *dnaK1* and *dnaK2*) in its genome (Mathé *et al.*, 2005). When *C. psychrerythraea* cells grown at 8°C were exposed to 18°C for 30 min, the level of mRNA of *groESL* operon markedly increased and reached a maximum of approximately 6-fold (Yamauchi *et al.*, 2012). Furthermore, transcriptions of both *dnaK1* and *dnaK2* were remarkably induced by a high temperature above 16°C (S. Fukata, unpublished). These results suggest that the heat shock response of *C. psychrerythraea* occurs at approximately 16°C.

Pseudoalteromonas haloplanktis TAC125 was isolated from Antarctic seawater, and its optimum growth temperature is about 20°C. In this bacterium, heat shock response is supposed to occur at 30°C on the basis of the observation that the synthesis of potential Hsp family members, including approximately 65-, 55-, 50-, 15- and 14-kDa proteins, was markedly induced at 30°C. Tosco and coworkers identified that the 55-kDa protein was GroEL, and revealed that the level of its mRNA was increased at 27°C by primer extension analysis (Tosco *et al.*, 2003).

The σ^{32}-type promoter was found in upstream regions of the *groESL* and *dnaK* of both psychrophilic bacteria (Table 7.1), suggesting that their expression is regulated by the σ^{32} system as in *E. coli* and *C. maris*. As described above, the specific features of the *rpoH*, such as the differences in degradation rate and in the secondary structure of *rpoH* mRNA, were optimized to ensure the expression of Hsp at relatively low temperature in *C. maris* (Yamauchi *et al.*, 2006). Further study is needed to reveal whether these features are also involved in the heat shock response of *C. psychrerythraea* and *P. haloplanktis*.

Hsp expression in psychrophilic yeasts

Deegenaars (1998) have reported the heat shock response of four psychrophilic yeasts from Antarctica: *Mrakia stokesii*, *Mrakia frigida*, *Mrakia gelida*, and *Leucosporidium antarcticum*. In mesophilic yeast *Saccharomyces cerevisiae*, the synthesis of about 80 proteins including major Hsp families, such as Hsp104, Hsp90, Hsp70, and Hsp60, was transiently induced after temperature increase from 25°C to 37°C (Miller *et al.*, 1982; Deegenaars and Watson, 1997). For these psychrophilic yeasts, 35°C is a severe high temperature, producing less than 10% survivors after treatment for 60 min. However, these cells were protected from 35°C treatment when they were exposed to a prior heat shock at 25°C for 3 h. Induction of the expression of potential Hsp family members as well as trehalose accumulation was observed in the cells of these psychrophilic yeasts after pre-treatment at 25°C for 3 h, suggesting that the psychrophilic yeasts possess the system for heat shock response similar to the case in *S. cerevisiae*.

The actual composition of the Hsp family in psychrophilic yeasts is still unclear. No signals were detected in each cellular extract prepared from four psychrophilic yeasts by Western blot analysis using antibodies against Hsp60 and Hsp104 of *S. cerevisiae*. These results imply that these psychrophilic yeasts do not possess protein homologues corresponding to the Hsp60 and Hsp104 families. The counterpart of Hsp104 was also found to be missing in another psychrophilic yeast, *Candida psychrophila* (Deegenaars and Watson, 1997). In three *Mrakia* species, *Mrakia stokesii*, *M. frigida* and *M. gelida*, marked induction of the 110-kDa protein was found by treatment at 25°C for 3 h. The authors have speculated that the function of Hsp104 is replaced by the 110-kDa protein in the psychrophilic yeasts. These results indicate that the psychrophilic yeasts have a different Hsp profile from *S. cerevisiae*. The relationship between different Hsp profiles and thermotolerance in psychrophilic yeasts remains to be elucidated.

Psychrophilic microorganisms in which the heat shock response has degenerated

Currently, the heat shock response produced by inducing the synthesis of Hsp is recognized as a universal mechanism in all three domains of life (prokaryotes, eukaryotes, and archaea). However, it has been reported that several psychrophilic organisms living in Antarctica have lost their heat shock response. For example, in a psychrophilic bacterium, *Oleispira antarctica* RB8, which grows optimally at 4°C, the amount of GroEL did not change in the temperature range from 4°C to 25°C (Ferrer *et al.*, 2004). In *Euplotes focardii*, which is the Antarctic species of the ciliate, no induction of Hsp70 expression was found in cells heat-treated at 20°C (La Terza *et al.*, 2001). It has also been reported that stress-induced expression of Hsp is absent in the Antarctic fish *Trematomus bernacchii* (Hofmann *et al.*, 2000). The loss of heat shock response in these organisms has been attributed to an ancient adaptation to the thermally stable waters of coastal Antarctica (where temperatures range from +0.3°C to −1.9°C) (Eastman, 1993). Since the Hsps are constitutively expressed in these psychrophiles, their major function is to assist the folding and maturation of newly translated polypeptides.

The habitat-correlated diversity of the heat shock response was observed in three *Hydra* species. *Hydra vulgaris* and *Hydra magnipapillata* induce the synthesis of several Hsps in response to heat stress and acquire thermotolerance by exposure to a mildly high temperature (Bosch *et al.*, 1988). In contrast, *Hydra oligactis* cannot synthesize detectable levels of Hsp or acquire thermotolerance after pretreatment with a mild heat shock (Bosch *et al.*, 1988). Such difference of expression patterns of Hsp is correlated with differences in the growing environments for each *Hydra* species; namely, *H. vulgaris* and *H. magnipapillata* can live in a wide temperature range, whereas *H. oligactis* is extremely sensitive to thermal stress and prefers a thermostable environment characterized by low temperatures. One of the factors involved in the lack of Hsp expression in *H. oligactis* is reduction of the stability of Hsp mRNA. The half-life of *hsp70* mRNA in heat-shocked *H. oligactis* is much shorter than that of *hsp70* mRNA in *H. magnipapillata* (Brennecke *et al.*, 1998). Taken together, these findings suggest that the heat shock response has degenerated in some psychrophilic species living in a thermostable environment.

Features of Hsp derived from psychrophilic microorganisms

The Hsps play important roles as molecular chaperones and/or proteases not only under heat stress conditions but also under physiological conditions. For example, the DnaK system is a negative regulator of the heat shock response in *E. coli*. DnaK directly associates with σ^{32}, and mediates its degradation with proteases (Rodriguez *et al.*, 2008). Thus, the amount of σ^{32} is very low at physiological temperature, and the expression of Hsp is maintained at a constant level. In *Bacillus subtilis*, GroEL interacts with a repressor protein, called HrcA, for transcription of the Hsp genes under physiological conditions, and activates its ability to bind DNA (Mogk *et al.*, 1997). Thus, the activated HrcA binds to the operator region of Hsp genes and represses their transcription. Such a modulation system with Hsps is broadly conserved from prokaryotes to eukaryotes (Abravaya *et al.*, 1992; Baler *et al.*, 1996; Kim and Schöffl, 2002; Schulz-Raffelt *et al.*, 2007).

Another role of Hsps under physiological conditions is assisting in the proper folding of newly translated proteins. Three molecular chaperones, namely trigger factor, DnaK (and its co-factors DnaJ and GrpE), and GroEL (and its co-factor GroES), are related to this function. Trigger factor is a ribosome-bound chaperone, and is the first chaperone interacting with all newly translated polypeptides (Hesterkamp *et al.*, 1996; Stoller *et al.*, 1995). Although 70% of the total proteins can fold spontaneously, the remaining 30% of the proteins need further assistance from Hsps, such as DnaK and GroEL. A subset of cellular proteins, at least 13 essential proteins, are strictly dependent on GroEL and GroES for proper folding, which is consistent with the fact that both GroEL and GroES are essential in the growth of *E. coli* (Kerner *et al.*, 2005; Fayet *et al.*, 1989). Furthermore, the combined absence of trigger factor and DnaK is lethal at physiological temperatures in *E. coli* (Deuerling *et al.*, 1999), suggesting the crucial functions of Hsps under physiological conditions.

If the psychrophilic Hsps also play important functions under physiological conditions, it is expected that they possess specific features that allow them to function efficiently at low temperature. Currently, a few groups including our own have performed the characterization of psychrophilic Hsps. In this section, we summarize the function and character of psychrophilic Hsps, focusing mostly on four Hsp families derived from psychrophilic microorganisms, namely, GroEL (the Hsp60 family), DnaK (the Hsp70 family), HtpG (the Hsp90 family), and the small Hsp family.

The GroEL chaperone system

The GroEL chaperone system is the most studied group of molecular chaperones. This system consists of the Hsp60 homologue, GroEL, which generally forms an ~800-kDa cylindrical complex consisting of two heptameric rings of 57-kDa subunits, and the heptameric oligomer of Hsp10 homologue GroES. Each subunit of GroEL is composed of three domains, namely, an apical domain containing the binding sites for non-native proteins and for GroES, an equatorial domain containing the ATP-binding site, and an intermediate domain, which connects the apical and equatorial domains (Xu *et al.*, 1997). In cooperation with GroES, GroEL assists in the proper folding of newly translated proteins and of unfolded proteins through an ATP-dependent process (Horwich *et al.*, 2006).

The GroEL-assisted folding pathway established so far is as follows (Fig. 7.6):

1. A non-native protein binds to the apical domain of GroEL.
2. Binding of ATP to GroEL leads to interaction with GroES, and a ternary complex containing GroEL, GroES, and the non-native protein is formed.
3. The non-native protein is encapsulated into the central cavity of GroEL, where proper folding is performed during ATP hydrolysis. The reaction of ATP hydrolysis in the *cis*-ring serves as a signal for initiation of the next round of folding reaction, and a non-native protein binds to the opposite ring as in the case of (1).
4. Both GroES and the folded protein are released from the *cis*-ring of GroEL. At this time, the ternary complex is formed at the opposite ring as in the case of (2).

In this system, transduction of an allosteric signal occurs between the two rings during the folding cycle; for example, releasing of GroES and folded proteins from one ring requires binding of GroES and ATP to another ring (Weissman *et al.*, 1996; Rye *et al.*, 1997). Thus, the double-ring assembly of GroEL is essential for the sequential cycle of GroEL-assisted folding reaction.

Although the GroEL system has been characterized mostly at mesophilic temperatures, Mendoza and coworkers (2000) have described the effects of low temperatures on the chaperone function of the *E. coli* GroEL system. Significant refolding activity was found above 19°C in the *E. coli* GroEL system, whereas the refolding activity was markedly decreased at 10°C. Furthermore, the ATPase activity of *E. coli* GroEL at 10°C was much lower than that at 25°C. The authors have concluded that the reduced ATPase activity is one of the factors for the diminished refolding activity at 10°C. These findings suggest that the mesophilic

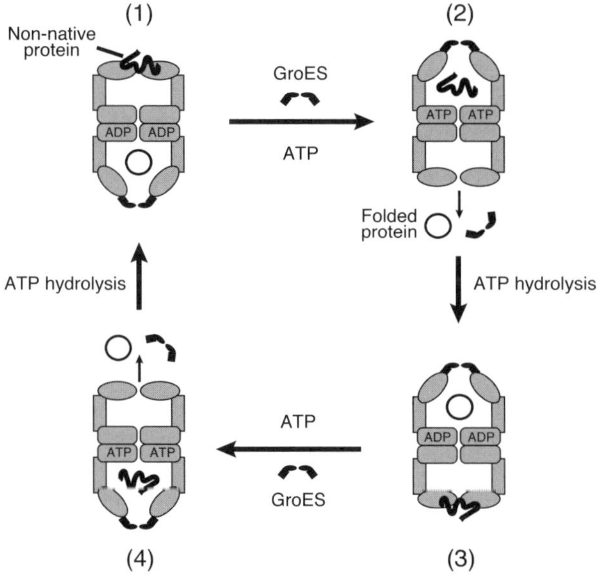

Figure 7.6 The reaction cycle for protein folding by the *E. coli* GroEL system. The cycle is mainly divided into four steps. (1) Non-native proteins interact with the *cis*-ring of GroEL. (2) GroES binds to the *cis*-ring of GroEL in the presence of ATP, and the non-native protein is encapsulated into a central cavity of the *cis*-ring of GroEL. (3) The encapsulated protein is folded in the central cavity during a reaction of ATP hydrolysis. After ATP hydrolysis, the second round of folding cycles is initiated in the *trans*-ring of GroEL. (4) The GroES and folded protein are released from the *cis*-ring of GroEL by binding ATP to the *trans*-ring of GroEL. It should be noted that the second round of reaction cycles proceeds at the opposite ring in the order (3), (4), (1) and (2).

GroEL fails to function at low temperatures at which the psychrophilic microorganisms grow optimally.

Features of the GroEL system of psychrophilic microorganisms

Here, we describe the features of the GroEL system derived from four psychrophilic bacteria, *Oleispira antarctica* RB8, *Pseudoalteromonas* sp. PS1M3, *Colwellia psychrerythraea* 34H, and *Pseudoalteromonas haloplanktis* TAC125. Although the amino acid sequences of these GroELs are highly homologous to that of *E. coli* GroEL (more than 75% identity), each psychrophilic GroEL showed distinct features.

The cold-active GroEL system

The GroEL system of *O. antarctica* (*Oa*GroEL and *Oa*GroES) attracted attention as a cold-active chaperone system based on the observation that *Oa*GroEL could support the growth of *E. coli* at 4°C (Ferrer *et al.*, 2003). Maximum refolding capacity of the *Oa*GroEL system was obtained at 4–18°C and this capacity drastically decreased over 20°C (Ferrer *et al.*, 2004). The maximum temperature of ATPase activity of *Oa*GroEL is 24–28°C (Ferrer *et al.*, 2004), which is 25°C lower than that of *E. coli* GroEL (Mendoza *et al.*, 1996). The cold-active *Oa*GroEL system should assist in the folding and maturation of various cellular proteins at low temperatures, resulting in the adaptation of *E. coli* cells to low temperatures. One of the interesting features of *Oa*GroEL is that its oligomeric state changes depending on the temperature. *Oa*GroEL exists as a single-ring heptamer at 4–10°C, whereas it converts into a double-ring tetradecamer over 10°C (Fig. 7.7A). These findings suggest that the single-ring state of *Oa*GroEL is a functional chaperone state, especially under low-temperature conditions, unlike *E. coli* GroEL for which the double-ring state is essential for chaperone function. Two amino acids, 468-K and 471-S of *Oa*GroEL (T and G in *E. coli* GroEL, respectively), have been identified as key factors for the temperature-dependent control of the oligomeric states of *Oa*GroEL (Ferrer *et al.*, 2004).

Another example of the cold-active GroEL system is the *Pseudoalteromonas* sp. PS1M3 GroEL (*Ps*GroEL), which showed significant activity of protein refolding at 10°C (Nakamura *et al.*, 2004). The recombinant *E. coli* GroEL, whose C-terminal region is replaced (residues 502–548) with that of *Ps*GroEL, showed effective refolding ability at 10°C, suggesting that this region is related to the temperature dependence of GroEL activity. Importance of the C-terminal region for temperature dependence of refolding activity has also been reported in archaeal Hsp60. In this case, charged residues in the C-terminal region are a key determinant of Hsp60 thermostability (Luo and Robb, 2011).

Energy-saving mechanism of the psychrophilic GroEL system

The double-ring assembly is essential for the reaction cycle of GroEL-assisted protein folding, since transduction of an allosteric signal occurring between the two rings plays important roles in the processes of ATPase reaction as well as release of GroES and refolded proteins (Weissman *et al.*, 1996; Rye *et al.*, 1997). Therefore, the double-ring GroEL will continue consuming ATP even in the absence of non-native proteins in cells. The GroEL of *O. antarctica* (*Oa*GroEL) exists as a single ring at low temperatures, and its ATPase activity was strongly inhibited in the presence of *Oa*GroES (Fig. 7.7A) (Ferrer *et al.*, 2004). Similar inhibition was observed in a single-ring mutant of *E. coli* GroEL called SR1, which is a functional impairment (Weissman *et al.*, 1995, 1996). The single-ring *Oa*GroEL is an active molecular chaperone below 10°C, and it enables the avoidance of extensive consumption of ATP by binding to *Oa*GroES when no denatured proteins exist in the cells (Fig. 7.7A). On the basis of these findings, it has been proposed that an ATP-saving mechanism in the *Oa*GroEL system overcomes severe growth conditions, such as cold environment and lack of nutrients.

The GroEL system of *C. psychrerythraea* (*Cp*GroEL and *Cp*GroES) showed a marked difference in nucleotide selectivity for protein folding reaction (Yamauchi *et al.*, 2012). Generally, molecular chaperones including GroEL are thought to use ATP preferentially for their functions, and these observations are reasonable because ATP is present at higher concentrations than other nucleotides in cells. However, in the case of the *Cp*GroEL system, ATP is not the optimum energy source for its refolding function. Recovery of chemically denatured malate dehydrogenase is more efficient in the presence of CTP (cytidine-5′-triphosphate) than ATP. Further analyses revealed that the differences of the strength of *Cp*GroEL–*Cp*GroES interaction contribute to the difference of nucleotide selectivity (Fig. 7.7B). The interaction of *Cp*GroEL-*Cp*GroES mediated by CTP binding is weaker than that mediated by ATP binding, resulting in refolding function of the *Cp*GroEL system by CTP. We consider that such a difference in nucleotide selectivity is another type of ATP-saving

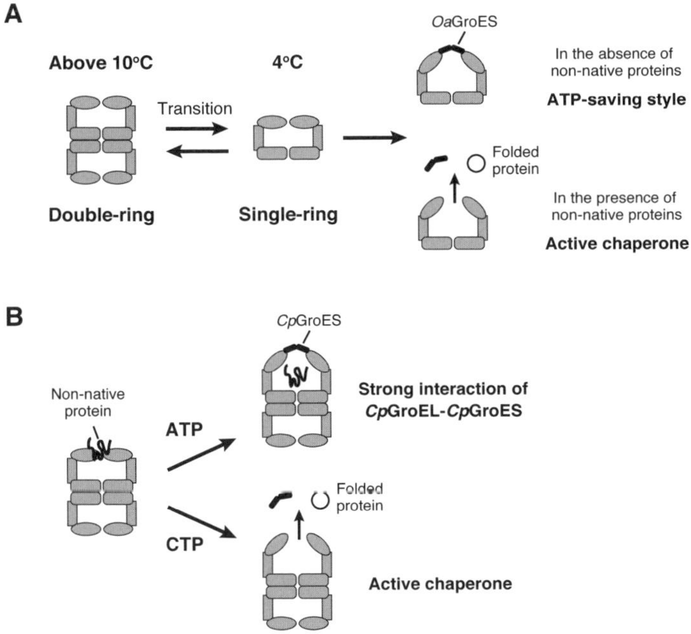

Figure 7.7 Features of the GroEL system of *O. antarctica* (*Oa*GroEL and *Oa*GroES) (A) and *C. psychrerythraea* (*Cp*GroEL and *Cp*GroES) (B). Transition of oligomer states of *Oa*GroEL depending on environmental temperatures and the difference in nucleotide selectivity of *Cp*GroEL are summarized. See text for more detailed information.

mechanism similar to the *Oa*GroEL system for avoiding excessive consumption of ATP under cold environmental conditions.

The psychrophilic GroEL system that shows the same feature as mesophilic system

In contrast to the psychrophilic GroELs described above, the GroEL system of *Pseudoalteromonas haloplanktis* TAC125 (*Ph*GroEL and *Ph*GroES) showed the same features as the mesophilic system (Tosco *et al.*, 2003). Although the maximum temperature of ATPase activity of *Ph*GroEL was slightly lower than that of *E. coli* GroEL (*Ec*GroEL), no significant difference in refolding activity was found between *Ph*GroEL and *Ec*GroEL systems. Interestingly, the *Ph*GroEL system shows low activity for protein refolding below 15°C, despite optimum growth temperatures of this bacterium. These findings suggest that the *Ph*GroEL system is not involved in cellular functions under physiological conditions, such as assisting proper folding of newly translated polypeptides. Recently, it has been reported that the trigger factor is a primary chaperone in *P. haloplanktis*, and is sufficient to facilitate protein folding at low temperatures (Piette *et al.*, 2010). It should be noted that the trigger factor of a psychrophilic bacterium, *Psychrobacter frigidicola*, could also assist in the proper folding of newly translated polypeptides without further assistance by other molecular chaperones (Robin *et al.*, 2009). These findings led to the proposal that the *Ph*GroEL system does not need to function under physiological conditions.

The DnaK chaperone system

The DnaK chaperone system is composed of DnaK (a bacterial homologue of the Hsp70 family) and two cofactors, DnaJ (a bacterial homologue of the Hsp40 family) and GrpE (a nucleotide exchange factor). DnaK consists of three domains, namely an N-terminal ATPase domain, a substrate-binding domain and a C-terminal domain (Mayer and Bukau, 2005). A model for the cycle of protein folding by the DnaK system is shown in Fig. 7.8. DnaJ binds to the non-native protein and delivers it to DnaK (1), and then DnaJ associates with the ATPase domain of DnaK and stimulates its ATPase activity (2) (Szabo *et al.*, 1994). The state of nucleotide binding to DnaK is closely related to the binding to non-native proteins. The ATP-bound DnaK shows low affinity for unfolded proteins, whereas the ADP-bound DnaK binds with high affinity to unfolded proteins (3) (Genevaux *et al.*, 2007). GrpE accelerates the release of ADP from DnaK (4), resulting in dissociation of non-native proteins from DnaK (5). Since non-native proteins interacting with DnaK are protected from further denaturation and aggregation, the possibility that they are folded spontaneously increases. Although the main functions of the DnaK system are to assist proper folding of newly translated proteins and to prevent the aggregation of proteins, this system can refold even aggregated proteins in cooperation with the Hsp104 family (called ClpB in *E. coli*) (Goloubinoff *et al.*, 1999). Therefore, the DnaK system plays important roles not only under stress conditions but also under physiological conditions. Importance of the DnaK chaperone system *in vivo* has been shown from the observation of phenotypes of a *dnaK*-null mutant of *E. coli* (Δ*dnaK* strain of *E. coli*), which lacks the ability to grow at both the far ranges of growth temperature, such as 16°C and 42°C (Bukau and Walker, 1989; Paek and Walker, 1987). Additionally, an *E. coli* mutant lacking both DnaK and trigger factor showed a lethal phenotype under physiological conditions (Deuerling *et al.*, 1999). Considering such important functions, it is expected that the

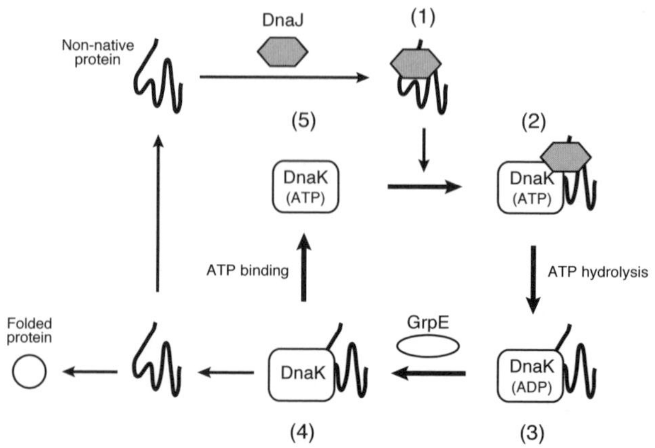

Figure 7.8 The reaction cycle for protein folding by the DnaK system. (1) Non-native proteins interact with DnaJ. (2) DnaJ delivers the non-native protein to DnaK, and stimulates its ATPase activity. (3) The ADP-bound type of DnaK tightly binds to the non-native protein. (4) GrpE promotes the dissociation of ADP from DnaK, and then the non-native protein is released from DnaK. Released non-native protein may fold to its native state spontaneously, or may be trapped again by the DnaK system. (5) The next round of reaction cycles is initiated by binding of ATP to DnaK again.

functional activity of the DnaK system in psychrophilic microorganisms is adapted to low temperatures.

Features of the DnaK system of psychrophilic microorganisms

To date, the DnaK systems of a psychrotrophic bacterium, *Shewanella* sp. Ac10, and a psychrophilic bacterium, *Colwellia psychrerythraea* 34H, have been characterized (Yoshimune *et al.*, 2005; S. Fukata, unpublished).

The optimum growth temperature of *Shewanella* sp. Ac10 is about 20°C, and this bacterium grows well at 4°C (Kulakova *et al.*, 1999). The heat shock response of *Shewanella* sp. Ac10 occurs at 24°C, given the observation that the level of DnaK protein increased about 6-fold after temperature increase from 4°C to 24°C for 30 min. The deduced amino acid sequence of DnaK of *Shewanella* sp. Ac10 (*She*DnaK) was highly homologous to that of mesophilic bacteria, such as *Shewanella oneidensis* and *E. coli* (93% and 77%, respectively). Nevertheless, *She*DnaK showed high ATPase activity compared with *E. coli* DnaK at temperatures ranging from 4°C to 28°C. The maximum temperature of ATPase activity of *She*DnaK is 35°C, which is slightly higher than the optimum growth temperature of *Shewanella* sp. Ac10. Similar temperature dependence of ATPase activity was also observed in the DnaKs from mesophilic and thermophilic bacteria. For example, a mesophilic bacterium, *E. coli*, optimally grows at 37°C, and the maximum temperature of ATPase activity of DnaK is approximately 52°C, while a thermophilic bacterium, *Thermus thermophilus*, optimally grows at 75°C and the maximum temperature of ATPase activity of DnaK is 90°C (Osipiuk and Joachimiak, 1997). Complementation analysis with the Δ*dnaK* strain of *E. coli*, which cannot grow at both high and low temperatures, revealed that *She*DnaK could support its growth only at a low temperature. This result suggests that *She*DnaK is not active at the high temperature range as in *E. coli*. The effect of cofactors, such as DnaJ and GrpE, on *She*DnaK function at low temperatures still remains to be elucidated.

Characterization of the DnaK system of *C. psychrerythraea* is somewhat complicated because of the existence of two *dnaK* genes (*dnaK1* and *dnaK2*) in its genome (Mathé *et al.*, 2005). Deduced amino acid sequences of DnaK1 and DnaK2 are highly homologous to *E. coli* DnaK (66% and 79% identity, respectively). To date, we have revealed three features associated with the difference in functions between DnaK1 and DnaK2 (S. Fukata, unpublished). The first feature is the difference in transcriptional patterns of two *dnaK* at physiological temperatures. Although transcription of both *dnaK1* and *dnaK2* was induced by heat stress, their transcriptional patterns were quite different at physiological temperatures, that is, a significant level of *dnaK2* mRNA was found at 8°C, whereas the transcription of *dnaK1* was completely inhibited. The second feature is the difference in effects of co-factors, DnaJ and GrpE, on ATPase activity of two DnaKs. Generally, the ATPase activity of DnaK is stimulated by the presence of DnaJ and GrpE (Liberek *et al.*, 1991). However, in the case of *C. psychrerythraea* DnaKs, DnaJ and GrpE stimulated the ATPase activity of DnaK2, whereas a little stimulation was found in DnaK1. Thus, it was speculated that DnaK2 is a common type of DnaK, as reported in other bacteria. The third feature is the difference in ability to complement the growth of Δ*dnaK* strain of *E. coli*. DnaK1 supported the growth only at high temperatures, whereas DnaK2 supported the growth only at low temperatures. These results suggest that *C. psychrerythraea* uses two types of DnaK depending on the temperature.

HtpG

HtpG, which is a prokaryotic homologue of the Hsp90 family, is present at high concentrations in cells under physiological conditions and is further induced by heat stress. In eukaryotes, Hsp90 interacts with various client proteins, such as kinases and steroid hormone receptors, many of which participate in signal transduction pathways. Since Hsp90 modulates stabilities and activities of these client proteins, it is an essential protein for viability under all conditions in eukaryotes (Wayne *et al.*, 2011). In contrast to the eukaryotic Hsp90, little is known about the correct function of the prokaryotic homologue HtpG. The *E. coli* mutant lacking *htpG* was viable but had a slight growth disadvantage under high-temperature conditions (Bardwell and Craig, 1988). It has been reported that the HtpG plays important roles in not only heat stress management but also cold acclimation in a mesophilic cyanobacterium, *Synechococcus elongatus* sp. PCC 7942 (Tanaka and Nakamoto, 1999; Hossain and Nakamoto, 2002). These results suggest that a major function of HtpG is the protection of cells from damage caused by temperature stresses.

García-Descalzo and coworkers (2011) have reported comparative data of HtpG-associated proteins among a mesophilic bacterium, *Shewanella oneidensis*, and two psychrophilic bacteria, *Shewanella frigidimarina* and *Psychrobacter frigidicola*. The class of HtpG-associated proteins is quite different between mesophilic and psychrophilic *Shewanella* species, whereas significant more common proteins were found in two psychrophilic bacteria, *Shewanella frigidimarina* and *Psychrobacter frigidicola*, in spite of their different genera. Both psychrophilic HtpG proteins interacted with the proteins involved in energy metabolism, such as isocitrate lyase and malate dehydrogenase, and those involved in translational machinery, such as elongation factor G and translation initiation factors. This commonality in HtpG-associated proteins between two psychrophilic bacteria may have contributed to the adaptation of metabolism and translational machinery to cold environments. Only two proteins, DnaK and GroEL, are common HtpG-associated proteins in three bacteria, suggesting the cooperative function of three Hsp families: HtpG, DnaK, and GroEL. Functional cooperation of these three Hsp families has also been reported in cyanobacteria (Hosaain and Nakamoto, 2002).

Small heat shock protein

Small heat shock proteins (sHsp) are the smallest Hsps, ranging from 15 to 30 kDa, and are ubiquitous proteins found in all kingdoms. They commonly contain a domain similar to the vertebrate eye lens α-crystalline, and form oligomeric complexes that interact with a variety of denatured proteins to prevent their aggregation (Kappé *et al.*, 2002; Haslbeck *et al.*, 2005; Nakamoto and Vigh, 2007). Currently, the sHsp family is considered to function as a storage depot for denatured proteins and deliver them to ATP-dependent chaperones (Mogk *et al.*, 2003; Ratajczak *et al.*, 2009).

As for the psychrophilic sHsp, sHsp of the archaea *Methanococcoides burtonii* (*Mbs*Hsp) has been characterized (Laksanalamai *et al.*, 2009). Although *M. burtonii* inhabits lake water in Antarctica where the temperature is always below 2°C, *Mbs*Hsp is fully functional at temperatures far above its optimum growth temperature. *Mbs*Hsp could prevent not only protein aggregation at 60°C *in vivo* but also loss of enzymatic activity of bovine glutamate dehydrogenase at 50°C *in vitro*. Our group has demonstrated that *E. coli* cells heterologously expressing *C. psychrerythraea* sHsp (designated as IbpA) enhanced their thermotolerance against treatment at 47°C, which is ca. 40°C higher than the optimum growth temperature of *C. psychrerythraea* (S. Yamauchi, unpublished). These findings suggest that psychrophilic sHsps are capable of functioning as molecular chaperones over a wide temperature range, but the biological meaning of sHsp function at high temperature remains to be elucidated.

Concluding remarks and future perspectives

Most psychrophilic microorganisms live in the hydrosphere, where the temperature is rather low and more stable than that in the atmosphere. This might lead us to imagine that their heat shock response is not so vital and poorly developed, but this is not the case. Actually, psychrophilic microorganisms induce the expression of typical Hsps that help them to survive under elevated temperature conditions in most cases, and the specific features of their response to high temperature are as follows:

1. The response to high temperature is the most sensitive among all living things because it starts working even at 16°C.
2. These organisms should have energy-saving systems to respond to high temperature because psychrophilic microorganisms must survive under conditions where the rate of chemical reactions is lower than that at regular temperatures.

Although the heat shock response starts working at relatively low temperatures in psychrophilic microorganisms, its system is the same as the mesophilic one. In the case of *C. maris*, the expression of Hsp is regulated by the σ^{32} system like in *E. coli*. However, the secondary structure of *rpoH* mRNA has been customized to induce the synthesis of σ^{32} at low temperatures (Yamauchi *et al.*, 2006). These findings suggest that the psychrophilic microorganisms have developed common mechanisms regulating Hsp expression for cold environments over the course of evolution. What is the other type of regulatory systems like, namely, negative regulation by repressor proteins? Is the thermostability of repressor proteins different from that of mesophilic ones? Further studies are needed to understand fully the heat shock response of psychrophilic microorganisms. Additionally, it is assumed that they might also have developed a cold shock response suitable for cold environments. In fact, the levels of mRNA of *cspA* and *cspG*, which are major cold shock gene homologues in *E. coli*, were found to be increased by a decrease of the temperature from 8°C to 1°C in the psychrophilic bacterium *Shewanella violacea* DSS12 (Fujii *et al.*, 1999). Recently, it has been demonstrated that the synthesis of *E. coli* CspA is also regulated at the translational stage by the secondary structure of its mRNA (Giuliodori *et al.*, 2010). Thus, comparisons of the structural information of these mRNAs between psychrophilic and mesophilic microorganisms make it possible to design the molecular switches with mRNA controlling gene expression on demand from various changes of temperature (Kortmann *et al.*, 2011).

Although psychrophilic enzymes exhibit highly specific activities at low temperature, they also display pronounced thermolability compared with their mesophilic and thermophilic counterparts. Thus, a number of denatured proteins start appearing in the cells at relatively low temperatures, and this causes major damage by high temperature in psychrophilic microorganisms. From genome information, it has been revealed that the psychrophilic microorganisms possess genes encoding a set of Hsps, just as the mesophilic and thermophilic microorganisms do. This is quite reasonable because Hsps play critical roles for their growth if we consider their functions as molecular chaperones as well as proteases for stress conditions. It should be noted that several psychrophilic Hsps were found to support the growth of *E. coli* at low temperatures (Ferrer *et al.*, 2003; Yoshimune *et al.*, 2005; S. Fukata, unpublished). However, it is also known that, upon cold shock in *E. coli*, Hsp expression is strongly repressed (Taura *et al.*, 1989). When the Hsps were heterologously expressed in *E. coli* at low temperatures, cellular survival was lost much faster than that in wild-type cells (Kandror and Goldberg, 1997). Thus, the psychrophilic Hsps, if not all, possess special features optimized to cold environments, for example, assisting the proper folding of proteins efficiently in the cold. Taking advantage of these features of psychrophilic Hsps, the cold-active GroEL system of *O. antarctica* has already been applied to an overexpression system for protein engineering to improve the yield of recombinant proteins at low temperature. In contrast, non-cold-adapted Hsps, such as GroEL of *P. haloplanktis* (Tosco *et al.*, 2003), would be specialized to function under heat stress conditions. Several psychrophilic microorganisms possess two or more sets of genes encoding the same Hsp family, and express them differently between physiological and heat stress conditions. This suggests that they might distinctly use duplicated Hsps depending on the environmental temperatures.

Acknowledgements

We greatly thank past and present students in our laboratory for their contributions to the study of the heat shock response in *Colwellia maris* and *C. psychrerythraea*. We thank Dr H. Okuyama for *C. maris* cells and for valuable comments on our study. We also thank Dr E. H. Morita and Dr Y. Nishiyama for critical discussions and for help in our study. This work was supported in part by grants from the Ministry of Education, Culture, Science, Sports and Technology, Japan.

References

Abravaya, K., Myers, M.P., Murphy, S.P., and Morimoto, R.I. (1992). The human heat shock protein hsp70 interacts with HUSASF, the transcription factor that regulates heat shock gene expression. Genes Dev. *6*, 1153–1164.

Aguilar, P.S., Hernandez-Arriaga, A.M., Cybulski, L.E., Erazo, A.C., and de Mendoza, D. (2001). Molecular basis of thermosensing: a two-component signal transduction thermometer in *Bacillus subtilis*. EMBO J. *20*, 1681–1691.

Allen, M.A. *et al.* (2009). The genome sequence of the psychrophilic archaeon, *Methanococcoides burtonii*: the role of genome evolution in cold adaptation. ISME J. *3*, 1012–1035.

Aono, E. *et al.* (2010). Complete genome sequence and comparative analysis of *Shewanella violacea*, a psychrophilic and piezophilic bacterium from deep sea floor sediments. Mol. BioSyst. *6*, 1216–1226.

Arsène, F., Tomoyasu, T., and Bukau, B. (2000). The heat shock response of *Escherichia coli*. Int. J. Food Microbiol. *55*, 3–9.

Baler, R., Zou, J., and Voellmy, R. (1996). Evidence for a role of Hsp70 in the regulation of the heat shock response in mammalian cells. Cell Stress Chaperones *1*, 33–39.

Bardwell, J.C.A., and Craig, E.A. (1988). Ancient heat shock gene is dispensable. J. Bacteriol. *170*, 2977–2983.

Bosch, T.C.G., Krylow, S.M., Bode, H.R., and Steele, R.E. (1988). Thermotolerance and synthesis of heat shock proteins: these responses are present in *Hydra attenuata* but absent in *Hydra oligactis*. Proc. Natl. Acad. Sci. U.S.A. *85*, 7927–7931.

Brennecke, T., Gellner, K., and Bosch, T.C.G. (1998). The lack of a stress response in *Hydra oligactis* is due to reduced *hsp70* mRNA stability. Eur. J. Biochem. *255*, 703–709.

Bukau, B., and Walker, G.C. (1989). Cellular defects caused by deletion of the *Escherichia coli dnaK* gene indicate roles for heat shock protein in normal metabolism. J. Bacteriol. *171*, 2337–2346.

Collins, T., Meuwis, M.A., Gerday, C., and Feller, G. (2003). Activity, stability and flexibility in glycosidases adapted to extreme thermal environments. J. Mol. Biol. *328*, 419–428.

Coote, P.J., Cole, M.B., and Jones, M.V. (1991). Induction of increased thermotolerance in *Saccharomyces cerevisiae* may be triggered by a mechanism involving intracellular pH. J. Gen. Microbiol. *137*, 1701–1708.

D'Amico, S., Collins, T., Marx, J.C., Feller, G., and Gerday, C. (2006). Psychrophilic microorganisms: challenges for life. EMBO Rep. *7*, 385–389.

D'Amico, S., Marx, J.C., Gerday, C., and Feller, G. (2003). Activity–stability relationships in extremophilic enzymes. J. Biol. Chem. *278*, 7891–7896.

Deegenaars, M.L., and Watson, K. (1997). Stress proteins and stress tolerance in an Antarctic, psychrophilic yeast, *Candida psychrophila*. FEMS Microbiol. Lett. *151*, 191–196.

Deegenaars, M.L., and Watson, K. (1998). Heat shock response in psychrophilic and psychrotrophic yeast from Antarctica. Extremophiles 2, 41–49.

Deuerling, E., Schulze-Specking, A., Tomoyasu, T., Mogk, A., and Bukau, B. (1999). Trigger factor and DnaK cooperate in folding of newly synthesized proteins. Nature *400*, 693–696.

Eastman, J.T. (1993). Antarctic Fish Biology: Evolution in a Unique Environment (New York: Academic Press).

Erickson, J.W., Vaughn, V., Walter, W.A., Neidhardt, F.C., and Gross, C.A. (1987). Regulation of the promoters and transcripts of *rpoH*, the *Escherichia coli* heat shock regulatory gene. Genes Dev. *1*, 419–432.

Fayet, O., Ziegelhoffer, T., and Georgopoulos, C. (1989). The *groES* and *groEL* heat shock gene products of *Escherichia coli* are essential for bacterial growth at all temperatures. J. Bacteriol. *171*, 1379–1385.

Feller, G., and Gerday, C. (1997). Psychrophilic enzymes: molecular basis of cold adaptation. Cell. Mol. Life. Sci. *53*, 830–841.

Ferrer, M., Chemikova, T.N., Yakimov, M.M., Golyshin, P.N., and Timmis, K.N. (2003). Chaperonins govern growth of *Escherichia coli* at low temperatures. Nat. Biotechnol. *21*, 1266–1267.

Ferrer, M., Lünsdorf, H., Chernikova, T.N., Yakimov, M., Timmis, K.N., and Golyshin, P.N. (2004). Functional consequences of single:double ring transitions in chaperonins: life in the cold. Mol. Microbiol. *53*, 167–182.

Fields, P.A., and Somero, G.N. (1998). Hot spots in cold adaptation: Localized increases in conformational flexibility in lactate dehydrogenase A_4 orthologs of Antarctic notothenioid fishes. Proc. Natl. Acad. Sci. U.S.A. *95*, 11476–11481.

Frydman, J., and Hartl, F.U. (1994). Molecular chaperone functions of hsp70 and hsp60 in protein folding. In the biology of heat shock proteins and molecular chaperones, R.I. Morimoto, A. Tissières and C. Georgopoulos, eds. (Cold Spring Harbor Laboratory Press), pp. 251–283.

Fujii, S., Nakasone, K., and Horikoshi, K. (1999). Cloning of two cold shock genes, *cspA* and *cspG*, from the deep-sea psychrophilic bacterium *Shewanella violacea* strain DSS12. FEMS Microbiol. Lett. *178*, 123–128.

García-Descalzo, L., Alcazar, A., Baquero, F., and Cid, C. (2011). Identification of in vivo HSP90-interacting proteins reveals modularity of HSP90 complexes is dependent on the environment in psychrophilic bacteria. Cell Stress Chaperones *16*, 203–218.

Genevaux, P., Georgopoulos, C., and Kelley, W.L. (2007). The Hsp70 chaperone machines of *Escherichia coli*: a paradigm for the repartition of chaperone functions. Mol. Microbiol. *66*, 840–857.

Georlette, D., Damien, B., Blaise, V., Depiereux, E., Uversky, V.N., Gerday, C., and Feller, G. (2003). Structural and functional adaptations to extreme temperatures in psychrophilic, mesophilic, and thermophilic DNA ligases. J. Biol. Chem. *278*, 37015–37023.

Gerike, U., Danson, M.J., and Hough, D.W. (2001). Cold-active citrate synthase: mutagenesis of active-site residues. Protein Eng. *14*, 655–661.

Gerike, U., Danson, M.J., Russell, N.J., and Hough, D.W. (1997). Sequencing and expression of the gene encoding a cold-active citrate synthase from an Antarctic bacterium, strain DS2–3R. Eur. J. Biochem. *248*, 49–57.

Giuliodori, A.M., Di Pietro, F., Marzi, S., Masquida, B., Wagner, R., Romby, P., Gualerzi, C.O., and Pon, C.L. (2010). The *cspA* mRNA is a thermosensor that modulates translation of the cold shock protein CspA. Mol. Cell 37, 21–33.

Goloubinoff, P., Mogk, A., Zvi, A.P.B., Tomoyasu, T., and Bukau, B. (1999). Sequential mechanism of solubilization and refolding of stable protein aggregates by a bichaperone network. Proc. Natl. Acad. Sci. U.S.A. *96*, 13732–13737.

Haslbeck, M., Franzmann, T., Weinfurtner, D., and Buchner, J. (2005). Some like it hot: the structure and function of small heat-shock proteins. Nat. Struct. Mol. Biol. *12*, 842–846.

Hawley, D.K., and McClure, W.R. (1993). Compilation and analysis of *E. coli* promoter DNA sequences. Nucleic Acids Res. *11*, 2237–2255.

Hengge, R., and Bukau, B. (2003). Proteolysis in prokaryotes: protein quality control and regulatory principles. Mol. Microbiol. *49*, 1451–1462.

Hesterkamp, T., Hauser, S., Lütcke, H., and Bukau, B. (1996). *Escherichia coli* trigger factor is a prolyl isomerase that associates with nascent polypeptide chains. Proc. Natl. Acad. Sci. U.S.A. *93*, 4437- 4441.

Hofmann, G.E., Buckley, B.A., Airaksinen, S., Keen, J.E., and Somero, G.N. (2000). Heat-shock protein expression is absent in the Antarctic fish *Trematomus bernacchii* (family Nototheniisae). J. Exp. Biol. *203*, 2331–2339.

Horwich, A.L., Farr, G.W., and Fenton, W.A. (2006). GroEL-GroES-mediated protein folding. Chem. Rev. *106*, 1917–1930.

Hossain, M.M., and Nakamoto, H. (2002). HtpG plays a role in cold acclimation in cyanobacteria. Curr. Microbiol. *44*, 291–296.

Houry, W.A. (2001). Chaperone-assisted protein folding in the cell cytoplasm. Curr. Protein Pept. Sci. *2*, 227–244.

Huston, A.L., Krieger-Brockett, B.B., and Deming, J.W. (2000). Remarkably low temperature optima for extracellular enzyme activity from Arctic bacteria and sea ice. Environ. Microbiol. *2*, 383–388.

Ishii, A., Suzuki, M., Sahara, T., Takada, Y., Sasaki, S., and Fukunaga, N. (1993). Genes encoding two isocitrate dehydrogenase isozymes of a psychrophilic bacterium, *Vibrio* sp. strain ABE-1. J. Bacteriol. *175*, 6873–6880.

Jiang, W., Hou, Y., and Inouye, M. (1997). CspA, the major cold shock protein of *Escherichia coli*, is an RNA chaperone. J. Biol. Chem. *272*, 196–202.

Kandror, O., and Goldberg, A.L. (1997). Trigger factor is induced upon cold shock and enhances viability of *Escherichia coli* at low temperatures. Proc. Natl. Acad. Sci. U.S.A. *94*, 4978–4981.

Kanemori, M., Nishihara, K., Yanagi, H., and Yura, T. (1997). Synergistic roles of HslVU and other ATP-dependent proteases in controlling in vivo turnover of σ^{32} and abnormal proteins in *Escherichia coli*. J. Bacteriol. *179*, 7219–7225.

Kappé, G., Leunissen, J.A., and de Jong, W.W. (2002). Evolution and diversity of prokaryotic small heat shock proteins. Prog. Mol. Subcell. Biol. *28*, 1–17.

Kerner, M.J., Naylor, D.J., Ishihama, Y., Maier, T., Chang, H.C., Stines, A.P., Georgopoulos, C., Frishman, D., Hayer-Hartl, M., Mann, M., and Hartl, F.U. (2005). Proteome-wide analysis of chaperonin-dependent protein folding *Escherichia coli*. Cell *122*, 209–220.

Kim, B.H., and Schöffl, F. (2002). Interaction between *Arabidopsis* heat shock transcription factor 1 and 70 kDa heat shock proteins. J. Exp. Bot. *53*, 371–375.

Kim, Y.H., Han, K.Y., Lee, K., and Lee, J. (2005). Proteome response of *Escherichia coli* fed-batch culture to temperature downshift. Appl. Microbiol. Biotechnol. *68*, 786–793.

Kortmann, J., Sczodrok, S., Rinnenthal, J., Schwalbe, H., and Narberhaus, F. (2011). Translation on demand by a simple RNA-based thermosensor. Nucleic acids Res. *39*, 2855–2868.

Kruuv, J., Glofcheski, D., Cheng, K.H., Campbell, S.D., Al-Qvsi, H.M., Nolan, W.T., and Lepock, J.R. (1983). Factors influencing survival and growth of mammalian cells exposed to hypothermia. l. Effects of temperature and membrane lipid perturbers. J. Cell Physiol. *115*, 179–185.

Kulakova, L., Galkin, A., Kurihara, T., Yoshimura, T., and Esaki, N. (1999). Cold-active serine alkaline protease from the psychrotrophic bacterium *Shewanella* strain Ac10: gene cloning and enzyme purification and characterization. Appl. Environ. Microbiol. *65*, 611–617.

Laksanalamai, P., Narayan, S., Luo, H., and Robb, F.T. (2009). Chaperone action of a versatile small heat shock protein from *Methanococcoides burtonii*, a cold adapted archaeon. Proteins *75*, 275–281.

La Terza, A., Papa, G., Miceli, C., and Luporini, P. (2001). Divergence between two Antarctic species of the ciliate *Euplotes*, *E. focardii* and *E. nobilii*, in the expression of heat-shock protein 70 genes. Mol. Ecol. *10*, 1061–1067.

Liberek, K., Marszalek, J., Ang, D., Georgopoulos, C., and Zylicz, M. (1991). *Escherichia coli* DnaJ and GrpE heat shock proteins jointly stimulate ATPase activity of DnaK. Proc. Natl. Acad. Sci. U.S.A. *88*, 2874–2878.

Luo, H., and Robb, F.T. (2011). A modulator domain controlling thermal stability in the Group II chaperonins of Archaea. Arch. Biochem. Biophys. *512*, 111–118.

Manzanera, M., Aranda-Olmedo, I., Ramos, J.L., and Marqués, S. (2001). Molecular characterization of *Pseudomonas putida* KT2440 *rpoH* gene regulation. Microbiology *147*, 1323–1330.

Mathé, B.A. *et al.* (2005). The psychrophilic lifestyle as revealed by the genome sequence of *Colwellia psychrerythraea* 34H through genomic and proteomic analyses. Proc. Natl. Acad. Sci. U.S.A. *102*, 10913–10918.

Mayer, M.P., and Bukau, B. (2005). Hsp70 chaperones: Cellular functions and molecular mechanism. Cell. Mol. Life Sci. *62*, 670–684.

Médigue, C. *et al.* (2005). Coping with cold: The genome of the versatile marine Antarctica bacterium *Pseudoalteromonas haloplanktis* TAC125. Genome Res. *15*, 1325–1335.

Mendoza, J.A., Dulin, P., and Warren, T. (2000). The lower hydrolysis of ATP by the stress protein GroEL is a major factor responsible for the diminished chaperonin activity at low temperature. Cryobiology *41*, 319–323.

Mendoza, J.A., Warren, T., and Dullin, P. (1996). The ATPase activity of chaperonin GroEL is highly stimulated at elevated temperatures. Biochem. Biophys. Res. Commun. *229*, 271–274.

Miller, M.J., Xuong, N.H., and Geiduschek, E.P. (1982). Quantitative analysis of the heat shock response of *Saccharomyces cerevisiae*. J. Bacteriol. *151*, 311–327.

Mogk, A., Deuerling, E., Vorderwülbecke, S., Vierling, E., and Bukau, B. (2003). Small heat shock proteins, ClpB and the DnaK system form a functional triade in reversing protein aggregation. Mol. Microbiol. *50*, 585–595.

Mogk, A., Homuth, G., Scholz, C., Kim, L., Schmid, F.X., and Schumann, W. (1997). The GroE chaperonin machine is a major modulator of the CIRCE heat shock regulon of *Bacillus subtilis*. EMBO J. *16*, 4579–4590.

Morimoto, R.I. (1998). Regulation of the heat shock transcriptional response: cross talk between a family of heat shock factors, molecular chaperones, and negative regulators. Genes Dev. *12*, 3788–3796.

Morita, M., Kanemori, M., Yanagi, H., and Yura, T. (1999). Heat-induced synthesis of σ^{32} in *Escherichia coli*: structural and functional dissection of *rpoH* mRNA secondary structure. J. Bacteriol. *181*, 401–410.

Morita, R.Y. (1975). Psychrophilic bacteria. Bacteriol. Rev. *39*, 144–167.

Nagai, H., Yuzawa, H., and Yura, T. (1991). Interplay of two cis-acting mRNA regions in translational control of σ^{32} synthesis during the heat shock response of *Escherichia coli*. Proc. Natl. Acad. Sci. U.S.A.. *88*, 10515–10519.

Nakamoto, H., and Vigh, L. (2007). The small heat shock proteins and their clients. Cell. Mol. Life Sci. *64*, 294–306.

Nakamura, T., Tanaka, M., Maruyama, A., Higashi, Y., and Kurusu, Y. (2004). A nonconserved carboxy-terminal segment of GroEL contributes to reaction temperature. Biosci. Biotechnol. Biochem. *68*, 2498–2504.

Narberhaus, F., Waldminghaus, T., and Chowdhury, S. (2006). RNA thermometers. FEMS Microbiol. Rev. *30*, 3–16.

Oikawa, T., Yamamoto, N., Shimoke, K., Uesato, S., Ikeuchi, T., and Fujioka, T. (2005). Purification, characterization, and overexpression of psychrophilic and thermolabile malate dehydrogenase of a novel Antarctic psychrotolerant, *Flavobacterium frigidimaris* KUC-1. Biosci. Biotechnol. Biochem. *69*, 2146–2154.

Osipiuk, J., and Joachimiak, A. (1997). Cloning, sequencing, and expression of *dnaK*-operon proteins from the thermophilic bacterium *Thermus thermophilus*. Biochim. Biophys. Acta *1353*, 253–265.

Paek, K.H., and Walker, G.C. (1987). *Escherichia coli dnaK* null mutants are inviable at high temperature. J. Bacteriol. *169*, 283–290.

Phadtare, S. (2004). Recent developments in bacterial cold shock response. Curr. Issues Mol. Biol. *6*, 125–136.

Piette, F., D'Amico, S., Struvay, C., Mazzucchelli, G., Renaut, J., Tutino, M.L., Danchin, A., Leprince, P., and Feller, G. (2010). Proteomics of life at low temperatures: trigger factor is the primary chaperone in the Antarctic bacterium *Pseudoalteromonas haloplanktis* TAC125. Mol. Microbiol. *76*, 120–132.

Ramírez-Santos, J., Collado-Vides, J., García-Varela, M., and Gómez-Eichelmann, M.C. (2001). Conserved regulatory elements of the promoter sequence of the gene *rpoH* of enteric bacteria. Nucleic Acids Res. *29*, 380–386.

Ratajczak, E., Zietkiewicz, S., and Liberek, K. (2009). Distinct activities of *Escherichia coli* small heat shock proteins IbpA and IbpB promote efficient protein disaggregation. J. Mol. Biol. *386*, 178–189.

Reisenauer, A., Mohr, C.D., and Shapiro, L. (1996). Regulation of a heat shock σ^{32} homolog in *Caulobacter crescentus*. J. Bacteriol. *178*, 1919–1927.

Richter, K., Haslbeck, M., and Buchner, J. (2010). The heat shock response: life on the verge of death. Mol. Cell *40*, 253–266.

Robin, S., Togashi, D.M., Ryder, A.G., and Wall, J.G. (2009). Trigger factor from the psychrophilic bacterium *Psychrobacter frigidicola* is a monomeric chaperone. J. Bacteriol. *191*, 1162–1168.

Rodriguez, F., Arsène-Ploetze, F., Rist, W., Rüdiger, S., Schneider-Mergener, J., Mayer, M.P., and Bukau, B. (2008). Molecular basis for regulation of the heat shock transcription factor σ^{32} by the DnaK and DnaJ chaperones. Mol. Cell *32*, 347–358.

Rye, H.S., Burston, S.G., Fenton, W.A., Beechem, J.M., Xu, Z., Sigler, P.B., and Horwich, A.L. (1997). Distinct actions of *cis* and *trans* ATP within the double ring of the chaperonin GroEL. Nature *388*, 792–798.

Sakamoto, T., and Bryant, D.A. (1997). Temperature-regulated mRNA accumulation and stabilization for fatty acid desaturase genes in the cyanobacterium *Synechococcus* sp. strain PCC 7002. Mol. Microbiol. *23*, 1281–1292.

Schulz-Raffelt, M., Lodha, M., and Schroda, M. (2007). Heat shock factor 1 is a key regulator of the stress response in *Chlamydomonas*. Plant J. *52*, 286–295.

Segal, G., and Ron, E.Z. (1996). Regulation and organization of the *groE* and *dnaK* operons in eubacteria. FEMS Microbiol. Lett. *138*, 1–10.

Stoller, G., Rücknagel, K.P., Nierhaus, K.H., Schmid, F.X., Fischer, G., and Rahfeld, J.U. (1995). A ribosome-associated peptidyl-prolyl *cis/trans* isomerase identified as the trigger factor. EMBO J. *14*, 4939–4948.

Szabo, A., Langer, T., Schröder, H., Flanagan, J., Bukau, B., and Hartl, F.U. (1994). The ATP hydrolysis-dependent reaction cycle of the *Escherichia coli* Hsp70 system-DnaK, DnaJ, and GrpE. Proc. Natl. Acad. Sci. U.S.A. *91*, 10345–10349.

Tanaka, N., and Nakamoto, H. (1999). HtpG is essential for the thermal stress management in cyanobacteria. FEBS Lett. *458*, 117–123.

Taura, T., Kusukawa, N., Yura, T., and Ito, K. (1989). Transient shut off of *Escherichia coli* heat shock protein synthesis upon temperature shift down. Biochem. Biophys. Res. Commun. *163*, 438–443.

Terao-Morita, M., Tanaka, Y., Kodama, T.S., Kyogoku, Y., Yanagi, H., and Yura, T. (1999). Translational induction of heat shock transcription factor σ^{32}: evidence for a built-in RNA thermosensor. Genes Dev. *13*, 655–665.

Tosco, A., Birolo, L., Madonna, S., Lolli, G., Sannia, G., and Marino, G. (2003). GroEL from the psychrophilic bacterium *Pseudoalteromonas haloplanktis* TAC 125: molecular characterization and gene cloning. Extremophiles *7*, 17–28.

Vigh, L., Horváth, I., Maresca, B., and Harwood, J.L. (2007). Can the stress protein response be controlled by 'membrane-lipid therapy'? Trends Biochem. Sci. *32*, 357–363.

Wada, A., and Suyama, A. (1986). Local stability of DNA and RNA secondary structure and its relation to biological functions. Prog. Biophys. Mol. Biol. *47*, 113–157.

Walter, S., and Buchner, J. (2002). Molecular chaperones-cellular machines for protein folding. Angew. Chem. Int. Ed. *41*, 1098–1113.

Watanabe, S., Yamaoka, N., Fukunaga, N., and Takada, Y. (2002a). Purification and characterization of a cold-adapted isocitrate lyase and expression analysis of the cold-inducible isocitrate lyase gene from the psychrophilic bacterium *Colwellia psychrerythraea*. Extremophiles *6*, 397–405.

Watanabe, S., Yamaoka, N., Takada, Y., and Fukunaga, N. (2002b). The cold-inducible *icl* gene encoding thermolabile isocitrate lyase of a psychrophilic bacterium, *Colwellia maris*. Microbiology *148*, 2579–2589.

Wayne, N., Mishra, P., and Bolon, D.N. (2011). Hsp90 and client protein maturation. Methods Mol. Biol. *787*, 33–44.

Weissman, J.S., Hohl, C.M., Kovalenko, O., Kashi, Y., Chen, S., Braig, K., Saibil, H.R., Fenton, W.A., and Horwich, A.L. (1995). Mechanism of GroEL action: productive release of polypeptide from a sequestered position under GroES. Cell *83*, 577–587.

Weissman, J.S., Rye, H.S., Fenton, W.A., Beechem, J.M., and Horwich, A.L. (1996). Characterization of the active intermediate of a GroEL-GroES-mediated protein folding reaction. Cell *84*, 481–490.

Xu, Z., Horwich, A.L., and Sigler, P.B. (1997). The crystal structure of the asymmetric GroEL-GroES-$(ADP)_7$ chaperonin complex. Nature *388*, 741–750.

Yamauchi, S., Okuyama, H., Morita, E.H., and Hayashi, H. (2003). Gene structure and transcriptional regulation specific to the *groESL* operon from the psychrophilic bacterium *Colwellia maris*. Arch. Microbiol. *180*, 272–278.

Yamauchi, S., Okuyama, H., Nishiyama, Y., and Hayashi, H. (2004). Gene structure and transcriptional regulation of *dnaK* and *dnaJ* genes from a psychrophilic bacterium, *Colwellia maris*. Extremophiles *8*, 283–290.

Yamauchi, S., Okuyama, H., Nishiyama, Y., and Hayashi, H. (2006). The *rpoH* gene encoding heat shock sigma factor σ^{32} of psychrophilic bacterium *Colwellia maris*. Extremophiles *10*, 149–158.
Yamauchi, S., Ueda, Y., Matsumoto, M., Inoue, U., and Hayashi, H. (2012). Distinct features of protein folding by the GroEL system from a psychrophilic bacterium, *Colwellia psychrerythraea* 34H. Extremophiles *16*, 871–882.
Yoshimune, K., Galkin, A., Kulakova, L., Yoshimura, T., and Esaki, N. (2005). Cold-active DnaK of an Antarctic psychrotroph *Shewanella* sp. Ac10 supporting the growth of *dnaK*-null mutant of *Escherichia coli* at cold temperatures. Extremophiles 9, 145–150.
Yumoto, I., Kawasaki, K., Iwata, H., Matsuyama, H., and Okuyama, H. (1998). Assignment of *Vibrio* sp. strain ABE-1 to *Colwellia maris* sp. nov., a new psychrophilic bacterium. Int. J. Syst. Bacteriol. *48*, 1357–1362.

Catalysis and Protein Folding in Psychrophiles

8

Charles Gerday

Abstract

Enzymes have the property to catalyse most of the chemical reactions occurring in living organisms and their efficiency is exponentially depending on temperature through the Arrhenius law. On the other hand psychrophiles are organisms that live in permanently cold habitats so, in the absence of adaptation, the rate of the chemical reactions would be to slow to sustain life. In fact, these organisms display, at the low temperature of their environment, metabolic fluxes close to that of their mesophilic counterparts. That means that their enzymes have been modified in order to compensate for the cooling effect on reaction rates. The analysis of the structure, catalytic properties and thermal stability of an already large number of psychrophilic enzymes has demonstrated that discrete changes in the amino acid sequence of these enzymes can confer a high catalytic efficiency at low and moderate temperatures. This is accompanied by a significant decrease of the thermal stability itself inducing a large increase of the flexibility of the molecular structure that enables the accommodation of the substrate even at temperatures lower than the freezing point of water. The folding process, which, in theory, is also rendered uneasy due to the large increase in viscosity, the low residual energy in the system and the depressive effect of low temperatures, notably on hydrophobic interactions, is also rescued, mainly, through the overexpression of peptidyl-prolyl *cis–trans* isomerases and of the trigger factor which is the first to take in charge the nascent polypeptide chain.

Introduction

In the absence of enzymes, the rate of chemical reactions occurring in living organisms would be, in general, too slow to sustain life and so enzymes appear as crucial components of the living cell. Their efficiency depends on several parameters, among which temperature appears as the most important one through the Arrhenius equation, initially proposed in 1889, that describes the thermodependence of the effect of acids on the velocity of the inversion of cane sugars (Arrhenius, 1889). This equation takes the form $d\ln k/dt = E_a/RT^2$ and on integration, assuming that E_a is constant within a limited range of temperature, that gives $k = Ae^{-Ea/RT}$, where k is the rate coefficient or rate constant; A is the pre-exponential term or frequency factor, which slightly depends on temperature, and from which one can also derive the value of the activation entropy of the reaction (Collins *et al.*, 2007); R is the gas constant (8.314 J/mol/K); and T is the temperature in kelvin. So the rate of a chemical or catalysed reaction strongly depends, in an exponential way, both on the activation energy, E_a, which is the energy of a fraction of the molecules that possess enough energy to react, and on temperature. On earth numerous species are living in cold environments which are widely spread on the surface of our planets; most of these organisms do not have any temperature regulation system and so their internal temperature fluctuates as a function of that of the surrounding environment. Depending on the cold habitat, the temperatures are rather diversified and, in most cases, vary from around 0°C, such as in sea waters for example, to about −20°C in permafrost zones or in frozen sea ices (Feller, 2010; Gerday, 2011). Within this range we will only consider organisms that display noticeable biological activities (Deming, 2002).Taking into account the Arrhenius law one can ask the question: How do these organisms succeed in displaying appropriate metabolic fluxes, comparable to those occurring in mesophilic counterpart? To fix the ideas, if one considers a chemical reaction having an activation energy, $E_{a,}$ equal to 50 kJ/mol, a temperature drop from 30°C to 20°C will decrease the rate of the reaction by a factor close to two. Indeed, in general, the activation energies of reactions occurring in mesophiles are around 50 kJ/mol and so the Q_{10}, which corresponds to the ratio between the rates of a reaction measured at an interval temperature of 10°C, is usually between 2 and 3. This high thermal dependence suggests that adaptations are necessary in organisms living in permanently cold habitats in order to keep reaction rates at a level compatible with life. From the Arrhenius equation, one way would be to decrease the activation energy of the numerous biochemical reactions occurring in living organisms. Enzymes have been designed to do so; in fact they do not really decrease the activation energy as such, but they modify the course of the reaction in order to find a more favourable way characterized by a lower activation energy. So, in psychrophiles, the survival is subordinated to a further decrease of the activation energy in order to render reaction rates appropriate at low temperatures. Loads of biological functions are affected by temperature (Margesin *et al.*, 2008) and especially the folding process of proteins that, in many cases, requires the assistance of other proteins known as chaperones. These proteins are ubiquitous, numerous and highly diversified, can act individually or in sequence from the beginning of the synthesis of the nascent polypeptide chain to its final destination in or outside the intracellular space; so they are also integral partners of the low temperature adaptation and part of this chapter will also focus on their role in the folding process of proteins from psychrophilic microorganisms.

Preliminary considerations

In many papers, the terms psychrotroph and psychrotolerant are alternatively used to designate microorganisms that resist temperatures around 30°C even so their usual habitat is permanently cold and close to the freezing point of water. These terms are based on the wrong concept that the temperature giving the shortest doubling time is close to the best temperature for these organisms. This idea does not resist a careful analysis of growing curves in terms of cell densities and enzymes production (Fig. 8.1). Indeed in cold-adapted organisms, exposed to more or less permanently cold habitats, the temperature giving the shortest doubling time also gives poor densities of cells indicating that at these temperatures, often erroneously called optimum growth temperatures, the organisms are under thermal stress that leads to unwanted metabolites or unappropriated concentrations (Feller *et al.*, 1994; Margesin, 2009). This is induced by the differential effect of temperature on the rate of catalysed reactions via the respective values of their activation energy. Short doubling

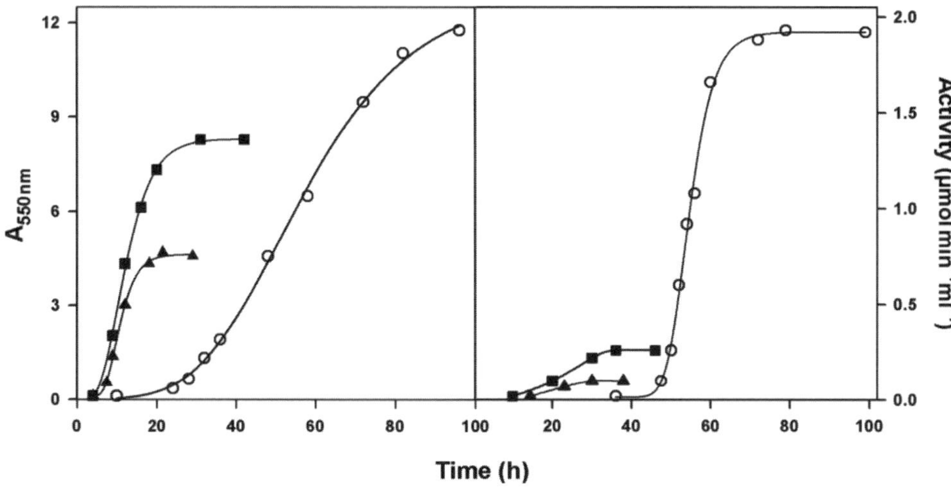

Figure 8.1 Growth (left) of the Antarctic strain, *Pseudoalteromonas haloplanktis* and α-amylase production (right) as a function of the temperature of the culture medium: open circle (4°C); black square (18°C) and black triangle, (25°C). Although the doubling time is shorter at 18°C and 25°C, the enzyme production is much lower than at 4°C and is nearly abolished at 25°C, a temperature close to the apparent optimum activity of the enzyme. (Adapted from Feller *et al*., 1994.)

times derive from kinetic effects; an increase in temperature increases reaction rates and forces the bacterium to divide at higher speed but this has nothing to do with the well-being of the microorganism, the temperature of which is on the contrary usually close to that of the environment. That does not mean that all organisms living in permanently cold habitats are perfectly adapted to their environment in terms of temperature. This probably depends on their initial capacity of temperature adaptation and specific evolution history. We therefore believe that there is a continuum in the adaptation to low temperatures that explains why some microorganisms cannot sustain temperatures above 10–15°C while other resist to temperatures close to 40°C. Therefore, the terms psychrotolerant and psychrotroph are inappropriate to designate organisms living in permanently cold environments and displaying efficient growth at low temperature, I mean that produce high density of cells close to the maximum that can be reached by these organisms. Therefore, in this review we will essentially use the term psychrophile to designate these organisms.

Often also, in the literature, the term optimum temperature is used to designate the maximum activity recorded when the activity of an enzyme is measured as a function of temperature. One should rather write « apparent optimum temperature » since this maximum is obtained by the crossing of two curves, one, derives from the Arrhenius equation and expresses the exponential dependence of reaction rates on temperature while the other corresponds to the inactivation, through unfolding, of the enzyme as a function of temperature. In consequence the apparent optimum temperature corresponds in fact to a mixture of native and unfolded enzyme and one cannot keep the enzymes for a long time at this temperature. The real optimum for an enzyme is close to the temperature that associates, over a long time, a good activity to an adequate stability, conditions that are essential for the survival of a living organism. It is necessarily situated on the left side of the activity curve.

Another common approximation, also found in many papers, is related to the evaluation of the thermal stability of an enzyme. In too many cases this has been achieved through the evaluation of the residual activity. Following this technique, an enzyme solution is exposed to various temperatures for a certain time, and then the residual activity is measured at a fixed temperature often close to room temperature. This can give erroneous data since in some cases and especially when cold-adapted enzymes are investigated, the thermal unfolding is reversible over cooling, and that leads to the overestimation of the thermal stability. When using this technique one has first to make sure that the unfolding is fully irreversible. So direct techniques such as spectrophotometric methods, circular dichroism and fluorescence, or micro calorimetry that can follow the sequential unfolding of the different calorimetric domains of a protein are by far preferable.

Enzyme activity under low temperature

Cold-adapted enzymes display activity curves shifted towards low temperatures when compared to their mesophilic counterparts. This is clearly illustrated in Fig. 8.2 in which the activity of a cold-adapted α-amylase is compared to the activity of the homologous thermophilic α-amylase from *Bacillus amyloliquefaciens* and this as a function of temperature. Several features are noticeable for the cold-adapted enzyme: first the apparent optimum is shifted towards low temperature by more than 30°C; second, the specific activity is much higher than that of the thermophilic counterpart over a temperature range from 0°C to 30°C; third, the activity rapidly drops for temperatures exceeding 25°C illustrating a higher thermal sensitivity. These properties characterize all cold-adapted enzymes although the amplitude of the differences varies according to the origin and type of the enzyme. That again supports the idea that there is a continuum in the adaptation to cold. What is the origin of the higher specific activity of cold-adapted enzymes at low and moderate temperatures? From

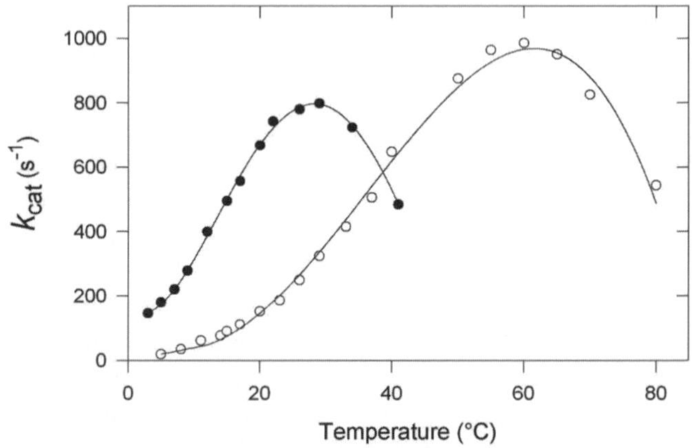

Figure 8.2 Activity of the α-amylase from the Antarctic sea strain *Pseudoalteromonas haloplanktis* (closed circles) as a function of temperature compared to that of the homologous α-amylase from *Bacillus amyloliquefaciens* (open circles). Noteworthy is the shift of the apparent optimum of the cold-adapted enzyme towards low temperatures and the much higher specific activity of the cold- adapted enzymes at low and moderate temperatures. (From Feller *et al.*, 1992.)

the Arrhenius equation, the most obvious way to increase the specific activity would be to decrease the activation energy E_a. The activation energy is equal to $\Delta H^* + RT$ and therefore a decrease of the activation enthalpy will favour reaction rate and will render the activity less dependent of the temperature since for $E_a = 0$, the exponential term will be equal to 1 and reaction rates will only depend on temperature through the pre-exponential term A. This term also depends on the activation entropy (Collins *et al.*, 2007), so to be complete, the full energy change associated with the formation of the activated complex ES* is equal to $\Delta G^* = \Delta H^* - T\Delta S^*$, and to render the catalysed reaction as much as possible fast, ΔG^* should be as low as possible. The activation energy of a few cold-adapted enzymes has been determined and compared to the activation parameters obtained for homologous mesophilic enzymes. The data are presented in Table 8.1. Most of these enzymes originate from bacteria exception being for arginine kinase isolated from deep-sea clam; lactate dehydrogenase (LDH), isolated from Antarctic fish; lysozyme isolated from an insect; and a xylanase purified from Antarctic yeast. The analysis of these data leads to several interesting

Table 8.1 Activation parameters of a few cold-adapted enzymes compared with those of mesophilic counterparts

Enzyme	Type	*T* (°C)	K_{cat} (s^{-1})	ΔG^* (kJ/mol)	ΔH^* (kJ/mol)	$T\Delta S^*$ (kJ/mol)	Reference
Amylase	Psy	10	294.0	57.7	34.7	–23.0	D'Amico *et al.* (2003)
	Mes		97.0	58.5	46.4	–12.1	
Arginine kinase	Psy	25	3.3	69.4	18.8	–50.6	Suzuki *et al.* (2011)
	Mes		13.4	66.6	41.9	–24.7	
Cellulase	Psy	4	0.18	71.6	46.2	–25.4	Garsoux *et al.* (2004)
	Mes		0.01	78.2	65.8	–12.4	
Chitinase	Psy	15	1.7	69.2	60.2	–9.0	Lonhienne *et al.* (2001a)
	Mes		3.9	67.2	74.3	+7.1	
Chitobiase	Psy	15	3.8	59.5	44.7	–14.8	Lonhienne *et al.* (2001b)
	Mes		0.9	63.5	71.5	+8.0	
Citrate synthase	Psy				7.4	ΔS^*= 22.7	Bjelic *et al.* (2008)
	Mes				11.5	ΔS^*= 9.7	
Endonuclease	Psy	5	9.41	62.8	33.4	–29.4	Altermark *et al.* (2007)
	Mes		1.03	67.9	74.0	+6.1	
LDH	Psy	0	250.0	75.0	22.0	–53.0	Coquelle *et al.* (2007)
	Mes		72.0	75.0	45.0	–30.0	
Lysozyme	Psy			45.1	31.9	–13.2	Sotelo-Mundo *et al.* (2007)
	Mes			46.2	49.4	+ 3.2	
Subtilisin	Psy	15	25.4	62.0	36.0	–26.5	Davail *et al.* (1994)
	Mes		5.4	66.0	46.0	–20.2	
Xylanase (bact)	Psy	10	515.5	54.0	21.0	–33.0	Collins *et al.* (2003)
	Mes		59.5	60.0	58.0	–2.0	
Xylanase (yeast)	Psy	5	14.8	52.3	45.3	–7.0	Petrescu *et al.* (2000)
	Mes		4.9	54.6	49.9	–4.7	

features. First, with the exception of arginine kinase and chitinase, the specific activity of the cold-adapted enzymes are much higher than that of mesophilic counterparts. In the case of the cold-adapted chitinase it has been proposed (Collins *et al.*, 2007) that a difference in substrate specificity can explain this unexpected data as chitins from different species are also structurally quite different. The case of the cold-adapted arginine kinase is more puzzling especially because the activation parameters strictly follows the canonical characters of cold-adapted enzymes. This arginine kinase is issued from deep-sea clams and, as the activity of the psychrophilic and mesophilic counterparts has been measured at 25°C, it is possible that the cold-adapted enzyme is partially inactivated at this temperature. This is corroborated by the fact that at 10°C the specific activity of the cold-adapted enzyme is three fold as high as that measured at 25°C. Second, the activation energy ΔG^* of the psychrophilic enzyme is systematically lower than that of the mesophilic counterpart, again with the exception of arginine kinase and chitinase, whereas for LDH the values are equal for unknown reasons. The lower activation energy of cold-adapted enzymes results from a drastic decrease of the activation enthalpy that renders the catalysed reactions much less dependent on temperature variations. The very large difference observed should normally lead to much lower values of the activation energy, ΔG^*, but the enthalpic term is significantly compensated by a large decrease of the activation entropy limiting the importance of the low activation enthalpy on ΔG^*. Enzyme activation is associated with a rearrangement of the enzyme structure in order to accommodate, in the most favourable way, the substrate and also to enable the release of the products. This process is often accompanied by a rearrangement of the water molecules that interact with the enzyme or that actively participate in the catalysed chemical reaction. The rearrangement of the enzyme stucture necessitates the breaking of some weak bonds and the respective values of the activation enthalpies suggest that less bonds have to be broken during the activation of cold-adapted enzymes when compared to their mesophilic counterparts. From the data presented in Table 8.1, one can see that, in most cases, the activation entropy is negative meaning that the activated state presents a lower disorder than the ground state. In a few cases however, mesophilic chitinase, chitobiase, endonuclease and lysozyme, the activation entropy is positive, indicating that the overall activated state presents, on the contrary, a more ordered organization. One has to keep in mind that, on activation, not only the enzyme structure changes but also, as mentioned above, the organization of water molecules. So, the positive activation entropies recorded in the case of some mesophilic enzymes could be due to additional disorders create by the movement and possible release of water molecules. We can see that, as far as the cold-adapted enzymes are concerned, the movement of water molecules, if any, does not compensate the reordering of the enzyme structure since activation entropies changes remain negative and the reordering appears to be much higher than in the case of mesophilic enzymes. This suggest that the ground state of cold-adapted enzymes displays higher local or overall disorder when compared to their mesophilic counterparts and the question is, is there any relation between the high specific activity of these enzymes and the suggested disorder?

Stability and flexibility

Some investigations have in a few cases indicated that some cold-adapted enzymes could have a thermal stability similar to that of mesophilic counterparts (Fedoy *et al.*, 2007;

Merlino *et al.*, 2010) but from an extremely large number of other investigations carried out on psychrophilic enzymes, both the thermal and chemical stability of these enzymes are much lower than that of their mesophilic counterparts (Feller, 2010; Collins *et al.*, 2007). The few discrepancies observed could arise from inappropriate techniques used to determine stability such as for example the determination of residual activity (see above). In support of this, it has been demonstrated that many cold-adapted enzymes show reversible thermal unfolding. In other cases (Merlino *et al.*, 2010), the thermal stability has been determined by CD measurements at 222 nm. This can also be inappropriate since this technique only evaluates the loss of secondary structure as a function of temperature but not the deleterious effect of heat on tertiary and quaternary structures. In the same paper, however, chemical denaturants, such as urea and GdCl, were also tested and they undoubtedly illustrate the much lower stability of cold-adapted enzymes. In the case of the isocitrate dehydrogenase from *Desulfotalea psychrophila,* which display an optimum growth at 10°C, the cold-adapted enzyme shows an overall thermal stability indeed higher than that of the mesophilic counterparts, as measured by differential scanning calorimetry (DSC). However, its specific activity is also lower than that of the mesophilic counterparts even at low temperature (5°C) in accordance with the higher value of the activation energy ΔG^* (Fedoy *et al.*, 2007). This is due to a lower activation enthalpy (ΔH^*) however unfavourably compensated by a more negative value of the activation entropy (ΔS^*). Some examples of the stability of cold-adapted enzymes are shown in Fig. 8.3 in association with activity curves recorded for psychrophilic, mesophilic and thermophilic α-amylases (A) and glycoside hydrolases (B). Again, the activity curves illustrate the important and progressive shift of the apparent optima towards low temperatures as a function of the environmental temperature of the respective organisms as well as a difference in enzyme inactivation. The thermal stability of these enzymes has been determined by fluorescence transition in the case of the α-amylases (C) and by differential scanning calorimetry in the case of the glycoside hydrolases, xylanase and cellulase (D). Fig. 8.3 illustrates one interesting feature of cold-adapted enzymes: contrary to mesophilic and thermophilic homologues, the thermal inactivation of the cold-adapted enzymes investigated precedes any significant structural changes as followed by fluorescence spectroscopy or differential calorimetry whereas for mesophilic and thermophilic enzymes the mid-point thermal transition matches the apparent optimum of activity. This means either that the active site of cold-adapted enzymes is more heat labile than the whole enzyme structure or that the ES complex dissociates before any changes in enzyme structure. This suggests highly flexible active sites or enzyme structures that can be easily but discretely altered by moderate increases of temperature. Indeed, a flexible structure can counteract the freezing effect of low temperatures on enzyme structures while inducing a higher thermal instability due to the need to weaken in a way or another the weak interactions that stabilize the structure.

In unfavourable environments, such as aqueous solvents, a nascent polypeptide chain that exposes numerous hydrophobic groups characterized by a poor solubility in water, tends to limit the contact with water molecules in burying most of these groups within a folded structure that is further stabilize by other weak interactions such as hydrogen bonds, van der Waals interactions, salt bridges and possibly disulfide bonds. Physical factors that limit the difference between the entropy of the unfolded and folded states, such as the glycine and proline contents, also significantly contribute to the stability of a protein structure. If one assumes a reversible equilibrium between the native and the unfolded form of the protein

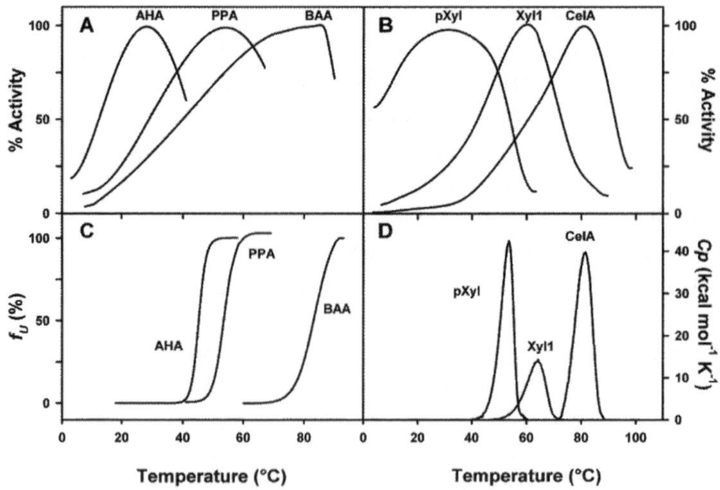

Figure 8.3 Enzyme activity of psychrophilic (AHA), mesophilic (PPA), and thermophilic α-amylases (BAA) (A) and of psychrophilic xylanase (pXyl) from Antarctic *Pseudoalteromonas haloplanktis*, mesophilic xylanase (Xyl11) from *Streptomyces* sp. S38, and thermophilic endoglucanase (CelA) from *Clostridium thermocellum* (B) and their corresponding thermal unfolding curves, as recorded by fluorescence spectroscopy for amylases (C) and by microcalorimetry for xylanases and endoglucanase (D). Note that for the two cold-adapted enzymes and contrary to what happens for mesophilic and thermophilic homologues the enzyme inactivation precedes any changes in the three-dimensional structure. (Taken with permission from Wiley from Georlette *et al*., 2004.)

one can have access to the thermodynamic stability, $\Delta G_{N\text{-}U}$, also known as conformational stability or Gibbs free energy of unfolding that corresponds to the energy required to unfold the protein. It is therefore a positive value that can be evaluated through the use of a combination between spectroscopic methods (CD, fluorescence) and chemical denaturants such as urea or guanidinium chloride (GdCl), or by a more convenient method, microcalorimetry (Feller, 2010), that gives a direct access to the enthalpy of unfolding, $\Delta H_{N\text{-}U}$. Stability curves, which express the value of the conformational stability as a function of temperature, can be constructed and examples are shown in Fig. 8.4. The three curves represent the evolution of the stabilization energy as a function of temperature for psychrophilic (AHA), mesophilic (PPA) and thermophilic (BAA) α-amylases.

These enzymes are characterized by quite different apparent optimum temperatures but their maximum stability is rather close and is around 25°C. This is due to the hydrophobic effect, which is maximum around room temperature (Kumar *et al.*, 2002). The maximum resistance to temperature, in the case of mesophilic and thermophilic α-amylase, appears to be reached by simply lifting the stability curve of the psychrophilic enzyme, and this is achieved through an increase in the stabilization enthalpy $\Delta H_{N\text{-}U}$, which also corresponds to an increase in the strength or number of weak bonds stabilizing the structure.

Two other models that account for the higher stability of thermophilic proteins have been described. The curves shown in Fig. 8.4 can be flattened to further shift the T_m (melting temperature where $\Delta G_{N\text{-}U} = 0$) towards high temperature. According to this model, the lesser dependence of $\Delta G_{N\text{-}U}$ on temperature would result from a lower magnitude of $\Delta H_{N\text{-}U}$,

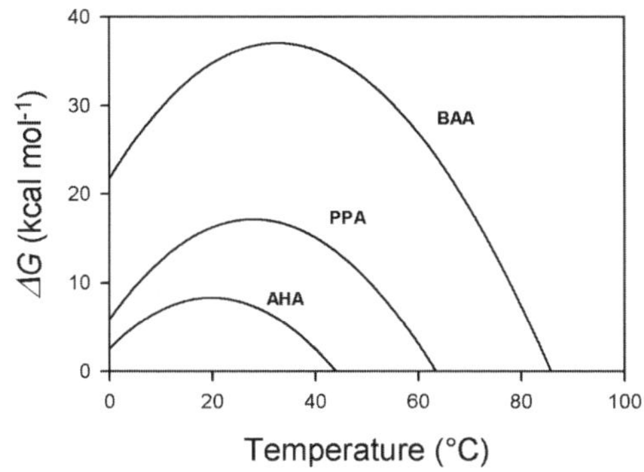

Figure 8.4 Thermodynamic stability curves of a psychrophilic α-amylase from Antarctic bacterium *Pseudoalteromonas haloplanktis* (AHA), of a mesophilic enzyme from pig pancreas (PPA), and of the mesophilic-thermophilic counterpart from *Bacillus amyloliquefaciens* (BBA) plotted as a function of temperature. For the three enzymes the maximum stability is reached near ordinary temperatures but the cold-adapted enzymes (AHA) is the most heat and cold sensitive. (Adapted from D'Amico *et al.*, 2003.)

of ΔS_{N-U}, and of ΔC_p, the difference between the specific heat of the unfolded and native forms of the protein that depends on the relative exposure of hydrophobic groups during unfolding (Beadle *et al.*, 1999). In a third model, the amplitude of ΔG_{N-U} of the thermophilic protein remains close to that of the mesophilic enzyme but its maximal value and the whole curve is shifted towards higher temperature. This is induced by a reduced loss in entropy on folding that often results from a decrease of the entropy of the unfolded state. That means that residual structures do exist in the unfolded state that can notably be induced by a higher proportion of proline residues and a lower proportion of glycine residues. A recent study carried out on 116 proteins from mesophiles and thermophiles (Sawle and Ghosh, 2011) favour the idea that there is a strong correlation between high T_m and reduced entropy change upon folding.

Another remarkable observation is that none of these α-amylase works, in natural environments, at the temperature corresponding to the maximal stability. For the cold-adapted α-amylase isolated from a microorganism living in Antarctic sea water, the difference is particularly acute. Indeed, the maximal stability is observed around room temperature whereas the environmental temperature is around 2°C. For the mesophilic and thermophilic enzymes one can see that the environmental temperatures are on the right side of the curves. The cold-adapted enzyme display therefore a marginal thermal stability at the temperature of the environment and by extrapolation one can expect that the enzyme will unfold at low temperatures, around −10°C. So, undoubtedly, cold denaturation should be taken into account when evaluating the survival of organisms at low temperatures. If the dissipative effect of heat on protein structure can explain heat denaturation through an excessive increase of ΔS_{N-U}, it is more difficult to explain cold denaturation but we know that some intramolecular bonds are weakened at low temperature, it is the case of hydrophobic bonds through unfavourable enthalpic effects that prevent the dehydration (de-encagement) of individual groups, an essential process that should precede any association. This is also true for salt bridges, and for a similar reasons, since it is necessary to desolvate any individual ion *prior* to form an ion pair. In this case the dielectric constant also plays an important role since it increases with a decrease in temperature and negatively affect the interaction between the ions through Coulomb's law. The low stability of cold-adapted enzymes, which becomes marginal around their usual environmental temperature, has been associated with the necessity to increase the flexibility of the molecular edifice in order to compensate for the freezing effect of low temperatures on the structure, a process that could prevent the interaction of the enzyme with the substrate.

A large consensus does exist around this idea. The determination of the local or global flexibility of enzymes is rather complex since flexibility has to be considered both in terms of amplitude of movements and movement's frequency. The flexibility of cold-adapted enzymes has been investigated by various techniques. We found that fluorescence quenching using acrylamide as a quencher is a rather convenient and straightforward technique that enables the direct evaluation of the penetrability of the proteins displaying a fluorescence signal at the level of buried tryptophan residues. On quenching, one observes a decrease in the amplitude of the fluorescence signal that is due to the diffusive collision between the quencher and tryptophan residues. This methodology has been largely used in our laboratory and has given clear evidence that psychrophilic enzymes are more easily penetrated than their mesophilic and thermophilic counterparts.

An example of the data is shown in Fig. 8.5; it concerns fluorescence quenching experiments carried out on psychrophilic α-amylase from Antarctic bacteria (AHA), mesophilic α-amylase from pig pancreas (PPA) and homologous enzyme from thermophilic *Bacillus amyloliquefaciens* (BAA). One can clearly see that the quenching of the fluorescence of tryptophan residues (F_0/F) is much more pronounced in the psychrophilic enzyme and that the effect significantly decreases from psychrophilic to thermophilic enzymes demonstrating the correlation that exists between the flexibility of the molecular edifice, the stability of these enzymes and their relative adaptation to low temperatures, a concept already developed long time ago by Hochachka and Somero (1984), re-evaluated by Somero(1995)

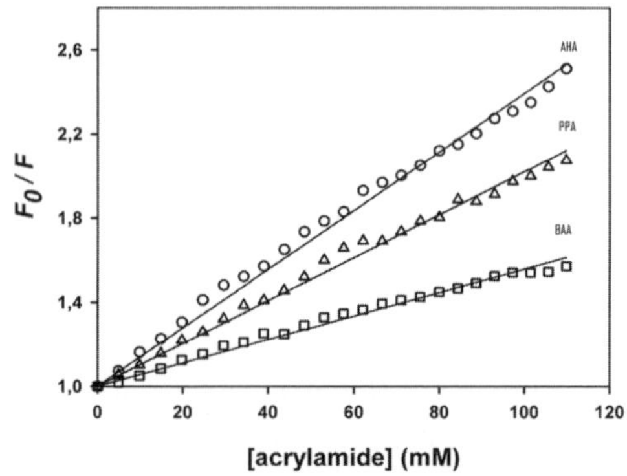

Figure 8.5 Fluorescence quenching using acrylamide of psychrophilic (open circles), mesophilic (open triangles), and thermophilic (open squares) α-amylases. The ratio F_0/F represents the ratio of the fluorescence signal in the absence of quenching agent and the fluorescence signal obtained at various concentrations of acrylamide. The relative steepness of the slope is related to the ability of the quencher to penetrate the structure. One can see that, by far, the psychrophilic enzyme is much more permeable than the mesophilic and thermophilic counterparts due to his higher flexibility.

and in the second and last edition of the initial book entitled *Biochemical Adaptation* by Hochachka and Somero (2002). Recent investigations based on the calculation of relative B-factors in psychrophilic and mesophilic iron superoxide dismutases (Merlino *et al.*, 2010) have also shown that the psychrophilic enzymes are characterized by an increase flexibility of the active site whereas several area of higher local flexibility have also been observed in psychrophilic hydroxylase with apparently, and in contrast with what is generally observed, no marked incidence on thermostability (Leiros *et al.*, 2007). The analysis of the stability, flexibility and kinetics of folding and refolding of multiple mutants of psychrophilic α-amylases have recently provide strong evidence that the weakening of intramolecular interactions in psychrophilic enzymes significantly increase the concerted motion of catalytic residues essential for high activity at low temperatures (Cipolla *et al.*, 2011).

Structural parameters involved in low stability

It has been shown that in most cases there is a correlation between the high specific activity of cold-adapted enzymes and their high flexibility itself deriving from an increase in structural instability. Now the question is 'Do these enzymes adopt a general strategy to reach the appropriate instability or is the strategy enzyme specific?' The molecular determinants possibly involved in cold adaptation have been enumerated in a large variety of cold-adapted enzymes and in particular in those of known three-dimensional structure. A relatively large number of crystallographic structures of cold-adapted enzymes are now available (Table 8.2). From the analysis of these structures, a general feature emerges: the three dimensional structure of cold-adapted enzymes is highly conserved when compared to the structure of their mesophilic counterparts, especially at the level of the structure implicated in substrate binding and catalysis and elsewhere the changes are discrete. Some important changes for the adaptation can easily escape the detection because long distance effects are not rare and can influence the substrate binding capacity, specificity and performance of the active site. One has to look for structural changes that, in principle, tend to decrease the stability (for an extensive discussion on thermal stability of proteins, see Smalas *et al.*, 2000; Vieille and Zeikus, 2001; and Kumar and Nussinov, 2001). Some modifications are detectable through the analysis of the amino acid composition in comparison with that of mesophilic and thermophilic homologues: a lower proline content increases the entropy of the unfolded state and destabilize the native state, a higher number of glycine residues can also increase the entropy of the unfolded state and stabilize this form, arginine residues and in particular high Lys/Arg + Lys ratios can also decrease the stability because Arg residues are able to form more stabilizing intramolecular bonds than Lys residues. Methionine has also been found to be implicated in a limited number of cases through the propensity of its lateral chain to create local disorder, this has been initially suggested by Gudmundsdottir *et al.* (1996) in cold-adapted elastase from Atlantic cod. Other molecular determinants can be repertoriated through the analysis of the three dimensional structure of the enzyme again as compared with the structure of mesophilic and thermophilic counterparts. One can mentioned: less hydrophobic and hydrogen interactions; less salt bridges; reduced strength of van der Waals interactions through the replacement, in the core of the protein, of bulky residues by less voluminous residues (Ile → Val for example); higher number of hydrophobic residues at the surface of the protein that reduces the entropy of the native state and therefore increases its free energy; higher number of charged (same charge) and polar residues at the surface of the

Table 8.2 Crystallographic structures of cold-adapted enzymes

Enzyme	Host organism	Reference
Adenylate kinase	*Bacillus globisporus*	Bae and Phillips (2004)
Adenylate kinase	*Marinibacillus marinus*	Davlieva and Shamoo (2009)
Alkaline metalloprotease	*Pseudomonas* sp.	Aghajari *et al*. (2003)
Alkaline phosphatase	*Pandalus borealis*	de backer *et al*. (2002)
Alkaline phosphatase	*Bacterial strain* TAB5	Wang *et al*. (2007)
Anionic trypsin	*Salmo salar*	Helland *et al*. (1998)
Alpha-amylase	*Pseudoalteromonas haloplanktis*	Aghajari *et al*. (1998)
Aspartate carbamoyltransferase	*Moritella profunda*	De Vos *et al*. (2007)
Beta-galactosidase	*Arthrobacter* sp. C2-2	Skalova *et al*. (2005)
Catalase	*Vibrio salmonicida*	Riise *et al*. (2007)
Cellulase	*Pseudoalteromonas haloplanktis*	Violot *et al*. (2005)
Citrate synthase	*Arthrobacter* sp. strain DS2-3R	Russell *et al*. (1998)
Elastase	*Salmo salar*	Berglund at al. (1995)
Endonuclease I	*Vibrio salmonicida*	Altermark *et al*. (2008)
Esterase	*Pseudoalteromonas* sp. 643A	Brzuszkiewicz *et al*. (2009)
Lactate dehydrogenase	*Champsocephalus gunnari*	Coquelle *et al*. (2007)
Lipase B	*Candida antarctica*	Uppenberg *et al*. (1994)
Malate dehydrogenase	*Aquaspirillium articum*	Kim *et al*. (1999)
Pepsin	*Gadus morhua*	Karlsen *et al*. (1998)
Phenylalanine hydroxylase	*Colwellia psychrerythraea*	Leiros *et al*. (2007)
Proteinase K-like	*Serratia* sp.	Helland *et al*. (2006)
Protein-tyrosine phosphatase	*Shewanella* sp.	Tsuruta *et al*. (2005)
S-formylglutathione hydrolase	*Pseudoalteromonas haloplanktis*	Alterio *et al*. (2010)
Subtilisin-like protease	*Vibrio* sp. PA-44	Arnorsdottir *et al*. (2005)
Subtilisin S41	*Bacillus* sp.	Almog *et al*. (2009)
Superoxide dismutase	*Aliivibrio salmonicida*	Pedersen *et al*. (2009)
Triose phosphate isomerase	*Vibrio marinus*	Alvarez *et al*. (1998)
Trypsin	*Oncorhynchus keta*	Toyota *et al*. (2002)
Uracil-DNA N-glycosylase	*Gadus morhua*	Leiros *et al*. (2003)
Xylanase	*Pseudoalteromonas* sp.	Van Petegem *et al*. (2003)

protein that, respectively, can weakened the compactness of the structure through repulsion effect or increase the interactions with the solvent. Are also concerned amino-aromatic and aromatic–aromatic interactions since the benzene ring is considered as a dipole with the face having a net negative charge and the C-H edges a positive charge; stability of α-helices through N- and C-capping of the helix dipole that displays a partially positive charge on the N-terminal side and a partially negative charge on the C-terminal side; increase loop lengths or additional loops that connect secondary structures and provide more amplitude to the relative movements between these structures.

All these parameters have been mentioned as possible determinants of the low stability of cold-adapted enzymes but the interesting thing is that the strategy is enzyme specific since each enzyme only utilizes an assortment of a rather limited number of these parameters. As examples and taking into account only rather recent reports one can mention that in cold-adapted LDH one observes a reduced number of ion pairs, a reduce size of ionic networks and accumulation of unfavourable interactions implicating charged residues (Coquelle *et al.*, 2007); in isocitrate dehydrogenase from the cold-adapted *Desulfotalea psychrophila*, a group of three methionine residues from each monomer formed a flexible zone at the level of the dimer interface; the number of ion pairs is also reduced and some charged residues are not compensated by opposite charges and one also observes a destabilization of the N-terminus end through the lack of charge–charge interactions; one has to note that the stabilization of the N-terminus is a characteristic feature of hyperthermophilic isocitrate dehydrogenases (Fedoy *et al.*, 2007). In subtilisin S41 originating from Antarctic bacteria (Davail *et al.*, 1994) the cold-adapted enzyme was found to have the largest surface area and the highest content of negatively charged residues (Almog *et al.*, 2009). In cold-adapted superoxide dismutase from *Aliivibrio salmonicida* an increased net negative charge of the surface of the protein was proposed to explain its lower stability when compared to the enzyme from *E. coli*. The crystal structure of S-formylglutathione hydrolase from the Antarctic strain *Pseudoalteromonas haloplanktis* TAC125 has been recently determined (Alterio *et al.*, 2010) and the analysis of the structure reveals that the number of arginine residues is significantly reduced as well as the number of ion pairs. Also, a large amount of the surface of the cold-adapted enzyme is occupied by polar residues especially by asparagine and glutamine residue. These factors were thought to confer the appropriate flexibility of the enzyme at low temperatures.

Folding at low temperature

The folding of proteins from mesophiles and thermophiles often needs assistance by various molecular tools that prepare the protein for proper folding, eliminate misfolded assemblies and prevent aggregation due to the exposure of hydrophobic groups of nascent polypeptide chains to an unfavourable aqueous environment. These proteins, known as molecular chaperones, force the synthesized proteins to adopt a favourable equilibrium between folded and unfolded form and in this way act as controllers of the protein concentration within the cell. They also play active roles in multimeric protein assembly, protein hydrolysis and protein transport. Some of these proteins are permanent hosts of the cell whereas others are inducible via heat shocks and cold shocks. They take in charge any non-native conformation up to their biologically active state and final destination. They can act as 'holdases' in stabilizing transient non-native conformation, as 'foldases' in assisting correct folding and as 'unfoldases' as they contribute to unfold misfolded proteins (Hoffmann *et al.*, 2010). It is worth mentioning that protein misfolding has been associated with more than 30 human diseases such as Alzheimer's, Parkinson's and Creutzfeldt–Jakob diseases (Tartaglia *et al.*, 2010). The most studied chaperones are the chaperonins, GroEL (Hsp 60) and its co-chaperone GroES (Hsp 10). mainly found in eubacteria and also in eukaryotic organelles such as mitochondria and chloroplasts. These chaperones form complex assemblies that can accommodate the polypeptide chain into a hydrophilic cavity that forces the hydrophobic residues to cluster to form the core of the native protein; the process is energetically driven by ATP binding and hydrolysis (Yebenes *et al.*, 2011). The DnaK (Hsp 70 in eukaryotes)/

DnaJ (Hsp 40)/GrpE system is also ATP dependent and is involved in many cellular processes: it prevents protein aggregation, assists in protein disaggregation, in defolding and refolding of improperly folded proteins and in targeting aggregated proteins to proteases. The third most important chaperone is the so-called trigger factor (48 kDa in *E. coli*), the only chaperone associated with ribosomes that is the first to be in contact with the nascent polypeptide chain and which forms a 1:1 stoichiometric complex with the ribosome at the level of the exit of the polypeptide chain. It is also found in the cytoplasm in rather high concentrations mainly under the form of a dimer that interacts with folded intermediates to be rescued by the DnaK/DnaJ/GrpE system. The cytoplasmic form of the trigger factor has been indeed found to be associated with a large diversity of full-length proteins and as such plays a role of independent chaperone that stabilizes native-like structures before their incorporation into multimeric assemblies (Martinez-Hackert *et al.*, 2009). The head of its dragon like structure has an activity of peptidyl-prolyl *cis/trans* isomerase, a very important enzyme necessary for the proper folding of proteins containing *cis*-proline configurations. It is commonly accepted that nascent polypeptide chains first interact with the ribosome-bound trigger factor and then, upon release, some proteins spontaneously fold whereas some other need further assistance by the GroEL and DnaK systems.

The heat-induced unfolding of proteins occurs with a positive modification of the enthalpy (endothermic process) associated with a large increase in entropy. Cold unfolding is somehow less understood and is believed to be induced by the rehydration of polar and non-polar groups and by a decrease of the strength of hydrophobic interactions. This results from the difficulty to disturb, at low temperature, the water molecules network encaging each hydrophobic group. As illustrated in Fig. 8.4, cold unfolding should occur, in most cases, at temperatures well below the freezing point of water and one can see that, in contrast to what was initially expected, psychrophilic enzymes are much more sensitive to low temperatures than mesophilic and thermophilic counterparts. So it is almost impossible to directly follow the process of cold unfolding of protein. One can artificially circumvent the difficulty by the addition of moderate concentrations of chemical denaturants such as urea and guanidinium chloride in order to shift the stability curve in the range of positive temperatures. It is clear, however, that these artifices render the environment quite different from physiological conditions.

The interpretation of data is not straightforward because of secondary effects of these chemicals that not only increase the temperature of cold denaturation but can also strongly modify the rate of unfolding. A recent report indeed suggests that cold denaturation is kinetically controlled and that kinetic stabilization could play an important role in the adaptation to cold of psychrophilic proteins (Romero-Romero *et al.*, 2011). An interesting case is that of Yfh 1, the yeast counterpart of frataxin, a protein causing in human beings a neurodegenerative disease, Friedreich's ataxia. This protein has the particularity to unfold at temperature higher than O°C and the unfolded state was studied by NMR spectroscopy (Adrover *et al.*, 2010). They demonstrate, first, that cold unfolding was highly cooperative leading to mostly unstructured state; second, that some duplicate chemical shifts were due to proline *cis/trans* isomerization; third, that solvation effects were clearly identified (this support the initial hypothesis suggesting that cold unfolding is due to a preferential hydration, at low temperatures, of polar and non polar groups); fourth, that although the structure appears to be close to an unfolded state, some residual secondary structures were detected with, however, no detectable similarity with any structure of the folded state; and, fifth, that the differences in chemical shifts indicate that the cold unfolded structure is different from the random coil obtained at 25°C (heat denaturation) probably partly due to hydration effects. From a comparative study of the thermodynamic stability of psychrophilic, mesophilic, and thermophilic α-amylases (Feller *et al.*, 1992; D'Amico *et al.*, 2003) it was shown that contrarily to high temperatures, at low temperatures the enzymes are entropically stabilized and enthalpically destabilized. The negative stabilization enthalpy is obviously due to the hydration process of individual groups, whereas the negative value of the stabilization entropy is the stabilization factor at temperatures close to that of the usual environment of psychrophiles. As at low temperatures psychrophilic enzymes also show a marginal stability their folding can also be impaired by the preferential hydration of groups that can form intramolecular interactions, by the decrease of the strength of hydrophobic and ionic interactions, by the difficulty to transform proline *trans* configuration into *cis* configuration and possibly by the high viscosity, in addition to macromolecular crowding (Gershenson and Gierasch, 2011), that characterizes intracellular aqueous environments at low temperatures.

Folding at low temperature can also be complicated by the possible incomplete adaptation of the proteins participating in the folding process. As in the case of thermophiles, psychrophiles have also the potential to express all type of chaperones. This fact is derived from the analysis of a significant number of genome sequences of psychrophiles (Casanueva *et al.*, 2010) but do they really express all of them or, depending on the species, only an assortment of these chaperones at low temperatures ? Only a few studies have been up to now addressing this question. They mainly focused on the expression of these chaperones as a function of growth temperatures of psychrophilic microorganisms. They concern *Shewanella* sp. SiB1 (Suzuki *et al.*, 2004), *Methanococcoides burtonii* (Goodchild *et al.*, 2004); *Exiguobacterium sibiricum* (Qiu *et al.*, 2006; Rodrigues *et al.*, 2008); *Psychrobacter arcticus* (Zheng *et al.*, 2007); *Shewanella livingstonensis* (Kawamoto *et al.*, 2007), *Spingopyxis alaskensis* (Ting *et al.*, 2010); *Pseudoalteromonas haloplanktis* (Piette *et al.*, 2010); *Shewanella* SIB1 FKBP22 (Budiman *et al.*, 2011), and *Acidithiobacillus ferroxidans* (Mykytczuk *et al.*, 2011). From this limited number of data one can hardly detect any uniform strategy in the process of assisting protein folding at low temperatures. Depending on the species, some chaperones are overexpressed or down regulated at low temperature. Table 8.3 summarizes the data obtained in the case of nine psychrophilic microorganisms. Striking is the systematic enhancement at low temperatures of peptidyl-prolyl *cis/trans* isomerase activity either through PPIase or TF that embark a PPiase domain. This suggest that the prolyl *cis–trans* isomerization is a particularly limiting step at low temperature, at the kinetic level.

The role of the GroEL/GroES couple is rather contrasted since it is up-regulated in certain species and down-regulated in others. In *Psychrobacter arcticus* the up-regulation of GroEL is associated with low temperature only in the absence of salts whereas it is down regulated at low temperatures in the presence of salts encountered in the natural environment of the permafrost strain. In this case it is clear that the possible salt dependence complicates the analysis of the data. In the same context the putative peptidyl-prolyl *cis/trans* isomerases are down- and up-regulated in the presence and absence of salts respectively. Also one of the two temperatures selected for comparison, 22°C, known as the so-called optimum temperature, should induce a severe thermal stress rendering even more complicated the interpretation of the data. In *Sphingopyxis alaskensis* a specific DnaK/DnaJ/GrpE system is largely expressed at low temperature but not more than the homologous system expressed at 30°C contrarily to the specific GroEL/GroES system (Ting *et al.*, 2010) which is more abundant at low

Table 8.3 Main chaperones up-regulated (+) and down regulated (-) at low temperatures in various cold-adapted microorganisms.

Chaperone	*Shewanella* (1, 2)	*Methanococcus* (3)	*Exiguobacterium* (4)	*Psychrobacter* (5)	*Sphingopyxis* (6)	*Pseudoalteromonas* (7)	*Shewanella* (8)	*Acidithiobacillus* (9)
TF	+		+			+		+
PPiase	+	+		±	+	+	+	+
GroEL		-		±	+	–		+
GroES				Uncertain	+			+
GrpE		–						
CLpB		+		+	+			
DnaK		–	+	–		-		
DnaJ		–		–	–	–		
sHSp (HSP20)		+			–	–		
Dsb				+		–		
SurA				+				

1, *Shewanella* SLB1 (Suzuki *et al*., 2004); 2, *Shewanella livingstonensis* (Kawamoto *et al*., 2007); 3, *Methanococcus burtonii* (Goodrich *et al*., 2004); 4, *Exiguobacterium sibiricum* (Qiu et *al*., 2006); 5, *Psychrobacter arcticus* (Zheng *et al*., 2007); 6, *Sphingopyxis alaskensis* (Ting *et al*., 2010); 7, *Pseudoalteromonas haloplanktis* (Piette *et al*., 2010); 8, *Shewanella* SIB1 (Budiman *et al*., 2011); 9, *Acidithiobacillus ferrooxidans* (Mykytczuk *et al*., 2011).

temperature. Here also a very high upper range temperature (30°C) was selected for the psychrophilic bacterium and this can also induce difficulties in drawing conclusions from these data.

In some cases, the GroEL/GroES system is repressed at low temperature as well as the DnaK/DnaJ/GrpE system; these results may be related to the fact that in psychrophiles the hydrophobic clusters that constitute the core of the cold-adapted proteins are often weaker than in mesophilic counterparts, and as the hydrophobic interactions are weakened at low temperature the risk of aggregation and misfolding is lower. This idea is corroborated by the fact that cold-adapted proteins in psychrophiles often show a spontaneous reversible thermal unfolding (Feller, 2010). It has also been shown that the overexpression of the trigger factor in *E. coli* represses the expression of other chaperones when the mesophilic bacterium is grown at low temperatures and the overexpresssion of hsps represses the viability of the cells at low temperatures (Kandror and Goldberg, 1997).

Conclusions

Psychrophiles are organisms that happily thrive at temperatures that can be well below the freezing point of water. This required numerous molecular adaptations, especially at the level of the enzymes which, in the absence of adaptation, and due to the exponential dependence of reaction rates on temperature, would not allow the survival of the organisms due to excessively slow biochemical reactions. The molecular adaptations have been achieved through rather discrete modifications of the amino acid sequences leading to a decrease in thermal stability favouring a significant increase of the flexibility of the molecular structure. This increase in flexibility compensates for the freezing effect of low temperatures on structure, the stiffness of which would otherwise prevent the appropriate accommodation of the substrate at low temperatures. Cold-adapted enzymes are therefore characterized by a very high activity at low and moderate temperatures that renders them highly suitable for many biotechnological processes. These enzymes are however prone to cold denaturation and display, as thermophilic enzymes at their usual environmental temperatures, a marginal stability that provides the required flexibility of the molecular edifice. This flexibility can be general or local and can affect various sites of the structure. The degree of adaptation can vary from one enzyme to the other and also depends on the organism. Some cold-adapted enzymes can work over a rather large temperature range or have on the contrary an activity limited to temperatures close to that of the environment; we believe that there is a continuum in the adaptation and that subclassification (psychrotolerant, psychrotroph) is useless and can be misleading. Depending on their evolution history each cold-adapted organism develops its own adaptation strategy precluding any general view of the molecular adaptation although the common way, which is probably the cheapest way in terms of energy cost, is to decrease the stability in order to reach an appropriate flexibility. Now if cold-adapted enzymes have been successfully modified in the course of evolution they also need proper folding in a context that is not necessarily favourable due to the high viscosity of the intracellular space and to the weakening of electrostatic and hydrophobic interactions at low temperatures. Proteomic experiments based on the comparison of the expression of proteins at low and moderate temperatures have been carried out. They are, however, scarce and still suffer from the complexity of the methodology required to identify changes that really take place. The choice of the marginal temperatures is also crucial and in a certain number of cases the so-called optimum temperatures, based on the shortest doubling time, have been selected forgetting that at this temperature some proteins suffers from heat stress that induces their thermal unfolding. From the few data available one can see that the over-expresssion of peptidyl-prolyl *cis/trans* isomerase is needed at low temperatures as well as that of the trigger factor which the first factor, on the ribosome, to interact with the nascent polypeptide chain.

References

Adrover, M., Esposito, V., Martorell, G., Pastore, A., and Temussi, P.A. (2010). Understanding cold denaturation: the case study of Yfh1. J. Am. Chem. Soc. *132*, 16240–16246.

Aghajari, N., Feller, G., Gerday, C., and Haser, R. (1998). Structures of the psychrophilic *Alteromonas haloplanctis* alpha-amylase give insights into cold adaptation at a molecular level. Structure *6*, 1503–1516.

Aghajari, N., Van Petegem, F., Villeret, V., Chessa, J.P., Gerday, C., Haser, R., Van Beeumen, J. (2003). Crystal structures of a psychrophilic metalloprotease reveal new insights into catalysis by cold-adapted proteases. Proteins *50*, 636–647.

Almog, O., Gonzalez, A., Godin, N., de Leeuw, M., Mekel, M.J., Klein, D., Braun, S., Shoham, G., and Walter, R.L. (2009). The crystal structure of the psychrophilic subtilisin S41 and the mesophilic subtilisin Sph reveal the same calcium-loaded state. Proteins, *74*, 489–496.

Alterio, V., Aurilia, V., Romanelli, A., parracino, A., Saviano, M., D'Auria, S., and De Simone G. (2010). Crystal structure of an S-formyl glutathione hydrolase from *Pseudoalteromonas haloplanktis* TAC 125. Biopolymers *93*, 669–677.

Altermark, B., Niiranen, L., Willassen, N.P., Smalas, A.O., and Moe, E. (2007). Comparative studies of endonuclease I from cold-adapted *Vibrio salmonicida* and mesophilic *Vibrio cholerae.* FEBS J. *274*, 252–263.

Altermark, B., Helland, R., Moe, E., Willassen, N.P., and Smalas, A.O. (2008). Structural adaptation of endonuclease I from the cold-adapted and halophilic bacterium *Vibrio salmonicida.* Acta Crystallogr. D Biol. Crystallogr. *64*, 368–376.

Alvarez, M., Zeelen, J.P., Mainfroid, V., Rentier-Delrue, F., Martial, J.A., Wyns, L., Wierenga, R.K., and Maes, D. (1998). Triose-phosphate isomerase (TIM) of the psychrophilic bacterium *Vibrio marinus.* Kinetic and structural properties. J. Biol. Chem. *273*, 2199–2206.

Arnorsdottir, J., Kristjansson, M.M., and Ficner, R. (2005). Crystal structure of a subtilisin-like serine proteinase from a psychrotrophic *Vibrio* species reveals structural aspects of cold adaptation. FEBS J. *272*, 832–845.

Arrhenius, S. (1889). Uber die reaktiongeschwindigkeit bei der inversion von rohrzucker durch säuren. Z. Physik. Chem. *4*, 226–248.

Bae, E., and Phillips, G.N. (2004). Structure and analysis of highly homologous psychrophilic, mesophilic and thermophilic adenylate kinases. J. Biol. Chem. *279*, 28202–28208.

Beadle, B.M., Baase, W.A., Wilson, D.B., Gilkes, N.R., and Shoichet, B.K. (1999). Comparing the thermodynamic stabilities of a related thermophilic and mesophilic enzyme. Biochemistry *38*, 2570–2576.

Berglund, G.I., Willassen, N.P., Hordvik, A., and Smalas, A.O. (1995). Structure of native pancreatic elastase from North Atlantic salmon at 1.61 Å resolution. Acta Crystallogr. D Biol. Crystallogr. *51*, 925–937.

Bjelic, S., Bransdal, B.O., and Aqvist, J. (2008). Cold adaptation of enzyme reaction rates. Biochemistry *47*, 10049–10057.

Budiman, C., Koga, Y., Takano, K., and Kanaya, S. (2011). FK506-binding protein 22 from a psychrophilic bacterium, a cold shock-inducible peptidyl prolyl isomerase with the ability to assist in protein folding. Int. J. Mol. Sci. *12*, 5261–5284.

Brzuszkiewicz, A., Nowak, E., Dauter, Z., Dauter, M., Cieslinski, H., Dlugolecka, A., and Kur, J. (2009). Structure of EstA esterase from psychrotrophic *Pseudoalteromonas* sp. 643A covanently inhibited by monoethylphosphonate. Acta Crystallogr. Sect. F Struct. Biol. Cryst. Commun. *65*, 862–865.

Casanueva, A., Tuffin, M., Cary, C., and Cowan, D.A. (2010). Molecular adaptations to psychrophily: the impact of 'omic' technologies. Trends Microbiol. *18*, 374–381.

Cipolla, A., D'Amico, S., Barumandzadeh, R., Matagne, A., and Feller, G. (2011). Stepwiese adaptations to low temperature as revealed by multiple mutants of psychrophilic alpha-amylase from antarctic bacterium. J. Biol. Chem. *286*, 38348–38355.

Collins, T., Meuwis, M.-A., Gerday, C., and Feller, G. (2003). Activity, stability and flexibility in glycosidases adapted to extreme thermal environments. J. Mol. Biol. *328*, 419–428.

Collins, T., D'Amico, S., Marx, J.-C., Feller, G., and Gerday, C. (2007). Coldf-adapted enzymes. In Physiology and Biochemistry of extremophiles, C. Gerday and N. Glansdorff, eds. (Washington, DC, ASM Press), pp.165–179.

Coquelle, N., Fioravanti, E., Weik, M., Vellieux, F., and Madern, D. (2007). Activity, stability and structural studies of lactate dehydrogenase adapted to extreme thermal environments. J. Mol. Biol. *374*, 547–562.

D'Amico, S., Gerday, C., and Feller, G. (2003). Activity–stability relationships in extremophilic enzymes. J. Biol. Chem. *278*, 7891–7896.

Davail, S., Feller, G., Narinx, E., and Gerday, C. (1994). Cold adaptation of proteins. Purification, characterization, and sequence of the heat-labile subtilisin from the Antarctic psychrophile *Bacillus* TA41. J. Biol. Chem. *269*, 17448–17453.

Davlieva, M., and Shammo, Y. (2009). Structure and biochemical characterization of an adenylate kinase originating from the psychrophilic organism *Marinibacillus marinus*. Acta Crystallogr. Sect. F Struct. Biol. Crysdt. Commun. *65*, 751–756.

de Baker, M., McSweeney, S., Rasmussen, H.B., Riise, B.W., Lindley, P., and Hough, E. (2002). The 1.9 Å crystal structure of heat-labile shrimp alkaline phosphatase. J. Mol. Biol. *318*, 1265–1274.

Deming, J.W. (2002). Psychrophiles and polar regions. Curr. Opin. Microbiol. *5*, 301–309.

De Vos, D., Xu, Y., Hulpiau, P., Vergauwen, B., and Van Beeumen, J.J. (2007). Structural investigation of cold activity and regulation of aspartate carbamoyltransferase from the extreme psychrophilic bacterium *Moritella profunda*. J. Mol. Biol. *365*, 379–395.

Fedoy, A.E., Yang, N., Martinez, A., Leiros, H.K., and Steen, I.H. (2007). Structural and functional properties of isocitrate dehydrogenase from the psychrophilic bacterium *Desulfotalea psychrophila* reveal a cold-active enzyme with an unusual high thermal stability. J. Mol. Biol. *372*, 130–149.

Feller, G. (2010). Protein stability and enzyme activity at extreme biological temperatures. J. Phys.: Condens. Matter. *22*, 323101–323118.

Feller, G., Lonhienne, T., Deroanne, C., Libioulle, C., Van Beeumen, J., and Gerday, C. (1992). Purification, characterization and nucleotide sequence of the thermolabile α-amylase from the Antarctic psychrotroph *Alteromonas haloplanktis* A 23. J.Biol. Chem. *267*, 5217–5221.

Feller, G., Narinx, E., Arpigny, J.-L., Zekhnini, Z., and Gerday, C. (1994). Temperature dependence of growth, enzyme secretion and activity of psychrophilic Antarctic bacteria. Appl. Microbiol. Biotechnol. *41*, 477–479.

Garsoux, G., Jamotte-Brasseur, J., Gerday, C., and Feller, G. (2004). Kinetic and structural optimisation to catalysis at low temperatures in a psychrophilic cellulase from the Antarctic bacterium *Pseudoalteromonas haloplanktis*. Biochem. J. *384*, 247–253.

Georlette, D., Blaise, V., Collins, T., D'Amico, S., Gratia, E., Hoyoux, A., Marx, J.C., Sonan, G., Feller, G., and Gerday, C. (2004). Some like it cold: biocatalysis at low temperatures. FEMS Microbiol. Rev. *28*, 25–42.

Gerday, C. (2011). Life at the extremes of temperature. In Bacterial stress response, G. Storz and R. Hengge, eds. (Washington, DC, ASM press), pp425–444.

Gershenson, A., and GieraschL.M. (2011). Protein folding in the cell: challenges and progress. Curr. Opin. Struct. Biol. *21*, 32–41.

Goodchild, A., Saunders, N.F., Erlan, H., Raftery, M., Guilhaus, M., Curmi, P.M., and Cavicchioli, R. (2004). A proteomic determination of cold adaptation in the Antarctic archaeon, *Methanococcoides burtonii*. Mol. Microbiol. *53*, 309–321.

Gudmunsdottir, E., Spilliaert, R., yang, Q., Craik, C.S., Bjarnason, J.B., and Gudmundsdottir, A. (1996). Isolation and characterization of two cDNAs from Atlantic cod encoding two distinct psychrophilic elastases. Comp. Biochem. Physiol. *113*, 795–801.

Helland, R., Leiros, I., Berglund, G.I., Willassen, N.P., and Smalas, A.O. (1998). The crystral structure of anionic salmon trypsin in complex with bovine pancreatic trypsin inhibitor. Eur. J. Biochem. *256*, 317–324.

Helland. R., Larsen, A.N., Smalas, A.O., and Willassen, N.P. (2006). The 1.8 Å crystal structure of a proteinase K-like enzyme from a psychrotroph *Serratia* species. FEBS J. *273*, 61–71.

Hochachka, P.W., and Somero, G.N. (1984). Biochemical Adaptation (Princeton: Princeton University Presss).

Hochachka, P.W., and Somero, G.N. (2002). Biochemical Adaptation. Mechanism and Process in Physiological Evolution. (New-York: Oxford University press).

Hoffmann, A., Bukau, B., and Kramer, G. (2010). Structure and function of the molecular chaperone, trigger factor. Biochim. Biophys. Acta *1803*, 650–661.

Kandror, O., and Goldberg, A.L. (1997. Trigger factor is induced upon cold shock and enhances viability of *Escherichia coli* at low temperatures. Proc. Natl. Acad. Sci. U.S.A. *94*, 4978–4981.

Karlsen, S., Hough, E., and Olsen, R.L. (1998). Structure and proposed amino-acid sequence of a trypsin from Atlantic cod (*Gadus morhua*). Acta Crystallogr. D Biol. Crystallogr. *54*, 32–46.

Kawamoto, J., Kurihara, T., Kitagawa, M., Kato, I., and Esaki, N. (2007). Proteomic studies of an Antarctic cold-adapted bacterium, *Shewanella livingtonensis* Ac 10, for global identification of cold-inducible proteins. Extremophiles *10*, 819–826.

Kim, S.Y., Hwang, K.Y., Kim, S.H., Sung, H.C., Han, Y.S., and Cho, Y. (1999). Structural basis for cold adaptation. Sequence, biochemical properties, and crystal structure of malate dehydrogenase from a psychrophile *Aquaspirillium arcticum*. J. Biol. Chem. *274*, 11761–11767.

Kumar, S., and Nussinov, R. (2001). How do thermophilic proteins deal with heat ? Cell. Mol. Life Sci. *58*, 1216–1233.

Kumar, S., Tsai, C.J., and Nussinov, R. (2002). Maximal stabilities of reversible two-state proteins. Biochemistry *41*, 5359–5374.

Leiros, I., Moe, E., Lanes, O., Smalas, A.O., and Willassen, N.P. (2003). The structure of uracil-DNA glycosylase from Atlantic cod (*Gadus morhua*) reveals cold-adaptation features. Acta Crystallogr. D Biol. Crystallogr. *59*, 1357–1365.

Leiros, H.K., Pey, A.L., Innselset, M., Moe, E., Leiros, I., Steen, I.H., and Martinez, A. (2007). Structure of phenylalanine hydroxylase from *Colwellia psychrerythraea* 34H, a monomeric cold active enzyme with local flexibility around the active site and high overall stability. J. Biol. Chem. *282*, 21973–21986.

Lonhienne, T., Baise, E., Feller, G., Bouriotis, V., and Gerday, C. (2001a). Enzyme activity determination on macromolecular substrates by isothermal titration calorimetry: application to mesophilic and psychrophilic chitinases. Biochim. Biophys. Acta *1545*, 349–356.

Lonhienne, T., Zoidakis, J., Vorgias, E., Feller, G., Gerday, C., and Bouriotis, V. (2001b). Modular structure, local flexibility and cold-activity of a novel chitobiase from a psychrophilic Antarctic bacterium. J. Mol. Biol. *310*, 291–297.

Margesin, R. (2009). Effect of temperature on growth parameters of psychrophilic bacteria and yeasts. Extremophiles *13*, 257–262.

Margesin, R., Schinner, F., Marx, J.-C., and Gerday, C. (2008). Psychrophiles: From Biodiversity to Biotechnology (Berlin: Springer Verlag).

Martinez-Hackert, E., and Hendrickson, W.A. (2009). Promiscuous substrate recognition in folding and assembly activities of the trigger factor chaperone. Cell *138*, 923–934.

Merlino, A., Russo-Krauss, I., Castellano, I., De Venditis, E., Rossi, B., Conte, M., Vergara, A., and Sica, F. (2010). Structure and flexibility in cold-adapted iron superoxide dismutase: The case of the enzyme isolated from *Pseudoalteromonas haloplanktis*. J. Struct. Biol. *172*, 343–352.

Mykytczuk, N.C.S., Trevors, J.T., Foote, S.J., Leduc, L.G., Ferroni, G.D., and Twine, S.M. (2011). Proteomic insights into cold adaptation of psychrotrophic and mesophilic *Acidothiobacillus ferrooxidans* strains. Antonie van Leeuwenhoek *100*, 259–277.

Pedersen, H.L., Willassen, N.P., and Leiros, I. (2009). The first structure of a cold-adapted superoxide dismutase (SOD): biochemical ans structural characterization of iron SOD from *Aliivibrio samonicida*. Acta Crystallogr. Sect. F Biol. Cryst. Commun. *65*, 84–92.

Petrescu, I., Lamotte-Brasseur, J., Chessa, J.-P., Ntarima, P., Claeyssens, M., Devreese, B., Marino, G., and Gerday, C. (2000). Xylanase from psychrophilic yeast *Cryptococcus adeliae*. Extremophiles *4*, 137–144.

Piette, F., D'Amico, S., Struvay, C., Mazzuchelli, G., Renaut, J., Tutino, M.L., Danchin, A., Leprince, P., and Feller, G. (2010). Proteomics of life at low temperatures: trigger factor is the primary chaperone in the Antarctic bacterium *Pseudoalteromonas haloplanktis* TAC 125. Mol. Microbiol. *76*, 120–132.

Qiu, Y., Kathariou, S., and Lubman, D.M. (2006). Proteomic analysis of cold adaptation in a Siberian permafrost bacterium *Exiguobacterium sibiricum* 255–15 by two-dimensional liquid separation coupled with mass spectrometry. Proteomics *6*, 5221–5233.

Riise, E.K., Lorentzen, M.S., Helland, R., Smalas, A.O., Leoros, H.-K.S., and Willassen, N.P. (2007). The first structure of a cold-active catalase from *Vibrio salmo*nicida at 1.96 Å reveals structural aspects of cold adaptation. Acta Crystallogr. D Biol. Crystallogr. *63*, 135–148.

Rodrigues, D.F., Ivanova, N., He, Z., Huebner, M., Zhou, J., and Tiedje, J.M. (2008). Architecture of thermal adaptation in an *Exiguobacterium sibiricum* strain isolated from 3 million year old permafrost: a genome and transcriptome approach. BMC Genomics *9*, 547.

Romero-Romero, M.L., Inglés-Prieto, A., Ibarra-Molero, B., and Sanchez-Ruiz, J.M. (2001). Highly anomalous energetics of protein cold denaturation linked to folding-unfolding kinetics. PLos One *6*, 1–7.

Russell, R.J., Gerike, U., Danson, M.J., Hough, D.W., and Taylor, G.L. (1998). Structural adaptations of the cold-active citrate synthase from an Antarctic bacterium. Structure *6*, 351–361.

Sawle, L., and Gosh, K. (2011). How do thermophilic proteins and proteosomes withstand high temperature ? Biophys. J. *101*, 217–227.

Skalova, T., Dohnalek, J., Spiwok, V., Lipovova, P., Vondrackova, E., Petrokova, H., Duskova, J., Strnad, H., Kralova, B., and Hasek, J. (2005). Cold-active beta-galactosidase from *Arthrobacter* sp. C2–2 forms compact 660 kDa hexamers: crystal structure at 1.9 Å resolution. J. Mol. Biol. *353*, 282–294.

Smalas, A.O., Schroder-Leiros, H.K., Os, V., and Willassen, N.P. (2000). Cold adapted enzymes. Biotechnol. Annu. Rev. *6*, 1–57.

Somero, G.N. (1995). Proteins and temperature. Annu. Rev. Phys. *57*, 43–68.

Sotelo-Mundo, R.R., Lopez-Zavala, A.A., Garcia-Orozco, K.D., Arvizu-Flores, A.A., Velazquez-Contreras, E.F., Vaalenzuela-Soto, E.M., Rojo-Dominguez, A., and Kanost, M.R. (2007). The lysozyme from insect (*Manduca sexta*) is a cold-adapted enzyme. Protein Pept. Lett. *14*, 774–778.

Suzuki, Y., Haruki, M., Takano, K., Morikawa, M., and Kanaya, S. (2004). Possible involment of an FKBP family member protein from a psychrotrophic bacterium, *Shewanella* sp. SLB1 in cold adaptation. Eur. J. Biochem. *271*, 1372–1381.

Suzuki, T., Yamamoto, K., Tada, H., and Uda, K. (2011). Cold-adapted features of arginine kinase from the deep-sea *Calyptogena kaikoi*. Mar. Biotechnol. *14*, 294–303.

Tartaglia, G.G., Dobson, C.M., Hartl, F.U., and Vendruscolo, M. (2010). Physicochemical determinants of chaperone requirements. J. Mol. Biol. *400*, 579–588.

Ting, L., Williams, T.J., Cowley, M.J., lauro, F.M., Guilhaus, M., Raftery, M.J., and Cavicchioli, R. (2010). Cold adaptation in the marine bacterium, *Sphingopyxis alaskensis* assessed using quantitative proteomics. Environ. Microbiol. *12*, 2658–2676.

Toyota, E., Ng, K.K., Kuninaga, S., Sekizaki, H., Itoh, K., Tanizawa, K., and James, M.N. (2002). Crystal structure and nucleotide sequence of an anionic trypsin from chum salmon (*Oncorrhynchus keta*) in comparison with Atlantic salmon (*Salmo salar*) and bovine trypsin. J. Mol. Biol. *324*, 391–397.

Tsuruta, H., Mikami, B., and Aizono, Y. (2005). Crystal structure of cold-active protein-tyrosine phosphatase from a psychrophile, *Shewanella* sp. J. Biochem. (Tokyo) *137*, 69–77.

Uppenberg, J., Hansen, M.T., Paktar, S., and Jones, T.A. (1994). The sequence, crystal structure determination and refinement of two crystal forms of lipase B from *Candida antarctica*. Structure, *2*, 293–308.

Van Petegem, F., Collins, T., Meuwis, M.A., Gerday, C., Feller, G., and Van Beeumen, J. (2003). The structure of a cold-adapted family 8 xylanase at 1.3 Å resolution. Structural adaptations to cold and investigation of the active site. J. Biol. Chem. *278*, 7531–7539.

Vieille, C., and Zeikus, G.J. (2001). Hyperthermophilic enzymes: sources, uses and molecular mechanisms for thermostability. Microbiol. Mol. Rev. *65*, 1–43.

Violot, S., Aghajari, N., Czjzek, M., Feller, G., Sonan, G., Gouet, P., Gerday, C., Haser, R., and Receveur-Bréchot, V. (2005). Structure of a full length psychrophilic cellulase from *Pseudoalteromonas haloplanktis* revealed by X-ray diffraction and small angle X-ray scattering. J. Mol. Biol. *348*, 120–11–1224.

Wang, E., Koutsioulis, D., Leiros, H.K., Andersen, O.A., Bouriotis, V., Hough, E., and Heikinheimo, P. (2007). Crystal structure of alkaline phosphatase from the Antarctic bacterium TAB5. J. Mol. Biol. *366*, 1318–1331.

Yebenes, H., Mesa, P., Munoz, I.G., Montoya, G., and Valpuesta, J.M. (2011). Chaperonins: two rings for folding. Trends Biochem. Sci. *36*, 424–432.

Zheng, S., Ponder, M.A., Shih, J.Y., Tiedje, J.M., Thomashow, M.F., and Lubman, D.M. (2007). A proteomic analysis of *Psychrobacter arcticus* 273–4 adaptation to low temperature and salinity using a 2-D liquid mapping approach. Electrophoresis *28*, 467–488.

Part III

Biomolecules Related to Cold

Psychrotolerant H_2O_2-resistant Bacteria and Environmental Adaptation of their Catalases

9

Isao Yumoto and Isao Hara

Abstract

Three novel species of psychrotolerant H_2O_2-resistant bacteria have been isolated from drain pools of a fish egg processing plant that uses H_2O_2 as a bleaching agent. Among them *Vibrio rumoiensis* was isolated from downstream of the drain pool, whereas *Exiguobacterium oxidotolerans* and *Psychrobacter piscatorii* were isolated from upstream. The catalase activity in the cell extract of the former was lower than that in the latter, which reflected the H_2O_2 concentration in their niches. The VktA (catalase from *V. rumoiensis*) and PktA (catalase from *P. piscatorii*) belonged to clade III, and EktA (catalase from *E. oxidotolerans*) belonged to clade I. From the temperature profile of their activities, these catalases exhibited heat sensitivity at higher than 60°C. The cell extract of *V. rumoiensis* exhibited lower catalase content (1.8%) than that of *P. piscatorii* (10%), whereas the former produces VktA, which exhibited a higher catalytic efficiency than the PktA. EktA content in the cell extract of *E. oxidotolerans* was 6.5% and EktA exhibited the highest catalase activity in both cell extract and the purified form among the three catalases. EktA possesses a wider bottleneck structure in the main channel for accepting substrates, enable it to react efficient with organic peroxides larger than H_2O_2 as substrates. Considering the environmental adaptation and distribution of microorganisms in this unique niche with high concentrations of H_2O_2 and low temperatures, certain variations of unknown bacterial species exist with certain variations of environmental adaptation mechanisms (e.g. cellular localization, production rate, and catalytic efficiency of catalase) depending on the environmental H_2O_2 concentration and fragility of cells. Furthermore, the characteristics of these catalase molecules reflect the environmental conditions (low temperatures and high concentrations of H_2O_2) under which these bacteria survive.

Introduction

In aerobic and facultatively aerobic microorganisms, electrons flow in the respiratory chain to translocate protons across the membrane for production of energy potential and redox reactions terminate upon reduction of molecular oxygen. These oxygen-metabolizing microorganisms produce reactive oxygen species (ROS) such as superoxide ($O_2 \cdot^-$), hydrogen peroxide (H_2O_2) and hydroxyl radicals (OH·) (Bolton *et al.*, 2000). For example, during the course of oxygen the reduction in the respiratory chain, there are chances in the direct reaction with oxygen (O_2) in the respiratory complex to produce $O_2 \cdot^-$ (Treberg *et al.*, 2011, Quinlan *et al.*, 2011). Following the production of $O_2 \cdot^-$, the substrate enzymatically reacts with superoxide dismutase (SOD). The presence of H_2O_2 in the cells may lead to the generation of more harmful ROS, OH·, in the presence of Fe^{2+} through the Fenton reaction (Halliwell and Gutteridge, 1999). Therefore, rapid elimination of H_2O_2 is crucial in microorganisms to prevent the generation of OH·. Catalases are typical H_2O_2- eliminating enzymes in microorganisms. ROS damage cell components such as DNA, RNA, proteins and lipids owing to their strong reactivity (Imlay and Linn,1988; Halliwell and Gutteridge, 1999; Rowe *et al.*, 2008). It is considered that catalases have very important roles in the protection of bacterial cells from damage caused by intracellular produced ROS.

Symbiotic microorganisms such as *Vibrio fisheri* in squid and root-colonizing bacteria such as those in legumes produce catalases to eliminate H_2O_2 produced from host cells (Katsuwon and Anderson, 1992; Rocha *et al.*, 1996; Visick and Ruby, 1998). When pathogenic microorganisms in the human body encounter macrophages, they are engulfed into phagosomes. The catalases in the phagosomes react with 100 mM H_2O_2 (Hust and Barrette, 1989). In addition, microorganisms may encounter H_2O_2-producing microorganisms in their niches. Catalases have very important roles in the protection of these microorganisms from such H_2O_2-producing microorganisms. For example, certain fungi produce H_2O_2 to eliminate competitors in their niches (Kim *et al.*, 1978). On the other hand, *Pseudomonas aeruginosa* cells produce a greenish substance, pyocyanin, which is an antibiotic that generates H_2O_2, and thereby affects the environmental niches (Norman *et al.*, 2004). Thus, catalases are very important for microorganisms to protect themselves against extracellular H_2O_2 and obtain their favourable niches. Thus, catalases are important for microorganisms for eliminating ROS not only intracellularly but also extracellularly.

As described above, microorganisms may encounter H_2O_2 as they interact with other organisms. However, such interactions may be limited between certain organisms. There have been few reports on microorganisms that can survive in an environment with high concentrations of H_2O_2. Therefore, we have initiated the study on the bacteria that are adapted to an environment containing high concentrations of H_2O_2 and the key enzymes for their adaptation, i.e. catalases. In addition, this environment where bacterial strains were isolated was a low-temperature environment (ca. 8°C). These combined extreme environmental conditions, i.e. high concentrations of H_2O_2 and low temperatures may have a great impact on the microorganisms and their enzymatic molecules.

Classification of catalases

Catalases (EC 1.11.1.6) have been investigated for a longer period than any other enzyme and are among the most efficient enzymes. More than 2000 DNA sequences of catalases have been determined. Catalases from different microorganisms can be classified into three phylogenetically unrelated groups according to their function, prosthetic group, subunit structure and amino acid sequence (Loewen *et al.*, 2000; Klotz *et al.*, 1997, Klots and Loewen, 2003; Zamocky *et al.*, 2008).

The first of the three groups, the so-called typical catalases (monofunctional catalases), are most widespread in nature, are identified in the *Bacteria* and *Archaea* domains, plants, fungi and animals, and exhibit efficient catalytic reaction. This group of catalases can be subdivided as follows: small-subunit catalases (subunit molecular mass: 55–69 kDa) possessing haem *b* as the prosthetic group and large-subunit catalases (subunit molecular mass: 75–84 kDa) possessing mainly haem *d* as the prosthetic group and one haem per subunit.

Typically, these catalases consist of four identical subunit structures, but dimers, hexamers, and even an unusual heterotrimer structure (from *Pseudomonas aeruginosa*) are also found. Both their catalytic activities and K_m values are high (20–200 mM). These catalases have a broad pH optimum range, are specifically inhibited by 3-amino-1,2,4-triazole and are resistant to reduction by dithionite (Kim *et al.*, 1994; Nadler *et al.*, 1986). So far, the 3D structures of 16 typical catalases have been determined.

The second group of catalases is catalase–peroxidases. Catalase–peroxidases are the so-called bifunctional catalases: they exhibit not only catalase activities but also peroxidase activities. However, their maximum catalytic activity is two or three orders lower than those of typical catalases. These enzymes are widely distributed in the *Bacteria, Archaea* and *Eukarya* domains (although only one fungal source in *Eukarya*) and 'typically have a dimeric or tetrameric structure with a subunit molecular mass of approximately 80 kDa, containing only one or two haem *b* per molecule. The amino acid sequences of these enzymes and their three-dimensional (3D) structures are similar to those of plant and fungal peroxidases. These enzymes have a distinct pH optimum, are not inhibited by 3-amino-1,2,4-triazole and are reduced by dithionite (Hochman and Shemesh, 1987; Kengen *et al.*, 2001). So far, the 3D structures of six catalase-peroxidases have been determined.

The third group of catalases is manganese-catalases (Mn-catalases). These enzymes are the so-called pseudocatalases and are widely distributed in lactic acid bacteria and thermophilic bacteria. Mn-catalases do not contain haem as the prosthetic group; rather, these enzymes use manganese ions. Therefore, the activities of these enzymes are not inhibited by cyanide or azide, which inhibit haem-containing catalases. Mn-catalases are generally hexameric and the molecular masses of their subunits range from 28 kDa to 35 kDa (Kono and Fridovich, 1983; Allgood and Perry, 1986). They have the smallest number of members among the three groups of catalases. So far, the 3D structures of three Mn-catalases have been determined.

Phylogenetic analysis based on the primary structure of bacterial typical catalases has revealed their subdivision into three distinct groups (Klotz *et al.*, 1997, 2003). This phylogenetic relationship is not directly related to rRNA-base phylogeny (Klotz *et al.*, 1997). Bacterial clade 1 catalases are small-subunit catalases and are related to plant catalases. These catalases are found in non-pathogenic and opportunistic pathogenic bacteria that are widespread in nature such as *Pseudomonas* species. Bacterial clade 2 catalases share common characteristics with fungal group 2 catalases. These catalases are large-subunit catalases including those originating from both high- and low-GC Gram-positive bacteria and *Gammaproteobacteria*. KatE from *Escherichia coli* belongs to in this clade. The bacterial clade 3 catalase is the most abundant subfamily that includes small-subunit catalases derived from *Proteobacteria, Bacteroides* and Gram-positive bacteria, many of which lead a parasitic or symbiotic life in association with a host organism such as *Haemophilus influenzae, Neisseria gonorrhoeae* or *Vibrio fischeri*. These catalase gene families may evolve in response to environmental pressure from host cells.

Mechanism of catalase reaction and gene regulation of catalases

All catalases possessing haem as the prosthetic group degrade H_2O_2 through an intermediate state of reaction in the overall reaction, $2H_2O_2 \rightarrow 2H_2O + O_2$. In the former stage reaction, one H_2O_2 molecule oxidizes the haem in the enzyme in the resting state (Fe^{3+} Por) to ferryl porphyrin with a porphyrin π-cation radical (Fe^{4+}=O $Por^{+}\cdot$): the intermediate state is compound I. In the latter stage reaction, the compound I is reduced by the second H_2O_2 molecule to molecular O_2 and H_2O (eqns. 9.1 and 9.2; Deisseroth and Dounce, 1970; Schonbaum and Chance, 1976).

$$\text{Resting state } (Fe^{3+}\text{ Por}) + H_2O_2 \rightarrow \text{Compound I } (Fe^{4+}{=}O\text{ Por}^{+}\cdot) + H_2O \quad (9.1)$$

$$\text{Compound I } (Fe^{4+}{=}O\text{ Por}^{+}\cdot) + H_2O_2 \rightarrow \text{Resting state } (Fe^{3+}\text{ Por}) + H_2O + O_2 \quad (9.2)$$

Bacterial catalase expressions are stimulated by several factors: the presence of H_2O_2, stresses such as osmotic pressure or heat, and the entry of cells into the stationary growth phase. There are bacterial species that possess multiple catalase isozymes encoded by different genes; which are regulated differently depending on the growth phase and response to several stresses, suggesting that these catalases may have different physiological functions corresponding to various stimuli.

E. coli produces two types of catalase – periplasmic catalase–peroxidase [hydroperoxidase I (HP I)] and cytoplasmic typical catalases [HP II (clade 2)] – encoded by *KatG* and *KatE*, respectively. The expression of *KatE* is controlled by the sigma factor, σ^s, which is controlled by RpoS regulon, in the stationary phase (Storz and Zheng, 2000). This may have a role in protection in the slow or no growth. The expression of *KatG* is controlled by the positive regulator OxyR, which is a 34-kDa LysR-type transcriptional regulator, in response to H_2O_2 and by RpoS in an OxyR-independent manner (Mukhopadhyay and Schellhorn, 1994; Ivanova *et al.*, 1994). OxyR can switch rapidly between the reduced and oxidized states and only that in the oxidized state acts as a transcriptional activator (Aslund *et al.*, 1999; Christman *et al.*, 1985). The expression of catalases may be related to the adaptation of organisms to low temperatures, because cold shock proteins (Csps) are required for the maximum KatG expression (Phadtare *et al.*, 2006).

Bacillus subtilis expresses three kinds of typical functional catalases, KatA, KatE and KatX (Petersohn *et al.*, 1999; Engelmann and Hecker, 1996). The expression of *KatA* is induced by H_2O_2 or under metal limitation condition and is regulated by the Fur family paralogue perR. On the other hand, the expression of *KatE* is induced by salt, heat and ethanol stresses and by glucose starvation via the sigma factor σ^s. In addition, *KatX*, which encodes the major catalase in dormant spores, is controlled by σ^B and forespore-specific σ^F.

Niches of H_2O_2-resistant bacteria

Based on the above-mentioned background, there are many chances for bacteria to encounter ROS including H_2O_2 produced intercellularly and extracellularly, and also, there are many defence mechanisms to diminish ROS in bacteria. Among the diverse defence mechanisms, diverse enzyme molecules are involved. On the other hand, there have been few reports on microorganisms that can survive in high concentrations of H_2O_2. There is an artificial environment with high concentrations of H_2O_2. A herring egg processing plant uses H_2O_2 as the bleaching agent to produce brightly coloured herring eggs. The plant possesses a water treatment system using activated sludge, which treats sewage produced in the plant. The water treatment system consists of five segments from upstream to downstream and as follows: a

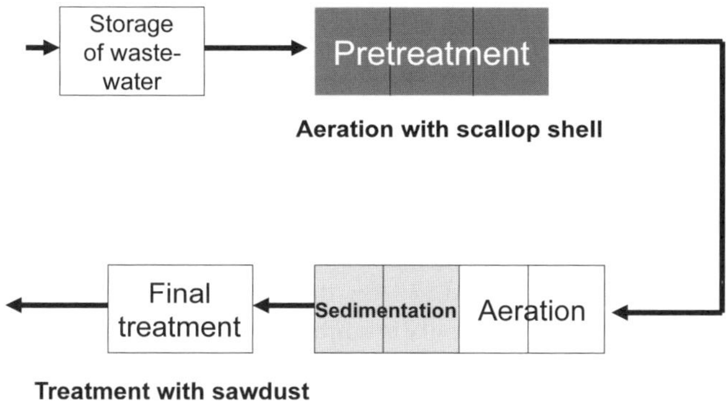

Figure 9.1 Water treatment system of fish processing factory. The water treatment system consists of five segments from upstream to downstream as follows: sewage storage tank (reservoir of sewage), pretreatment tank (aeration with scallop shells facilitates aeration efficiency to diminish concentrations of H_2O_2), ordinary aeration tank (degradation of organic substances by microorganisms), sedimentation tank [anaorobic treatment (e.g. denitrification) and separation of activated sludge] and aerobic treatment with sawdust.

sewage storage tank, a pretreatment tank (vigorous aeration with scallop shells to diminish concentration of H_2O_2), an ordinary aeration tank (treatment of organic substances by activated sludge containing microorganisms), a sedimentation tank (anaerobic treatment and separation of activated sludge) and final aerobic treatment with sawdust (containing bacterial consortia) (Fig. 9.1). The concentration of H_2O_2 is higher in the upstream tank than in the downstream tank. The water treatment system has high concentrations of H_2O_2 and low temperatures. It is quite intriguing to determine what kinds of microorganisms existing in such an environment and the H_2O_2-eliminating enzymes (catalases) involved. Therefore, we have initiated the studies on the isolation and identification of bacteria isolated from this H_2O_2-containing environment and we have also purified several catalases and characterized them.

Vibrio rumoiensis and its catalase

Vibrio rumoiensis S-1^T was firstly isolated as a psychrotolerant exhibiting high catalase activity from downstream (sedimentation tank) of the drain pool of a fish processing plant (Fig. 9.1) that uses H_2O_2 as a bleaching agent for brightening the colour of fish eggs (Yumoto *et al.*, 1998, 1999). The strain is a Gram-negative, non-motile, psychrotolerant bacterium. The temperature and H_2O_2 concentrations are around 8°C and 1.5–6 mM, respectively. The catalase activity of the cell extract (4000–8000 U/mg protein) is one or two orders of magnitude higher than those of *Alcaligenes faecalis, Corynebacterium glutamicum, Staphylococcus aureus* and *Pseudomonas fluorescens*. The strain seems to possess only one kind of catalase. Electron microscopic observation revealed that the surface of the bacterium was covered by blebs. Generally, *Vibrio* species possesses flagella; however, *V. rumoiensis* does not possess flagella. On the basis of the phylogenetic position, *V. rumoiensis* is classified into the genus *Vibrio*. In addition, *V. rumoiensis* exhibits a distinctive position among the genus *Vibrio* in the phylogenetic tree. When *V. rumoiensis* cells were inoculated into a medium containing either 5 mM H_2O_2 or no H_2O_2, they grew similarly. On the other hand, when *V. rumoiensis* cells were grown on plates containing various concentrations of H_2O_2, the colony-forming ability of individual *V. rumoiensis* cells was inferior to those of *Escherichia coli, Bacillus subtilis* and *Vibrio parahaemolyticus*. These findings suggest that although individual *V. rumoiensis* cells are sensitive to H_2O_2, a group of these cells can accumulate a sufficient amount of catalase that protects them from H_2O_2 (Ichise *et al.*, 1999).

Catalase of *V. rumoiensis* (VktA) has been purified and characterized (Yumoto *et al.*, 2000). Its molecular mass is 230 kDa, and it consists of four equimolar subunits of 57.3 kDa. The absorption spectrum of purified VktA exhibits a Soret band at 406 nm and additional minor peaks between 500–550 nm and 630 nm. Its isoelectric point (p*I*) is 5.5. The catalase is not reduced by dithionite and its activity is inhibited by cyanide, aside and 3-amino-1,2,4-triazole. The enzyme has a broad pH optimum range. The spectral shape of the enzyme is not altered by adding dithionite. On the basis of above findings, catalase from *V. rumoiensis* is classified into the typical catalase processing small subunits.

The kinetic analysis of VktA catalase activity shows an apparent V_{max} of 7.95×10^5 μmol H_2O_2/μmol haem/s and an apparent K_m of 35 mM for H_2O_2. The catalytic efficiency k_{cat}/K_m of VktA (2.3×10^7/s/M) was the highest among reported clade 3 catalases isolated from pathogenic or symbiotic bacteria owing to the low K_m. The pH dependence of VktA activity reveals a broad optimum pH range from pH 6 to 10. Although residual activity is observed at pH 4 and pH 5, the activity completely disappears at pH 3. VktA activity shows a broad pH range of stability from pH 6 to pH 10. While more than 80% of the activity remained at pH 5, the activity completely disappeared at pH 3.

Although it has been reported that the temperature dependence of catalase activity at higher than 70°C, an obvious temperature dependence between 10 and 70°C has not been reported until the report on VktA. A temperature dependence between 10 and 70°C with the optimum temperature at 40°C was observed for VktA (Fig. 9.2). Such an obvious temperature dependence of the activity is not observed in *Micrococcus luteus* and bovine liver catalases (Yumoto *et al.*, 2000). On the other hand, the thermostability of VktA was lower than those of *M. luteus* and bovine liver catalases. Furthermore, the catalytic activity of VktA was 527,500 U/mg protein at 40°C under standard reaction conditions (30 mM H_2O_2, pH 7) and the activity was 1.5 and 4.3 times higher than those of *M. luteus* and bovine liver catalases, respectively. Thus, these unique properties of VktA are in accordance with the environmental conditions under which the microorganism survives. This example is quite unique due to the environmental pressures elevate both catalytic efficiency and temperature instability.

The VktA gene (*vktA*) has been cloned and its sequence has been determined (Ichise *et al.*, 1999, 2000). The *vktA* consists of an open reading frame of 1530 bp encoding 508 amino acids with a calculated molecular mass of 57657.79 Da. Upstream of the start codon, putative promoter sequences were found. The promoter sequences are predicted to be recognized by RNA polymerase carrying the housekeeping sigma factor, σ^{70}. Another putative promoter binding site for OxyR exists overlapping at the −35 promoter site (Bin Haji Mohd Taha *et al.*, 2012). A comparison of the deduced amino acid sequence of VktA with those of other catalases revealed that the active sites (His^{61}, Thr^{100} and Asn^{134}), binding sites of the distal region of haem (Val^{102}, Thr^{124} and Phe^{139}) and the proximal sites of haem (Tyr^{244} and Arg^{351}) are well conserved. VktA may contain NADPH inside of its molecule because NADPH-binding sites (His^{180}, Arg^{189}, Val^{288} and Lys^{291}) were observed (Zámocký and Koller, 1999). The amino acid sequence of VktA reveals that it belongs to clade 3 (Fig. 9.3). It has been reported that pathogenic or symbiotic microorganisms possessing only a single

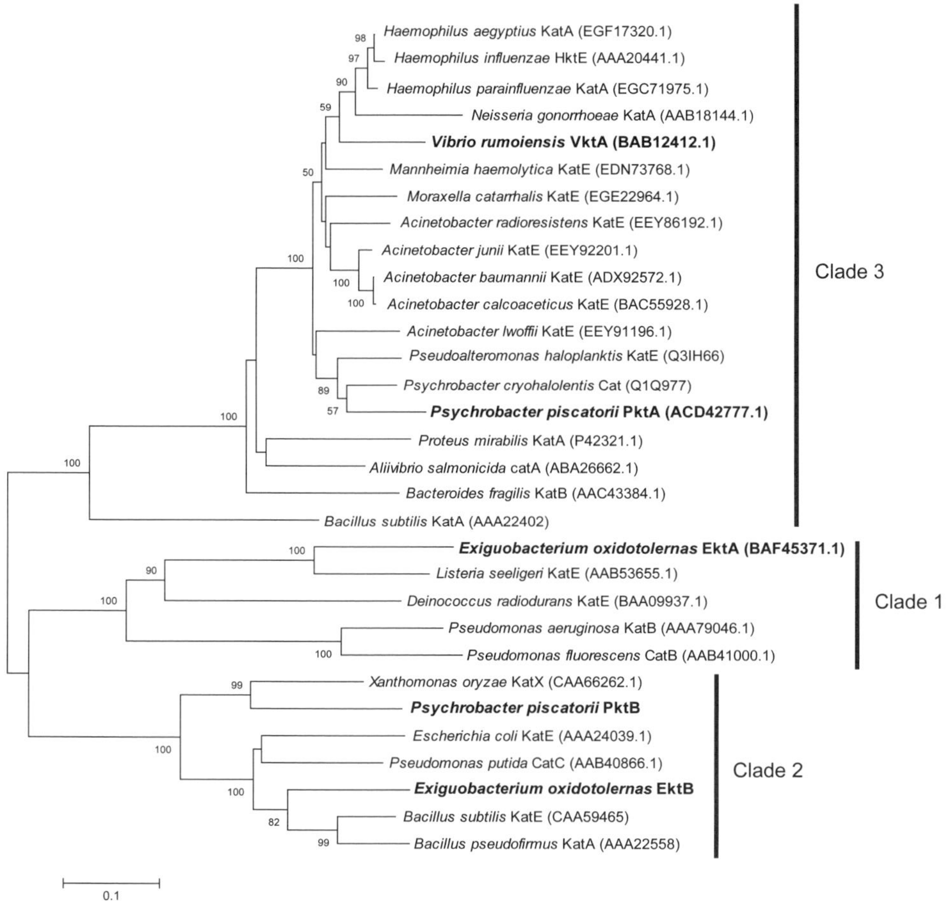

Figure 9.2 Phylogenetic positions of VktA, PktA, PktB, EktA and EktB. PktB and EktB do not seem to be expressed. The phylogenetic tree was constructed using the CLUSTAL W program (Thompson *et al.*, 1999) with multiple alignments using the neighbour-joining method (Saitou and Nei, 1987). Numbers at the branches are bootstrap values higher than 50. These percentages are based on 1000 replicates. Bar, 0.1 changes per amino acid position. The amino acid sequences of catalases used for the alignment are shown with their GenBank/DDBJ/EMBL accession numbers in parentheses.

isozyme belong to clade 3 catalase. Therefore, there is a possibility that *V. rumoiensis* was originally a symbiotic bacterium in marine organisms, e.g. herring, because the bacterium was isolated from a herring egg processing plant. It is considered that bacteria possessing only a single isozyme belonging to clade 3 catalase select the appropriate enzyme with catalytic efficiency for eliminating H_2O_2 produced from host cells.

It is considered that the localization of catalases is important for the effective elimination of either intracellular or extracellular H_2O_2 to prevent damage to cell components. The localization of VktA in the cytoplasmic, periplasmic and extracellular fractions is determined by estimation of catalase activity and Western blot analysis of fractionated cells cultivated for 21 h at 27°C (Yumoto *et al.*, 2000). The presence of VktA in cytoplasmic and

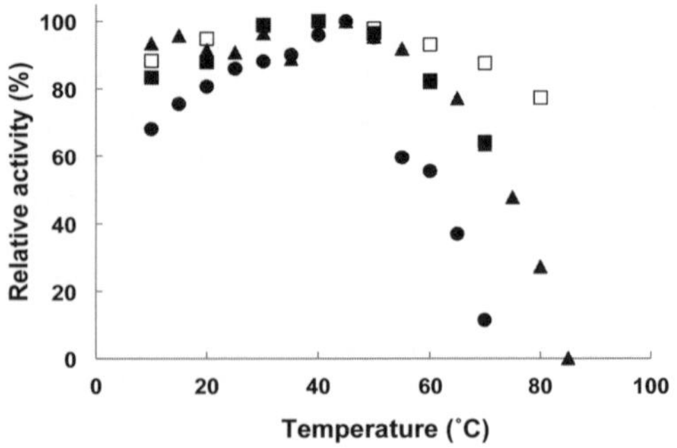

Figure 9.3 Temperature dependence of activities of catalases. Catalase activity is estimated under standard assay conditions in presence of 30mM H_2O_2 at pH 7 (50mM potassium buffer) at 25°C. Symbols: closed squares, Vkt; closed triangles, PktA; closed circles, EktA; open squares, *Micrococcus luteus* catalase.

periplasmic fractions is ascertained by Western blot analysis. The ratio of catalase activity in the culture extracellular, periplasmic and intracellular fractions was 1:7:202. Almost no growth hindrance of *V. rumoiensis* in the presence of 100 mM H_2O_2 in culture broth middle of experimental phase was observed. On the other hand, a spontaneous catalase-less mutant of *V. rumoiensis* is not able to tolerate H_2O_2 even at 5 mM in the medium. The contribution of catalase to H_2O_2 tolerance is also demonstrated in the comparison between catalase-deficient *E. coli* strain UM2 and the strain UM2 carrying *vktA* (Ichise *et al.*, 2008). It seems that H_2O_2 tolerance is attributed to the catalase activity in cells themselves. The presence of cell surface catalase was observed by immunoelectron micrography using an antibody to the purified intracellular catalase molecule (VktA) in the cells of *V. rumoiensis*. In conclusion, it is considered that the presence of cell surface catalase contributes to catalase activity in cells themselves. This will be important for the rapid elimination of extracellular H_2O_2 to prevent the damage to cell components.

Psychrobacter piscatorii and its catalase

Psychrobacter piscatorii T-3-2^T was isolated from upstream (pretreatment tank) of the drain pool of the fish processing plant (Fig. 9.1) that uses H_2O_2 as a bleaching agent (Yumoto *et al.*, 2010). The temperature was around 8°C and the H_2O_2 concentration fluctuated between 6 and 38 mM. The strain is a Gram-negative, non-motile, psychrotolerant bacterium. The growth temperature range of the strain is 0–30°C. Cells growth occurs in a medium supplemented with 0–12% NaCl. The strain hydrolyses tributyrin and Tweens 20, 40, 60 and 80 but not casein, gelatin, starch or DNA. The cell extract of the strain exhibits a catalase activity of 12,000 U/mg protein, which is much higher than that of the phylogenetic neighbours of other *Psychrobacter* spp. such as *Psychrobacter nivimaris* (15 U/mg protein), *Psychrobacter proteolyticus* (29 U/mg protein) and *Psychrobacter aquamaris* (1800 U/mg protein).

P. piscatorii T-3 was isolated from the same origin, and it exhibits a higher catalase activity than the strain T-3–2^T (Kimoto *et al.*, 2008, 2012). Its cell extract exhibits a catalase activity of 19,700 U/mg protein, which was higher than that of the cell extract from *M. luteus*, which is used for industrial catalase production. The catalase (PktA) is purified by one-step anion-exchange chromatography and one-step hydrophobic chromatography. A large amount of cells can be easily used for enzyme purification because the catalase content is high in the cell extract and the purification procedure does not require the gel filtration step. The purified catalase showed a final specific activity of 222,000 U/mg protein. The molecule exhibits a tetrameric structure consisting of a subunit of ca. 59 kDa. The absorption spectrum of the purified catalase shows a Soret band at 406 nm and additional minor peaks between 500–550 nm and 630 nm.

The kinetic analysis of PktA activity shows an apparent V_{max} of 2.4×10^5 µmol H_2O_2/µmol haem/s and an apparent K_m of 75 mM for H_2O_2. The catalytic efficiency, k_{cat}/K_m, of PktA (3.2×10^6/s/M) was higher than that of several other clade 3 catalases from pathogenic or symbiotic bacteria owing to the low K_m. This suggests that *P. piscatorii* cells themselves are not as resistant to H_2O_2 as other pathogenic or symbiotic bacteria possessing clade 3 catalases. It is considered that the bacterial cells protect themselves by producing catalase with low K_m because the cells may be fragile even at low concentrations of H_2O_2. The dependence of PktA activity on pH exhibits a broad optimum pH range from pH 5.0 to pH 10.0. The residual activity at pH values 4.0 and 3.0 was higher than that of *M. luteus* catalase (MLC). The pH stability of the catalase activity of PktA is highest at pH 8.0. However, the dependence of the activity of MLC and *E. coli* HP II on temperature is scarcely observed in the temperature range of 10–70°C (Switala *et al.*, 1999). On the other hand, PktA catalase activity exhibits a clear decrease at temperatures higher than 55°C.

The entire nucleotide sequence of the PktA (1533 bp) gene was determined. The deduced amino acid sequence of PktA showed that the catalase is composed of 510 amino acids with a calculated molecular mass of 58,313.19 Da, and the predicted p*I* is 5.45. Upstream of the PktA-coding region, the promoter sequences are predicted to be recognized by RNA polymerase carrying the housekeeping sigma factor. Another putative promoter binding site for OxyR exists beside the upstream of the −35 promoter site. The amino acid sequence shows the highest identity of 84.9% with *Psychrobacter cryohalolentis*, a psychrotolerant bacterium obtained from Siberian permafrost. The phylogenetic tree constructed on the basis of the amino acid sequence of PktA shows that it is a clade 3 catalase and shows the closest kinship with the catalase of *P. cryohalolentis* (Fig. 9.2). Although it has been reported that the phylogenetic tree constructed from the amino acid sequence of catalase is not related to that constructed on the taxonomic basis (e.g. 16S rRNA gene sequence) (Klotz *et al.*, 1997), PktA was shown to occupy a position related to the catalase from taxonomically related species. A comparison of PktA amino acid sequence with those of catalases carried out to gain a general understanding of the function of amino acid residues revealed that the active sites (His^{65}, Ser^{104} and Asn^{138}), the binding sites of the distal region of haem (Val^{106}, Thr^{128} and Phe^{143}) and the proximal sites of haem (Tyr^{348} and Arg^{355}) are conserved. PktA probably possesses NADPH owing to the conservation of amino acid residues of NADPH-binding sites (His^{184}, Arg^{193}, Val^{292} and Lys^{295}) (Zámocký *et al.*, 1999).

Cell extract from *P. piscatorii* exhibits higher catalase content (10%) than that from *V. rumoiensis* (1.8%), and the former produces catalase (PktA) exhibiting a lower catalytic efficiency [k_{cat}/K_m of PktA (3.2×10^6/s/M)] than the latter [k_{cat}/K_m of VktA (2.3×10^7/s/M)]. This catalytic efficiency may be related to the fragility of cells and the amount of catalase produced in *V. rumoiensis*. The higher production of catalase in *P. piscatorii* may related to the H_2O_2 concentration in the environment in which the bacterium lives.

PktA exhibits a relatively high catalytic efficiency and a very high production rate. These features are related to the environmental adaptation of the microorganism. On the other hand, the amino acid sequence of PktA shows the highest homology with the catalase of *P. cryohalolentis*, which was isolated from Siberian permafrost. These findings suggest that the molecular feature of PktA reflects not only the environmental pressure but also the taxonomic relationship.

Exiguibacterium oxidotolerans and its catalase

Exiguibacterium oxidotolerans T-2-2^T was isolated from upstream (pretreatment tank) of the drain pool of the fish processing plant (Fig. 9.1) that uses H_2O_2 as a bleaching agent where *P. piscatorii* was isolated (Yumoto *et al.*, 2004). The growth pH range of the strain is pH 7–10. The strain is a Gram-positive rod, facultatively aerobic and motile with peritrichous flagella. It grows in the presence of NaCl at 0–12% but not above 15% NaCl. It grows at 4–40°C (optimum 34°C). The microorganism is positive for β-galactosidase and acid phosphatase, but negative for reduction of NO_3^- to NO_2^-, and indole and H_2S production. It hydrolyses casein, gelatin, starch and DNA, but not chitin, alginic acid, and Tweens 20, 40, 60 and 80. The growth of the cells is not inhibited even in the presence of 100 mM H_2O_2. The cell extract of the strain exhibits a catalase activity of 28,000 U/mg protein, which is higher than those of *V. rumoiensis* and *P. piscatorii*.

Catalase of *E. oxidotolerans* (EktA) was purified in three steps consisting of two steps of anion-exchange chromatography and one step of gel filtration chromatography (Hara *et al.*, 2007). The purified EktA exhibits a specific activity of 430,000 U/mg protein under a standard reaction condition in the presence of 30 mM H_2O_2 at 25°C in 50 mM sodium phosphate (pH 7.0). The molecular mass of a subunit of the catalase was estimated to be 64 kDa by SDS-PAGE. The native molecular mass was estimated to be 240 kDa by gel filtration. The results indicate that the catalase is composed of four identical subunits. The absorption spectrum of purified catalase exhibits a Soret band at 408 nm and additional minor peaks between 500–550 nm and 627 nm.

A temperature dependence between 5°C and 70°C with the optimum temperature at 45°C was observed for EktA. Although the catalase activity decreased steadily from 45°C to 5°C (from 100% to 58%), it decreased drastically from 45 to 70°C (from 100% to 38%) (Fig. 9.3). The pH dependence between pH 2 and pH 10 with the optimum pH at 7 was observed. Catalase activity is completely lost at pH 3. These clear temperature and pH dependences are quite unique compared with other reported catalases.

When a reactive intermediate formation using EktA is estimated by stopped-flow experiments with methyl hydroperoxide in 50 mM phosphate buffer (pH 7.0) at 5°C, compound I formation is observed and most of compound I directly returns to the resting state without conversion to compound II, which is a one-electron reaction product of compound I (Hara *et al.*, 2007). The formation rate constant of compound I of EktA with H_2O_2 is slightly higher than that of *M. luteus* catalase (MLC) and 3.2 times higher than that of bovine liver catalase (BLC). The formation rate constant of compound I of EktA with methyl hydroperoxide (CH_3OOH) is 12.2 times higher than that of BLC and is 37.6 times higher than that of MLC. The formation rate constant of EktA with peracetic acid (CH_3CO_3H) is 76.9 times higher than that of BLC and is 1190 times higher than that of MLC. In addition, the formation rate constant of EktA with *tert.*-butyl hydroperoxide [$(CH_3)_3C\text{-}OOH$] is 11.0 times higher than that of BLC and 71.7 times higher than that of MLC.

The entire 1546 bp DNA sequence that includes the entire nucleotide sequence of EktA catalase was determined (Hara *et al.*, 2007). The deduced amino acid sequence shows that EktA is composed of 491 amino acids with a calculated molecular mass of 564887.70 Da, and the predicted p*I* is 5.18. Although the reason is not known, this value is inconsistent with the result of SDS-PAGE. The deduced amino acid sequence showed the highest identity (69.2%) with the catalase from *Listeria seeligeri*. The upstream of EktA the promoter sequence corresponding –10 and -35 sites, which encode the promoter sequences predicted to be recognized by RNA polymerase carrying the housekeeping sigma factor, was identified. Another putative promoter binding site for OxyR exists at an upper site 18 bp from the –35 site. This is consistent with the finding that continuous addition of H_2O_2 at an appropriate concentration in the culture medium results in a higher catalase activity than its control (Takebe *et al.*, 2007). A comparison of the deduced EktA amino acid sequence with those of other catalases revealed that the active site (His^{56}, Ser^{104} and Asn^{138}), the binding site of the distal region of haem (Val^{97}, Thr^{119} and Phe^{142}) and the proximal sites of haem (Tyr^{339} and Arg^{346}) are well conserved (Zámocký *et al.*, 1999). However, NADPH-binding sites are not found in the amino acid sequence of EktA. Thus, a comparison of amino acid sequences of catalases revealed that the most important residues for the activity are well conserved. EktA is a clade 1 catalase in the phylogenetic tree constructed on the basis of amino acid sequences of catalases and is related to *Listeria seeligeri* KatE (Fig. 9.2).

The crystal structure of EktA has been determined with refined to 2.4 Å resolutions (Hara *et al.*, 2007). As suggested by the amino acid sequence of EktA, the structure around the haem environment is similar to those of other catalases [*Pseudomonas syringae* catalase (PSCF), MLC and BLC]. PSCF was first determined its crystal structure in clade I catalase and EktA is the second example belonging to the same clade. Although the superimposed structures of EktA and PSCF exhibit that the tertiary structures of conserved amino acid residues around the haem are identical, the bottleneck of the substrate accessible to the main channel in EktA is larger than PSCF. The main channel defines the accessibility of the substrate and this underpins the high velocity of catalase activity. The narrow channel, that is, the lower part of the channel between protohaem and conserved Asp^{127}, comprises 14 amino acid residues: Val^{73}, His^{74}, Val^{115}, Asp^{127}, Pro^{128}, Asn^{147}, Phe^{152}, Phe^{153}, Phe^{160}, Phe^{163}, Ile^{164}, Gln^{167}, Trp^{185} and Leu^{198} in BLC numbering. Although Val^{73}, His^{74}, Val^{115}, Asp^{127}, Pro^{128}, Asn^{147}, Phe^{152}, Phe^{153} and Phe^{160} are highly conserved, Phe^{163}, Ile^{164}, Gln^{167}, Trp^{185} and Leu^{198} are less conserved among typical catalases. The latter combination in EktA was totally different from that in BLC as follows: Met^{145}, Val_{146} Lue^{149}, Met^{167} and Ile^{180} (in EktA numbering). In particular, Met^{167} and Ile^{180} are rarely located at the entrance of the narrow channel in the catalase sequence in the database. It can be predicted that the Trp residues (i.e. Trp^{185} in BLC and Trp^{171} in MLC) at the entrance of the narrow channel (bottleneck) decrease the diameter of the bottleneck of the main channel in BLC and MLC. On the other hand, Met^{167} is located in EktA instead of the Trp residue. This may make the bottleneck of EktA wider than those of BLC and MLC. The bottleneck exists at ~14–15 Å from protohaem. The diameters of the bottlenecks are 2.7 Å for EktA, 2.3 Å for BLC, and 2.1 Å for MLC. The rate constants of compound I formation using methyl hydroperoxide, paracetic acid and *tert.*-butyl hydroperoxide are directly proportional to the diameter of the bottleneck. The result indicates that EktA rapidly reacts with organic peroxide molecule, which is larger than the H_2O_2 molecule owing to the large bottleneck in the main channel.

Discussion

Although the occurrence of H_2O_2 is not very rare in natural environments, its continuous presence at high concentrations at low temperatures may be rare. Owing to the genetic diversity of bacteria, it is expect that unique bacteria exist in such an environment (low temperatures and high concentrations of H_2O_2). Three novel species of psychrotolerant H_2O_2-resistant bacteria have been isolated from drain pools of a fish egg processing plant that uses H_2O_2 as a bleaching agent. *P. piscatorii* and *E. oxidotolerans* were isolated from upstream (pretreatment tank) of the water treatment system (Fig. 9.1 and Table 9.1). *Psychrobacter* species and *Exiguobacterium* species have been isolated from Siberian permafrost soil. In addition, it has been demonstrated that *Exiguobacterium* species ubiquitously exist in permafrost soil. *Psychrobacter* species and *Exigiobacterium* species have also been isolated from several cold environments such as Antarctic soil and ice, the Himalayan mountains and saline lake in Romania (average water temperature: 10°C) (Borsodi *et al.*, 2010; Rodrigues *et al.*, 2007, 2009). Therefore, it is considered that these genera of bacteria are widely distributed in cold natural environments. It can be considered that these genera involve a variety of species and are genetically diverse. It can also be considered that among them, species that were able to tolerate high concentrations of H_2O_2 were selected in the waste water treatment system of the processing plant.

Identification of bacteria and characterization of catalases were carried out to determine the taxonomic positions of the bacteria adapted to the environment with high concentrations of H_2O_2 and the catalytic properties of catalases, respectively. *V. rumoiensis* was isolated from downstream (lower concentration of H_2O_2 than upstream) of the sewage treatment system, whereas *E. oxidotorelans* and *P. piscatorii* were isolated from upstream of the system. The concentration of H_2O_2 in the origin seems to affect the catalase activity of the cell extract. The catalase activity of the cell extract from *V. rumoiensis* is 7300 U/mg protein, whereas those from *E. oxidotorelans* and *P. piscatorii* are 24,600 U/mg protein and 19,700 U/mg protein, respectively. On the other hand, the activity of purified PktA was the lowest PktA, 222,000 U/mg protein; VktA, 395,000 U/mg protein; and EktA, 430,000 U/mg protein (Table 9.2). The catalase contents are 10% in the cell extract from *P. piscatorii*, 0.7% from *V. rumoiensis* and 6.5% from *E. oxidotolerans* (Table 9.1). These suggest that the environmental adaptation of bacteria to high concentrations of H_2O_2 is due to not only the high catalase activity in purified enzyme but also the high expression level of catalase.

It has been reported that symbiotic, parasitic and pathogenic microorganisms possess only one clade 3 catalase as a single isozyme. It is considered that catalases from microorganisms associated with hosts have evolved for their protection against H_2O_2 produced by their

Table 9.1 Summary of characteristics of isolates from drains of waste water treatment system containing H_2O_2

Strain	Origin in the water treatment system	Catalase	Catalase activity of cell extract (U/mg)	Rate of catalase content in cell extract	Fragility of cells
V. rumoiensis	Downstream	VktA	7300	1.8%	Fragile
P. piscatorii	Upstream	PktA	19,700	10%	Medium
E. oxidotolerans	Upstream	EktA	28,000	6.5%	Hard

Table 9.2 Summary of characteristics of catalase from isolates from drains of waste water treatment system containing H_2O_2

Catalase	Catalase activity (U/mg protein)[a]	k_{cat}/K_m (/s/M)	Temperature dependence higher than 60°C	p*I*
VktA	395,000	2.3×10^{7}[b]	Yes	5.5
PktA	222,000	3.2×10^{6}[c]	Yes	5.5[d]
EktA	430,000	Not determined	Yes	5.2[d]

[a]Standard reaction condition in the presence of 30mM H_2O_2 at pH 7.
[b]Determined by spectrophotometer.
[c]Determined by oxygen electrode.
[d]Predicted value based on the amino acid sequence.

hosts. Although *P. piscatorii* and *E. oxidotolerans* possess the clade 2 catalase genes, they express only clade 3 catalase under ordinary culture conditions (Fig. 9.2). This may indicate that the bacteria express selectively suitable catalase genes for their adaptation in environments with high concentrations of H_2O_2.

An analysis of the dependence of the activity of the three catalases, VktA, PktA and EktA on temperature, reveals that all these catalases exhibit an obvious decrease in activity at higher than 60°C with the optimum temperature in the range of 40–45°C (Fig. 9.3). Although there are differences in the temperature profiles of their activities, it is considered that the characteristic decrease in the activity at higher than 60°C with the optimum temperature in the range of 40–45°C is due to their being produced by psychrotolerant microorganisms. VktA is first found as a temperature-dependent catalase in the temperature range of 10–70°C with the obvious optimum temperature at 40°C. In 2006, catalase from *Aliivibrio* (*Vibrio*) *salmonicida* was found as the first psychrophilic catalase that exhibits an obvious decrease in activity at 20°C with the optimum temperature range of 0–10°C (Loewntzen *et al.*, 2006). On the other hand, the mesophilic catalase, *E. coli* HP II (clade 2 catalase) exhibits the obvious decrease in activity at 78°C with the optimum temperature range of 50–74°C (Switala *et al.*, 1999). On the other hand, another mesophilic catalase, MLC, exhibits the obvious decrease in activity at 80°C with the optimum temperature range of 10–70°C. It can be concluded that the three catalases exhibit high catalytic activity and are adapted to low temperatures and these characteristics reflect the environment in which these microorganisms survive. It is considered that the variations of productivity and catalytic properties among the three species depend on the characteristics of the species such as cell fragility and the genetic variations among the genera and species involved.

Future studies

In the series of studies described above, the existence of bacteria adapted to an environment with high concentrations of H_2O_2 and low temperatures has been demonstrated. In addition, catalases from three isolates are also adapted to high H_2O_2 concentrations and low temperatures. These microorganisms and catalases may have specifically evolved for their adaptation to such an environment because there is no natural environment similar to that of the water treatment system using H_2O_2. Consideration of the molecular basis of the mechanisms underlying the processes of selection and evolution is very important in biology. Discoveries of microorganisms similar to these isolated ones and comparison of whole-genome sequences may lead to a key to clarifying the molecular basis.

Although the history of studies of catalases is longest among various enzymes, it has not been know that catalase activity hardly exhibits a distinct temperature dependency at lower than 70°C. VktA is the firstly catalase found to exhibit temperature dependence. In addition, VktA exhibits high catalytic efficiency as determined on the basis of its low K_m and high V_{max}. This is probably related to the fragility of *V. rumoiensis* cells owing to the necessity to diminish ambient H_2O_2 concentration effectively. On the other hand, PktA exhibits a high similarity with the catalase from *P. cryohalolentis*. These findings described above suggest that the evolution of the catalase molecule is affected by environmental pressure and/or the taxonomic relation of the catalase-producing microorganism. Systematic comparative studies of catalases from extremophiles may lead to a key to clarifying the evolutional process of catalases in their functions.

Website resources

Website of catalase in The Online Macromolecular Museum Exhibits (http://www.callutheran.edu/Academic_Programs/Departments/BioDev/omm/catalase/cat1.htm)
PeroxiBase (http://peroxibase.toulouse.inra.fr/introduction.php)
Catalase website (http://www.catalase.com/index.htm)
EMBEL-EBI Catalase, mono-functional, haem-containing
(http://www.ebi.ac.uk/interpro/DisplayIproEntry?ac=IPR018028)
EMBEL-EBI Clade 1, 3 Catalase
(http://www.ebi.ac.uk/interpro/DisplayIproEntry?ac=IPR024711)

References

Allgood, G.S., and Perry, U.J. (1986). Characterization of a manganese-containing catalase from the obligate thermophile *Thermoleophilum album*. J. Bacteriol. *168*, 563–567.

Aslund, F., Zheng, M., Beckwith, J., and Storz, G. (1999). Regulation of OxyR transcription factor by hydrogen peroxide and the cellular thiol-disulfide status. Proc. Natl. Acad. Sci. U.S.A. *96*, 6161–6165.

Bin Haji Mohd Taha, A.I., Okuyama, H., Ohwada, T., Yumoto, I., and Orikasa, Y. (2012). Exogenous catalase gene expression as a tool for enhancing metabolic activity and production of biomaterials in host microorganisms. In: Innovations in Biotechnology, Agbo, E.C. ed., InTech, Rijeka, Croatia, pp. 251–274.

Bolton, J.L., Trush, M.A., Penning, T.M., Dryhurst, G., and Monks, T.J. (2000). Role of quinones in toxicology. Chem. Res. Toxicol. *13*, 135–160.

Borsodi, A.K., Kiss, R.I., Cech, G., Vajna, B., Tóta, E.M., and Márialigeti, K. (2010). Diversity and activity of cultivable aerobic planktonic bacteria of a saline lake located in Sovata, Romania. Folia Microbiol. *55*, 461–466.

Christman, M.F., Morgan, R.W., Jacobson, F.S., and Ames, B.N. (1985). Positive control of regulon for defence against oxidative stress and some heat-shock proteins in *Salmonella typhimurium*. Cell *41*, 753–762.

Deisseroth, A., and Dounce, A.L. (1970). Catalase: Physical and chemical properties, mechanism of catalase, and physiological role. Physiol. Rev. *50*, 319–375.

Engelmann, S., and Hecker, M. (1996). Impaired oxidative stress resistance of *Bacillus subtilis* sigB mutants and the role of katA and katE. FEMS Microbiol. Lett. 145: 63–69.

Halliwell, B., and Gutteridge, J.M.C. (1999). *Free radical in biology and medicine*, 3rd ed.; Clarendon Press: Oxford, United Kigdom.

Hara, I., Ichise, N., Kojima, K., Kondo, H., Ohgiya, S., Matsuyama, H., and Yumoto, I. (2007). Relationship between the size of the bottleneck 15 Å away from iron in the main channel and reactivity of catalase corresponding to the molecular size of substrates. Biochemistry *46*, 11–22.

Hust, J.K., and Barrette, W.C., Jr. (1989). Leukocytic oxygen activation and microbial oxidative toxins. CRC Crit. Rev. Biochem. Mol. Biol. *24*, 271–328.

Ichise, N., Morita, N., Hoshino, T., Kawasaki, K., Yumoto, I., and Okuyama, H. (1999). A mechanism of resistance to hydrogen peroxide in *Vibrio rumoiensis* S-1. Appl. Environ. Microbiol. *65*, 73–79.

Ichise, N., Morita, N., Kawasaki, K., Yumoto, I., and Okuyama, H. (2000). Gene cloning and expression of the catalase from the hydrogen peroxide-resistant bacterium *Vibrio rumoiensis* S-1 and its subcellular localization. J. Biosci. Bioeng. *90*, 530–534.

Ichise, N., Hirota, K., Ichihashi, D., Nodasaka, Y., Morita, N., Okuyama, H., and Yumoto, I. (2008). H_2O_2 tolerance of *Vibrio rumoiensis* S-1^T is attributable to the cellular catalase activity. J. Biosci. Bioeng. *106*, 39–45.

Imlay, J.A., and Linn, S. (1988). DNA damage and oxygen radical toxicity. Science *240*, 1302–1309.

Ivanova, A., Miller, C., Gilinsky, G., and Eisentark, A. (1994). Role of *rpoS* (*katF*) in *oxyR*-independent regulation of hydroperoxidase I in *Escherichia coli*. Mol. Microbiol. *12*, 571–578.

Katsuwon, J., and Anderson, A.J. (1992). Characterization of catalase activities in root colonizing isolates of *Pseudomonas putida*. Can. J. Microbiol. 38, 1026–1032.

Kengen, S.W., Bikker, F.J., Hagen, W.R., de Vos, W.M., and van der Oost, J. (2001). Characterization of catalase-peroxidase from the hyperthermophilic archaeon *Archaeoglobus fulgidus*. Extremophiles *5*, 323–332.

Kim, K.K., Fravel, D.R., and Papavizas, G.C. (1978). Identification of a metabolite produced by *Talaromyces flavus* as glucose oxidase and in the biocontrol of *Verticillium dahlia*. Phytopathology *78*, 488–492.

Kim, H., Lee, J.S., Hah, Y.C., and Roe, J.H. (1994). Characterization of major catalase from *Streptmyces coelicolor* ATCC 10147. Microbiology *140*, 3391–3397.

Kimoto, H., Matsuyama, H., Yumoto, I., and Yoshimune, K. (2008). Heme content of recombinant catalase from *Psychrobacter* sp. T-3 altered by host *Escherichia coli* growth conditions. Protein Expres. Purif. *59*, 357–359.

Kimoto, H., Yoshimune, K., Matsuyama, H., and Yumoto, I. (2012). Characterization of catalase from psychrotolerant *Psychrobacter piscatorii* T-3 exhibiting high catalase activity. Int. J. Mol. Sci. *13*, 1733–1746.

Klotz, M.G., Klassen, G.R., and Loewen, P.C. (1997). Phylogenetic relationships among prokaryotic and eukaryotic catalases. Mol. Biol. Evol. *14*, 951–958.

Klotz, M.G., and Loewen, P.C. (2003). The molecular evolution of catalase hydroperoxidase: evidence for multiple lateral transfer of gene between prokaryota and from bacteria into eukaryote. Mol. Biol. Evol. *20*, 1098–1112.

Kono, Y., and Fridovich, I. (1983). Isolation and characterization of pseudocatalase of *Lactobacillus plantarum*. J. Biol. Chem. *258*, 6015–6019.

Loewen, P.C., Klotz, M.G., and Hassett, D.J. (2000). Catalase – an 'old' enzyme that continues to surprise us. ASM News *66*, 76–82.

Loewntzen, M.S., Moe, E., Jouve, H.M., and Willassen, N.P. (2006). Cold adapted features of *Vibrio salmonicida* catalase: characterization and comparison to the mesophilic counterpart from *Proteus mirabilis*. Extremophiles *10*, 427–440.

Mukhopadhyay, S., and Schellhom, H.E. (1994). Induction of *Escherichia coli* hydroperoxidase I by acetate and other weak acids. J. Bacteriol. 176, 2300–2307.

Nadler, V.; Goldberg, I., and Hochman, A. (1986). A comparative study of bacterial catalases. Biochim. Biophys. Acta *882*, 234–241.

Norman, R.S., Moeller, P., McDonald, T.J., and Morris, P.J. (2004). Effect of pyocyanin on a crude-oil-degrading microbial community. Appl. Environ. Microbiol. *70*, 4004–4011.

Petersohn, A., Engelmann, S., Setlow, P., and Hecker, M. (1999). The katX gene of *Bacillus subtilis* is under dual control of sigmaB and sigmaF. Mol. Gen. Genet. *262*, 173–179.

Phadtare, S., Tadigotla, V., Shin, W.H., Senguupta, A., and Severinov, K. (2006). Analysis of *Escherichia coli* global gene expression profiles in response to overexpression and deletion of CspC and CspE. J. Bacteriol. *188*, 2521–2527.

Ponder, M.A., Gilmour, S.J., Bergholz, P.W., Mindock, C.A., Hollingsworth, R., Thomashow, M.F., and Tiedje, J.M. (2005). Characterization of potential stress responces in ancient Siberian permafrost psychroactive bacteria FEMS Microbiol. Ecol. *53*, 103–115.

Quinlan, C.L., Gerencser, A.A., Treberg, J.R., and Brand, M.D. (2011). The mechanism of superoxide production by the antimycin-ihnibited mitochondrial Q-cycle. J. Biol. Chem. *286*, 31361–31372.

Rocha, E.R., Selby, T., Coleman, J.P., and Smith, C.J. (1996). Oxidative stress response in an anaerobe, *Bacteroides fragilis*: a role for catalase in protection against hydrogen peroxide. J. Bacteriol. *178*, 6895–6903.

Rodrigues, D.F., da C Jesus E., Ayala-Del-Río H.L., Pellizari, V.H., Gilichinsky, D., Sepulveda-Torres, L., and Tiedje, J.M. (2009). Biogeography of two cold-adapted genera: *Psychrobacter* and *Exiguobacterium*. ISME J. *3*, 658–665.

Rodrigues, D.F., and Tiedji, J.M. (2007). Multi-locus real-time PCR for quantitation of bacteria in the environment reveals *Exiguobacterium* to be prevalent in permafrost. FEMS Microbiol. Ecol. *59*, 489–499.

Rowe, L.A., Degtyareva, N and Doetsch, P.W. (2008). DNA damage-induced reactive oxygen species (ROS) stress response in *Saccharomyces cerevisiae*. Free Radic. Biol. Med. *45*, 1167–1177.

Petersohn, A., Engelmann, S., Setlow, P., and Hecker, M. (1999). The kata gene of *Bacillus subtilis* is under dual control of sigmaB and sigmaF. Mol. Gen. Genet. *262*, 173–179.

Rodrigues, D.F., da C Jesus, E., Ayala-Del-Río, H.L., Pellizari, V.H., Gilichinsky, D., Sepulveda-Torres, L., and Tiedje, J.M. (2009). Biogeography of two cold-adapted genera: *Psychrobacter* and *Exiguobacterium*. ISME J. *3*, 658–665.

Saitou, N., and Nei, M. (1987). The neighbor-joining method: a new method for reconstructing phylogenetic trees. Mol. Biol. Evol. *4*, 406–425.

Schonbaum, G.R., and Chance, B. (1976). Catalase. In: The Enzyme. Boyer, P.D. ed., Academic Press, New York, USA, pp. 363–408.

Storz, G., and Zheng, M. (2000). Oxydative Stress. In: Bacterial Stress Response. Storz, G., and Hegge-Aronis, R eds. (American Society for Microbiology Press, Washington, DC, USA), pp. 47–59.

Switala, J., O'Neil, J.O., and Loewen, P.C. (1999). Catalase HPII from *Escherichia coli* exhibits enhanced resistance to denaturation. Biochemistry *38*, 3895–3901.

Takebe, F., Hara, I., Matsuyama, H., and Yumoto, I. (2007). Effect of H_2O_2 under low- and high-aeration-level conditions on growth and catalase activity in *Exiguobacterium oxidotolerans* T-2–2^T. J. Biosci. Bioeng. *104*, 464–469.

Thompson, J.D., Higgins, D.G., and Gibson, T.J. (1994). CLUSTAL W: improving the sensitivity of progressive multiple sequence alignment through sequence weighting, position-specific gap penalties and weight matrix choice. Nucleic Acid Res. 22: 4673–4680.

Treberg, J.R., Quinlan, C.L., and Brand, M.D. (2011). Evidence for two sites of superoxide production by mitochondrial NADH-ubiquinone oxidoreductase (Complex I). J. Biol. Chem. *286*, 27103–27110.

Visick, K.L., and Ruby, E.G. (1998). The periplasmic, group III catalase *Vibrio fischeri* is required for normal symbiotic competence and is induced both by oxidative stress and by approach to stationary phase. J. Bacteriol. *180*, 2087–2092.

Yumoto, I., Yamazaki, K., Kawasaki, K., Ichise, N., Morita, N., Hoshino, T., and Okuyama, H. (1998). Isolation of *Vibrio* sp. S-1 exhibiting extraordinarily high catalase activity. J. Ferment. Bioeng. *85*, 113–116.

Yumoto, I., Iwata, H., Sawabe, T., Ueno, K., Ichise, N., Matsuyama, H., Okuyama, H., and Kawasaki, K. (1999). Characterization of a facultatively psychrophilic bacterium, *Vibrio rumoiensis* sp. nov., that exhibits high catalase activity. Appl. Environ. Microbiol. *65*, 67–72.

Yumoto, I., Ichihashi, D., Iwata, H., Istokovics, A., Ichise, N., Matsuyama, H., Okuyama, H., and Kawasaki, K. (2000). Purification and characterization of a catalase from the facultative psychrophilic bacterium *Vibrio rumoiensis* S-1^T exhibiting high catalase activity. J. Bacteriol. *182*, 1903–1909.

Yumoto, I.; Hishinuma-Narisawa, M.; Hirota, K.; Shingyo, T.; Takebe, F.; Nodasaka, Y.; Matsuyama, H., and Hara, I. (2004). *Exiguobacterium oxidotolerans* sp. nov., a novel alkaliphile exhibiting high catalase activity. Int. J. Syst. Evol. Microbiol. *54*, 2013–2017.

Yumoto, I., Hirota, K., Kimoto, H., Nodasaka, Y., Matsuyama, H., and Yoshimune, K. (2010). *Psychrobacter piscatorii* sp. nov., a psychrotolerant bacterium exhibiting high catalase activity isolated from an oxidative environment. Int. J. Syst. Evol. Microbiol. *60*, 205–208.

Zámocký, M., and Koller, F. (1999). Understanding the structure and function of catalases: clues from molecular evolution and in vitro mutagenesis. Prog. Biophys. Mol. Biol. *72*, 19–66.

Zamocky, M., Furtmüller, P.G., and Obinger, C. (2008). Evolution of catalases from bacteria to humans. Antioxid Redox Signal *10*, 1527–1548.

10 Microorganisms in a Permafrost Ice Wedge and their Resuscitation-promoting Factors

Taiki Katayama and Michiko Tanaka

Abstract

Bacteria and fungi preserved for long periods at subzero temperatures in Alaskan and Siberian permafrost ice wedge were reactivated on agar by aerobic cultivation at 15°C. Culturable bacteria differed among ice samples, but several phylogenetic groups were closely related to those in other frozen environments. Incubation under controlled temperatures and the Arrhenius profiles of bacterial isolates from the Alaskan ice wedge indicated that they could grow at temperatures below 0°C without remarkable alterations in their cellular process. The novel ice wedge isolates, *Glaciibacter superstes* AHU1791^T and *Tomitella biformata* AHU1821^T, increased membrane fluidity at lower to subzero temperatures by modulating the fatty acid composition of the cytoplasmic membrane. Reactivating the non-culturable state of ice wedge isolates using resuscitation promoting factor (Rpf) and culturing melted ice wedge with Rpf provided indirect evidence that non-culturable bacteria exist within the ice wedge *in situ*. Bacteria in ice wedges can change their membrane fatty acid composition and/or structure to survive, but may then lose the ability to grow under laboratory conditions, as a final adaptation to long periods in a frozen natural environment.

Introduction

Permafrost ice wedge

Various types of ice have been formed in freezing and frozen ground in extremely cold areas of the world. The volume and frequency of wedge ice, pore ice, segregated ice and buried ice are the most significant among types of ground ice (Mackay, 1972; Harry, 1989). Ice wedges are linear, V-shaped bodies of nearly pure ice that exist within permafrost and are formed through the infiltration of surface water into thermal contraction cracks (Steven *et al.*, 2006). Ice wedge grows as a result of repeated cycles of frost cracking during winter followed by the infiltration of snow, meltwater, soil and other materials into the cracks. Material incorporated into the ice wedge quickly becomes frozen and the ice, as well as ice in soil and organic particles including microorganisms, can be preserved frozen for thousands of years until they are exposed and melted. The estimated age of such ice is 15,000 years BP (before present) (=15,000 years before 1950 AD; thus, the ice and incorporated materials has remained frozen for 15,000 years.

Although microorganisms have been identified in frozen environments, the preservation of ancient bacteria or fungi within ice wedge has not been extensively investigated.

Revival of microorganisms from ice wedge

The initial objective of our studies was to determine whether viable microorganisms could be isolated from ice wedge, and how long they had been preserved. We obtained several samples from the Fox permafrost tunnel in Alaska (N64°57.084, W147°37.250) (Johansen *et al.*, 1988), where numerous ice wedges in ice-rich syngenetic permafrost (yedoma) are exposed in the tunnel wall and preserved at a temperature of roughly −3°C by the Cold Regions Research and Engineering Laboratory (CRREL). Ice wedges in the tunnel exhibit numerous thin, vertical bands of sediments and ice veinlets characteristic of undisturbed ice wedges, as well as numerous small air bubbles, indicating that they have never thawed. In fact, the stable carbon isotopic ratio of methane derived from one ice wedge was −84.651‰, which differs from that of atmospheric methane. A radiocarbon date of 24,884 ± 139 years BP was obtained from the same sample, demonstrating that the ice wedge had remained frozen for about 25,000 years (Katayama *et al.*, 2007). Ice samples were also taken from an exposure on the left bank of the Aldan River at Mammoth Mountain in Eastern Siberia (N62°59, E133°58) under the project entitled, 'Extremophiles in lithosphere, their biodiversity, distribution and biomedical applications'. This ice was not radiocarbon dated because the age was estimated as being older than 10,000 years based on measurements of organic materials from the same horizon of ice, which had an age that ranged from 11,000 to about 40,000 years. The estimated annual temperature of the Siberian ice wedge is −4°C, which is similar to that of Alaskan ice (A. Brouchkov, unpublished). The sampled ice wedges were stored frozen during transportation, surface-sterilized, melted aseptically and spread on agar plates at our laboratory.

Adaptations of preserved microorganisms to sustain life in ice

We then investigated how viable bacteria have adapted and survived in ice at subzero temperatures for such long periods. We speculated that ice wedge isolates could adapt to the natural environment by entering a dormant, non-culturable state or a lower or quiescent metabolic state for over thousands of years. The activity of the microbial community in permafrost *in situ* has also been examined in several studies. Recent findings indicate that microorganisms are capable of both growth and metabolic activity at the subzero temperatures characteristic of permafrost (Steven *et al.*, 2006; Hinsa-Leasure *et al.*, 2010). The latest report described the biogeochemical characteristics of four ice wedges from the Fox permafrost tunnel and compared them with those measured in the Chena River, which flows within 12 km of the Fox permafrost tunnel (Douglas *et al.*, 2011). Ice wedges yield major constituent concentrations such as potassium, magnesium, calcium, sulfate, fluoride, chloride, and nitrate within an order of magnitude of Chena River values and contain slightly more dissolved organic carbon, and total dissolved nitrogen than the highest concentrations measured in the Chena River (Douglas *et al.*, 2011). Steven *et al.* (2006) described that liquid water exists as a very thin film surrounding soil and ice particles at subzero temperatures within permafrost. These films also allow for the mass transfer of nutrients and metabolic end products. However, pure ice is impermeable to potassium and chlorine ions and viable bacteria are rarely recovered from pure ice systems such as ice wedges (Gilichinsky *et al.*, 1995; Gilichinsky, 2002) or massive ground ice formations (Steven *et al.*, 2006).

Since the term 'non-culturable' was described by Xu *et al.* (1982), hundreds of studies have investigated viable, but non-culturable, states of bacteria. A lack of nutrients, low temperatures, high pressure, sharp changes in pH or salinity, and other environmental factors initiate a series of cellular events that includes changes in cellular morphology and the concentration and/or structure of major biopolymers, as well as cessation of the growth of organisms on solid or liquid laboratory media. Most Gram-negative heterotrophic, Gram-positive asporogenic and Gram-variable bacteria can enter a viable, but non-culturable state to adapt to the environment (Colwell and Grimes, 2000). Possible multistress conditions such as subzero temperature, low water activity, limited nutrient or gas availability for long periods in an ice wedge might also be appropriate for organisms to devise survival strategies such as entering into a non-culturable state.

This chapter describes the diversity of culturable microorganisms in ancient ice wedges and their characteristics. Resuscitation-promoting factors are also described and discussed based on the assumption that most ice wedge isolates have remained in a dormant, non-culturable state or a lower or quiescent metabolic state at subzero temperatures for thousands of years.

Microorganisms in Alaskan and Siberian permafrost ice wedge

Viable bacteria and fungi isolated by culture-based techniques

We examined the recovery of microbes by culturing melted ice in plates containing agar. A portion of ice wedge was surface sterilized by immersion in 70% ethanol and burned to remove contaminating surface ice and ethanol (Katayama *et al.*, 2007). After aerobic culture at 15°C, bacterial cells from melted Alaskan and Siberian ice formed colonies at densities up to 10^6 and 10^4 CFU per ml, respectively. We isolated 270 and 89 aerobic or facultative anaerobic heterotrophic bacterial strains from the respective samples. Based on a definition of operational taxonomic units (OTU) as strains sharing over 97% similarity in the 16S rRNA gene sequences, 41 and 30 OTUs were isolated from the Alaskan and Siberian samples, respectively. These strains were affiliated with the classes, *Alpha-*, *Beta-* and *Gammaproteobacteria, Bacilli* and *Actinobacteria* (Table 10.1). *Alpha-* and *Betaproteobacterial* OTUs were isolated only from the Siberian sample. Spore-forming bacteria, such as the genera *Bacillus, Promicromonospora* and *Nonomuraea,* were isolated from the Siberian ice, but not from the Alaskan ice. Most of the OTUs (85%) from the Alaskan ice wedge were non-spore-forming actinomycetes in the class *Actinobacteria* comprising the genera *Arthrobacter, Microbacterium, Leifsonia, Aeromicrobium, Brevibacterium, Cryobacterium, Salinibacterium* and others including the recently proposed genera, *Glaciibacter* and *Tomitella* (Katayama *et al.*, 2009, 2010). Many of the isolates were phylogenetically close to bacteria originating in other cold or frozen environments despite differences in culture media and conditions. Indeed, 24 (59% of total) and 8 (27%) OTUs from the Alaskan and Siberian samples, respectively (Table 10.1), shared over 99% similarity in 16S rRNA gene sequences with strains isolated from permafrost soils in north-western China, Siberia, northern Norway, the Canadian high Arctic (Bai *et al.*, 2006; Vishnivetskaya *et al.*, 2006; Hansen *et al.*, 2007; Wilhelm *et al.*, 2011), Antarctic sea ice, seawater and lakes (Bowman *et al.*, 1997; Christner *et al.*, 2001; Reddy *et al.*, 2002; Michaud *et al.*, 2004; Peeters *et al.*, 2011), an alpine ice cave in Austria (Margesin *et al.*, 2004) and a glacier ice from China (Xiang *et al.*, 2005). However, genetic coherence in 16S rRNA genes does not represent similarity in phenotypic characteristics. Only four OTUs affiliated with the genera *Arthrobacter, Citricoccus* and *Leifsonia* shared over 99% similarity between the Alaskan and Siberian ice samples. Bacterial community structures

Table 10.1 Phylogenetic distribution of bacteria isolated from Alaskan and Siberian ice wedges

Alaskan isolates			Siberian isolates		
Class	Genus	Number of OTUs	Class	Genus	Number of OTUs
Gammaproteobacteria			*Alphaproteobacteria*		
	*Lysobacter**	1		*Brevundimonas**	1
Bacilli				*Phyllobacterium*	1
	*Carnobacterium**	1		*Sphingomonas**	1
	*Planococcus**	1	*Betaproteobacteria*		
	*Planomicrobium**	2		*Massilia*	1
Actinobacteria			*Gammaproteobacteria*		
	Streptomyces	1		*Acinetobacter*	1
	Tessaracoccus	1		*Luteimonas*	1
	Nocardioides	1	*Bacilli*		
	*Aeromicrobium**	2		*Staphylococcus*	3
	*Rhodococcus**	1		*Bacillus**	3
	Tomitella	1		*Sprosarcina**	1
	*Dietzia**	1	*Actinobacteria*		
	*Janibacter**	1		*Streptomyces*	4
	*Brachybacterium**	1		*Nonomuraea*	1
	*Brevibacterium**	2		*Promicromonospora*	3
	*Micrococcus**	1		*Brevibacterium**	1
	Citricoccus	1		*Kocuria**	1
	*Arthrobacter**	8		*Micrococcus**	1
	Agrococcus	1		*Citricoccus*	1
	*Cryobacterium**	2		*Arthrobacter**	2
	Glaciibacter	1		*Microbacterium*	2
	Microbacterium	4		*Leifsonia**	1
	*Salinibacterium**	3			
	*Leifsonia**	3			

OTUs were classified based on the partial 16S rRNA gene sequence using RDP classifier (Wang *et al.*, 2007). The genera indicated by asterisks include OTUs that phylogenetically closely related to bacteria from other cold or frozen environments.

also differed, according to PCR-DGGE of the 16S rRNA gene, with each of four ice wedges collected from the Fox permafrost tunnel (T. Katayama, unpublished). We used complex medium with rich nutrition, diluted complex medium and modified minimal medium as previously described (Katayama *et al.*, 2007) to isolate diverse culturable microbes from Alaskan ice wedge. The numbers of CFU and OTUs normalized to the number of strains did not significantly differ (data not shown). However, among OTUs with <97% similarity with reported type strains, 22% and 19% were more frequently isolated from diluted and modified minimal medium, respectively, than the complex medium (10%). In fact, strains affiliated with *Tomitella biformata*, a recently proposed genus, and species in the suborder *Corynebacterineae* were isolated from the diluted and synthetic media.

Fungal colonies from both Alaskan and Siberian ice wedge also reproduced at a density of 10–10^3 CFU per ml. Based on analyses of large subunit D1/D2 region of nuclear rRNA gene and internal transcribed spacer (ITS) sequences as well as morphological observations, strains isolated from one sample of Alaskan ice wedge were closely related to *Geomyces* sp. and *Phaeococcomyces* sp., which are anamorphic states of ascomycotic fungi and yeast species, respectively. Basidiomycetous yeasts affiliated with *Rhodotorula* sp. and an anamorphic state of *Leucosporidium* sp. have also been isolated from other Alaskan ice wedges. The fungi recovered from the Siberian ice wedge were not identified. The genus *Geomyces*, which is widespread in permafrost (Ozerskaya *et al.*, 2009), was also the most frequently isolated from the five samples of Alaskan ice wedge. They were all phylogenetically related to the anamorphic state of genus *Pseudogymnoascus*; however, isolates from three ice samples were affiliated with species different from those in Siberian permafrost or Antarctica (Asano, 2010).

These results showed that the viable cell count on agar, the phylogeny of the isolates and DNA differed among ice samples, suggesting that the surrounding environment and accidental movements of materials that trapped organisms within ice affect microbial diversity in ice wedge.

Characteristics of growth at low temperature

To further characterize the ice wedge isolates, we examined cell growth at −5°C, 4°C, 15°C, 27°C and 37°C by measuring diameters of colonies. The tested bacterial and fungal isolates grew at 4°C and 20°C, but not at 37°C. Some of the bacterial isolates affiliated with *Arthrobacter, Microbacterium, Rhodococcus, Planococcus, Planomicrobium, Glaciibacter* and *Tomitella* and fungal isolates of *Geomyces* and *Leucosporidium* could also grow at −5°C, i.e. at the temperature in which they were found *in situ*. Consistent with the permafrost bacteria (Steven *et al.*, 2007, 2009), most were eurypsychrophiles that grew over a wide range of temperatures from below 0°C to above 20°C rather than stenopsychrophiles that are psychrophilic microorganisms with a restricted range of growth temperature (Feller and Gerday, 2003; Cavicchioli, 2006). Since categorization for eurypsychrophiles or stenopysychrophile does not indicate cold adaption (Cavicchioli, 2006), we evaluated the isolates using the modified Arrhenius equation (Mohr and Krawiec, 1980) that expresses the thermodependence of bacterial growth. The Arrhenius profiles (growth rate versus reciprocal temperature) of bacteria including psychrophiles, mesophiles and thermophiles are often characterized by a linear portion in suboptimal range and deviations below and above the range (Mohr and Krawiec, 1980; Reichardt and Morita, 1982; Berger *et al.*, 1996; Guillou and Guespin-Michel, 1996; Choma *et al.*, 2000; Bakermans and Nealson, 2004). The point at which the linear curve deviates from Arrhenius law represents the inability of cells to maximize growth rate due to high or low temperature stress (Guillou and Guespin-Michel, 1996). The slope for the growth of *Arthrobacter* sp. strain AHU1770 (a representative of the strain isolated most frequently from the Alaskan ice wedge) between −5°C and 15°C deviated between 20°C and 25°C in the Arrhenius plot (Fig. 10.1), indicating that cellular processes between −5°C and 15°C are only dependent on temperature and occurred consistently. *Arthrobacter* sp. strain AHU1770 can thus grow at temperatures below zero without significantly altering cellular processes as a response to low temperature stress. However, further characterization

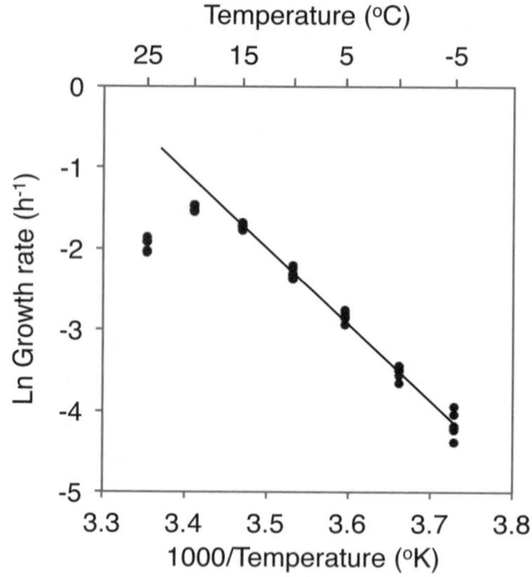

Figure 10.1 Growth of Alaskan ice wedge isolate, *Arthrobacter* sp. AHU1770 is temperature dependent. Natural logarithm of growth rates vs. temperature. Kelvin and Celsius temperatures are indicated at the bottom and top, respectively. The trend line fitting to the data point ranging from −5°C to 15°C is also shown.

such as proteomic analysis at various temperatures is required to demonstrate the theoretical interpretation of cold adaptation by *Arthrobacter* sp. strain AHU1770.

Cold adaptation by modulation of membrane fluidity

The growth and adaptation of microorganisms at low temperature depend on the ability of cells to restore the function of a stiffening cytoplasmic membrane by increasing its fluidity (Shivaji and Prakash, 2010). The ability of the ice wedge isolates *Glaciibacter superstes* strain AHU1791^T and *Tomitella biformata* strain AHU1821^T to modulate their membrane fatty acid composition in response to a shift to subzero temperatures was examined to assess membrane adaptation. Fatty acids found in cells of *G. superstes* were iso- and anteiso-branched, saturated and unsaturated with chains of 14–17 carbon atoms (Katayama *et al.*, 2009). When cultured at lower temperatures, the proportion of unsaturated fatty acids increased while that of saturated acids decreased. In addition, iso-$C_{15:0}$ significantly increased at the expense of iso-$C_{16:0}$ at −5°C, indicating that *G. superstes* modulates membrane fluidity at subzero temperatures by desaturating fatty acids and increasing the proportions of shorter chain fatty acids, both of which increase membrane fluidity. Such changes in fatty acid composition did not significantly correlate with Arrhenius profiles with respect to the growth of this strain (data not shown). The ratio of monounsaturated to saturated fatty acids in *T. biformata* cells increases in response to a decrease in temperature from 25°C to −5°C (Katayama *et al.*, 2010). More significantly, the proportion of *cis*- to *trans*-unsaturated acids increased depending on the extent of the decrease in temperature. Since an increased *cis*/*trans* ratio weakens the density and packing of the membrane (Heipieper *et al.*, 1992), *T. biformata* desaturated and *cis*–*trans*-isomerized fatty acids to compensate for membrane stiffening at low and subzero temperature.

Resuscitation promoting factor in ice wedge isolates

Resuscitation-promoting factor (Rpf) is a protein discovered in the actinomycetes *Micrococcus luteus* that reactivates and promotes the growth of viable, but non-culturable cells on solid or liquid culture medium. This factor also stimulates the growth of other high G + C Gram-positive bacteria including many species within the genera *Mycobacterium* (Mukamolova *et al.*, 1998; Shleeva *et al.*, 2004). Whole genomic sequencing indicated that *rpf*-like genes are widely distributed among Gram-positive bacteria with a high G + C content (Mukamolova *et al.*, 1998). The Rpf protein has a lysozyme-like active site with muralytic activity for the *M. luteus* cell wall (Mukamolova *et al.*, 2006), but whether or not Rpf protein is specific for bacterial adaptation to cold remains uncertain. Indeed, viable bacteria have been preserved for long period at subzero temperatures within ice wedge, where they have been exposed to stressors such as desiccation, starvation and local oxygenic or hypoxic conditions. The fatty acid composition of membranes changed so that novel ice wedge isolates could grow at controlled lower temperatures in our laboratory. We speculated that ice wedge isolates trapped in subzero ice for millennia cease to grow on solid or in liquid media and assume a viable, but non-culturable state as a final adaptive strategy to surviving in such extreme conditions. A non-culturable state is defined as a condition in which bacterial cells are alive, but unable to grow in standard solid or liquid laboratory media. Non-culturable cells are detected by comparing the number of living cells under epifluorescent microscopy with the number of colonies formed on plates containing media optimized for cell growth. The direct detection and discrimination of non-culturable cells within the natural environment *in situ* is quite difficult. We used reactivating factors in an effort to revive non-culturable cells or non-culturable state of isolates and provide experimental evidence of their existence in ice wedge. Several reports have described temperature, heat shock, organic nutrition and manipulating culture conditions *in vitro* as factors that can stimulate the cell division and growth of non-culturable Gram-negative bacteria (Nilsson *et al.*, 1991, Wai *et al.*, 1996, Rosak *et al.*, 1984).

Although the dominant taxon of 16S rRNA gene clones in Alaskan ice wedge was *Gammaproteobacteria* (93.1% of the total numbers of clones; Katayama *et al.*, 2007), most Alaskan ice wedge isolates were non-spore-forming *Actinobacteria*. In addition, some colonies on the initial isolation plate could not be subcultured in fresh media. We then considered adding Rpf, which reactivates the non-culturable *Actinobacteria*, as one of effective factors. Thus, we started screening and isolating the *rpf* gene from ice wedge isolates to prepare Rpf protein with which to revive hypothetical non-culturable *Actinobacteria* within ice wedge.

We searched for specific Rpf domain fragment sequences in *Actinobacteria* among ice wedge isolates to identify the *rpf* gene, express the protein and confirm that it can revive the non-culturable state of the source organism and other *Actinobacteria* that constitute the bacterial community within ice.

Degenerate primers specific for the Rpf domain fragment sequence in the *rpf* gene were constructed by aligning known (Rpf domain) amino acid sequences obtained from GenBank. By PCR, cloning and sequencing, gene fragments that encode a putative Rpf domain were identified in 17 of 23 Alaskan strains affiliated with the genera *Micrococcus*, *Citricoccus*, *Arthrobacter*, *Brachybacterium*, *Frigoribacterium*, *Cryobacterium*, *Agrococcus*, *Dietzia*, and *Tomitella* and in 6 of 19 Siberian strains affiliated with the genera *Streptomyces*, *Arthrobacter*, and *Kocuria*. The putative amino acid sequences were related to those of *M. luteus*, indicating that *rpf* gene homologues exist in genomic DNA of various genera of actinomycetes

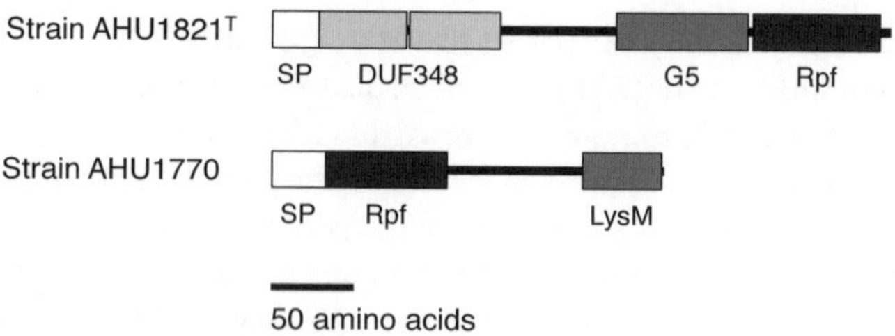

Figure 10.2 Putative domain structure of Rpf from *Tomitella biformata* AHU1821^T (372 aa) and *Arthrobacter* sp. AHU1770 (237 aa). SP; signal peptide, DUF348; domain of unknown function, G5; domain with putative cell-wall adhesive function, Rpf; Rpf domain that has lysozyme-like activity, LysM; peptidoglycan-binding motif.

isolates, especially of *Arthrobacter* from both Alaskan and Siberian ice wedge. The putative open reading frames of *rpf* genes from *Tomitella biformata* AHU1821^T and from *Arthrobacter* sp. AHU1770, which was the most frequent isolate, were determined by Southern hybridization using the PCR product of the Rpf domain as a probe and inverse PCR. The signal peptide and conserved domains within the putative *rpf* genes were then determined. The putative amino acid sequences of the ORF in *T. biformata* AHU1821^T and *Arthrobacter* sp. AHU1770 were 44.9% and 35.8% similar to RpfB in *Mycobacterium tuberculosis* H37Rv and to Rpf in *M. luteus*, respectively.

A possible active centre and the amino acids responsible for enzymatic activity within the Rpf domain (Cohen-Gonsaud *et al.*, 2004), were preserved both in strain AHU1821^T and in strain AHU1770. Fig. 10.2 shows the domain structures of the putative amino acid sequences of the ORFs in strain AHU1821^T and AHU1770, respectively. Both sequences contained signal peptides as well as the Rpf in *M. luteus*. The domain structures of strains AHU1821^T and AHU1770 differed. The sequences of strain AHU1821^T belonged to the RpfB subfamily defined by Ravagnani *et al.* (2005), which contains a DUF348 domain of unknown function and a G5 domain. The Rpf G5 domain might have a cell wall adhesive function as suggested by Ruggiero *et al.* (2009). The sequences of strain AHU1770 were affiliated with the LysM subfamily that has the peptidoglycan-binding motif, LysM (Bateman and Bycroft, 2000). These results indicated that muralytic enzymatic activities might be similar if these homologous genes are expressed.

A fragment of the *rpf* gene from *T. biformata* AHU1821^T was inserted into an expression vector and transformed into *Escherichia coli* and *Rhodococcus erythropolis*. Expressed protein purified by nickel affinity column chromatography was examined the activity in *T. biformata* cells. The lag phase in liquid culture was reduced and the CFUs significantly increased in low concentration of active cells when transferred to fresh medium containing the recombinant protein (rRpf) fraction, indicating that rRpf from *T. biformata* promotes the growth of active cells. Non-culturable *T. biformata* cells were induced by prolonged incubation in modified minimal medium under oxygen-limited conditions. After transfer to fresh liquid media with the rRpf fraction, these non-culturable cells started to grow and the rRpf fraction also promoted colony formation by non-culturable *T. biformata* on solid agar (Puspita *et al.*, 2013). The numbers of colonies other than *T. biformata* increased when melted ice wedge samples were incubated on agar with rRpf fraction (Puspita, I.D., Uehara, M., Katayama, T., Tanaka, M., Kitagawa, W., Nakatsu, C.H., Kamagata, Y., and Asano, K., unpublished). These results suggested that ice wedge isolates could enter a non-culturable or quiescent metabolic state and then resume cellular processes in the presence of Rpf protein. Furthermore, the ice wedge itself might have preserved some non-culturable actinomycetes at subzero temperatures for long periods.

Discussion and conclusive remarks

Living bacteria have been preserved in ice wedges for tens of thousands of years at subzero temperatures. Aerobic or facultative anaerobic heterotrophic bacteria isolated from Alaskan and Siberian ice wedges and grouped into over 30 operational taxonomic units (>97% similarity in 16S rRNA gene sequence) affiliated with five classes were culturable in the laboratory. Only a few groups of fungal colonies, namely *Geomyces* sp., *Phaeococcomyces* sp., *Rhodotorula* sp. and an anamorphic state of *Leucosporidium* sp., reproduced under laboratory conditions. Some bacterial ice wedge isolates adapted to grow under controlled lower to subzero temperatures by modulating cytoplasmic membrane fluidity. How yeasts adapt to frozen environments remains unclear, but some actinomycetes might lose their ability to grow on solid or liquid laboratory media and enter a viable but non-culturable state as a final series of cellular events within subzero ice. Rpf might be involved in resuming the cell division and growth of such non-culturable cells as one of reactivating factor.

Natural permafrost enrichment (NPS) consisting of thawing permafrost at 4°C and incubating the thawed material for up to 12 weeks before direct seeding on agar, increases the recovery of both the numbers and diversity of viable cells from most permafrost samples (Vishnivetskaya *et al.*, 2000). Such enrichment increased direct bacterial counts in some samples, suggesting increased cell recovery was due to bacterial reproduction within the samples, but in others, there was no increase in bacterial counts indicating that NPS facilitates the recovery of cells from a viable, but non-culturable state. Similarly, preliminary incubation in anaerobic and aerobic liquid media for up to 1 year before seeding on agar remarkably increased the recovery and diversity of organisms recovered from Greenland deep ice core samples (Miteva *et al.*, 2004). These findings also support the notion that many types of bacteria can survive in a non-culturable state in a frozen environment.

Future trends

Very little is presently known about how microorganisms remain alive and the non-dividing or lower metabolic states of bacteria in frozen environments such as permafrost, sea ice, glacier ice, alpine ice caves and ice wedge. Additionally, the mechanism of Rpf protein as a reactivating factor, which might work not only for frozen environmental bacteria, but also for soil and subsurface bacteria, has not been investigated in detail, although Rpf homologues are exclusively found within the phylum *Actinobacteria*. Recent reports indicate that Rpf works through several pathways (Mir *et al.*, 2011), such as mycobacterial peptidoglycan remodelling and the production of muropeptide signalling molecules. Bacterial species of microbiota in the natural environment respond to stimuli in various ways and some *Actinobacteria* might enter various types of non-culturable or quiescent metabolic states. Various types of factors might be necessary to resuscitate such non-culturable cells and promote the growth in the laboratory setting.

Thus, the exploitation of reactivating factors such as Rpf-like proteins and further functional research of Rpf will be needed to increase the colonization efficiency of environmental bacteria in a laboratory and to understand bacterial proliferation, dormancy and regrowth in natural environment.

Acknowledgements

These studies were supported by grant and donation from the Institute for Fermentation, Osaka, Japan. We are grateful to Drs T. Douglas, A. Brouchkov and M. Fukuda for sampling of ice wedge and insightful discussion.

References

Asano, K. (2010). Exhaustive isolation, identification, and preservation of microorganisms from a permafrost ice wedge. Inst. Ferment. Osaka Res. Commun. *24*, 7–16.

Bai, Y., Yang, D.Q., Wang, H., Xu, S., Wang, X.X., and An, L.Z. (2006). Phylogenetic diversity of culturable bacteria from alpine permafrost in the Tianshan Mountains, northwestern China. Res. Microbiol. *157*, 741–751.

Bakermans, C., and Nealson, K.H. (2004). Relationship of critical temperature to macromolecular synthesis and growth yield in *Psychrobacter cryopegella*. J. Bacteriol. *186*, 2340–2345.

Bateman, A., and Bycroft, M. (2000). The structure of a LysM domain from E-coli membrane-bound lytic murein transglycosylase D (MltD). J. Mol. Biol. *299*, 1113–1119.

Berger, F., Morellet, N., Menu, F., and Potier, P. (1996). Cold shock and cold acclimation proteins in the psychrotrophic bacterium *Arthrobacter globiformis* SI55. J. Bacteriol. *178*, 2999–3007.

Bowman, J.P., McCammon, S.A., Brown, M.V., Nichols, D.S., and McMeekin, T.A. (1997). Diversity and association of psychrophilic bacteria in Antarctic sea ice. Appl. Environ. Microb. *63*, 3068–3078.

Cavicchioli, R. (2006). Cold-adapted archaea. Nat. Rev. Microbiol. *4*, 331–343.

Choma, C., Guinebretiere, M.H., Carlin, F., Schmitt, P., Velge, P., Granum, P.E., and Nguyen-The, C. (2000). Prevalence, characterization and growth of *Bacillus cereus* in commercial cooked chilled foods containing vegetables. J. Appl. Microbiol. *88*, 617–625.

Christner, B.C., Mosley-Thompson, E., Thompson, L.G., and Reeve, J.N. (2001). Isolation of bacteria and 16S rDNAs from Lake Vostok accretion ice. Environ. Microbiol. *3*, 570–577.

Cohen-Gonsaud, M., Keep, N.H., Davies, A.P., Ward, J., Henderson, B., and Labesse, G. (2004). Resuscitation-promoting factors possess a lysozyme-like domain. Trends Biochem. Sci. *29*, 7–10.

Colwell, R.R., and Grimes, D.J. (2000). Semantics and strategies. In Nonculturable Microorganisms in the Environment, R.R. Colwell, and D.J. Grimes, eds. (Washington, DC: American Society for Microbiology), pp. 1–6.

Douglas, T.A., Fortier, D., Shur, Y.L., Kanevskiy, M.Z., Guo, L.D., Cai, Y.H., and Bray, M.T. (2011). Biogeochemical and geocryological characteristics of wedge and thermokarst-cave Ice in the CRREL permafrost tunnel, Alaska. Permafrost Periglac. *22*, 120–128.

Feller, G., and Gerday, C. (2003). Psychrophilic enzymes: hot topics in cold adaptation. Nat. Rev. Microbiol. *1*, 200–208.

Gilichinsky (2002). Pemafrost as a microbial habitat. In Encyclopedia of Environmental Microbiology, G. Briton, ed. (New York: John Wiley & Sons), pp. 932–956.

Gilichinsky, D.A., Wagener, S., and Vishnivetskaya, T.A. (1995). Permafrost microbiology. Permafrost Periglac. *6*, 281–291.

Guillou, C., and Guespin-Michel, J.F. (1996). Evidence for two domains of growth temperature for the psychrotrophic bacterium *Pseudomonas fluorescens* MF0. Appl. Environ. Microbiol. *62*, 3319–3324.

Hansen, A.A., Herbert, R.A., Mikkelsen, K., Jensen, L.L., Kristoffersen, T., Tiedje, J.M., Lomstein, B.A., and Finster, K.W. (2007). Viability, diversity and composition of the bacterial community in a high Arctic permafrost soil from Spitsbergen, Northern Norway. Environ. Microbiol. *9*, 2870–2884.

Harry, D.G. (1989). Ground ice and permafrost. In Advances in Periglacial Geomorphology, M.J. Clark, ed. (New York: Wiley), pp. 114–149.

Heipieper, H.J., Diefenbach, R., and Keweloh, H. (1992). Conversion of cis unsaturated fatty acids to trans, a possible mechanism for the protection of phenol-degrading *Pseudomonas putida* P8 from substrate toxicity. Appl. Environ. Microbiol. *58*, 1847–1852.

Hinsa-Leasure, S.M., Bhavaraju, L., Rodrigues, J.L., Bakermans, C., Gilichinsky, D.A., and Tiedje, J.M. (2010). Characterization of a bacterial community from a Northeast Siberian seacoast permafrost sample. FEMS Microbiol. Ecol. *74*, 103–113.

Johansen, N.I., Huang, S.L., and Aughenbaugh, N.B. (1988). Alaska's CRREL permafrost tunnel. Tunn Undergr Sp Tech *3*, 19–24.

Katayama, T., Kato, T., Tanaka, M., Douglas, T.A., Brouchkov, A., Abe, A., Sone, T., Fukuda, M., and Asano, K. (2010). *Tomitella biformata* gen. nov., sp. nov., a new member of the suborder *Corynebacterineae* isolated from a permafrost ice wedge. Int. J. Syst. Evol. Microbiol. *60*, 2803–2807.

Katayama, T., Kato, T., Tanaka, M., Douglas, T.A., Brouchkov, A., Fukuda, M., Tomita, F., and Asano, K. (2009). *Glaciibacter superstes* gen. nov., sp. nov., a novel member of the family *Microbacteriaceae* isolated from a permafrost ice wedge. Int. J. Syst. Evol. Microbiol. *59*, 482–486.

Katayama, T., Tanaka, M., Moriizumi, J., Nakamura, T., Brouchkov, A., Douglas, T.A., Fukuda, M., Tomita, F., and Asano, K. (2007). Phylogenetic analysis of bacteria preserved in a permafrost ice wedge for 25,000 years. Appl. Environ. Microbiol. *73*, 2360–2363.

Mackay, J.R. (1972). The world of underground ice. Ann. Associ. Am. Geograph. *62*, 1–22.

Margesin, R., Schumann, P., Sproer, C., and Gounot, A.M. (2004). *Arthrobacter psychrophenolicus* sp nov., isolated from an alpine ice cave. Int. J. Syst. Evol. Microbiol. *54*, 2067–2072.

Michaud, L., Di Cello, F., Brilli, M., Fani, R., Lo Giudice, A., and Bruni, V. (2004). Biodiversity of cultivable psychrotrophic marine bacteria isolated from Terra Nova Bay (Ross Sea, Antarctica). FEMS Microbiol. lett. *230*, 63–71.

Mir, M., Asong, J., Li, X., Cardot, J., Boons, G.J., and Husson, R.N. (2011). The extracytoplasmic domain of the *Mycobacterium tuberculosis* Ser/Thr kinase PknB binds specific muropeptides and is required for PknB localization. PLoS pathogens *7*, e1002182.

Miteva, V.I., Sheridan, P.P., and Brenchley, J.E. (2004). Phylogenetic and physiological diversity of microorganisms isolated from a deep greenland glacier ice core. Appl. Environ. Microbiol. *70*, 202–213.

Mohr, P.W., and Krawiec, S. (1980). Temperature characteristics and Arrhenius plots for nominal psychrophiles, mesophiles and thermophiles. J Gen Microbiol *121*, 311–317.

Mukamolova, G.V., Kaprelyants, A.S., Young, D.I., Young, M., and Kell, D.B. (1998). A bacterial cytokine. Proc. Natl. Acad. Sci. U.S.A. *95*, 8916–8921.

Mukamolova, G.V., Murzin, A.G., Salina, E.G., Demina, G.R., Kell, D.B., Kaprelyants, A.S., and Young, M. (2006). Muralytic activity of *Micrococcus luteus* Rpf and its relationship to physiological activity in promoting bacterial growth and resuscitation. Mol. Microbiol. *59*, 84–98.

Nilsson, L., Oliver, J.D., and Kjelleberg, S. (1991). Resuscitation of *Vibrio vulnificus* from the viable but non-culturable state. J. Bacteriol. *173*, 5054–5059.

Ozerskaya, S., Kochkina, G., Ivanushkina, N., and Gilichinsky, D.A. (2009). Fungi in permafrost. In Permafrost Soils, R. Margesin, ed. (Berlin, Germany: Springer-Verlag), pp. 85–95.

Puspita, I.D., Uehara, M., Katayama, T., Kikuchi, Y., Kitagawa, W., Kamagata, Y., Asano, K., Nakatsu, C.H., and Tanaka, M. (2013). Resuscitation promoting factor (Rpf) from *Tomitella biformata* AHU1821^{T} promotes growth and resuscitates non-dividing cells. Microb. Environ. *28*, 58–64.

Peeters, K., Hodgson, D.A., Convey, P., and Willems, A. (2011). Culturable diversity of heterotrophic bacteria in Forlidas Pond (Pensacola Mountains) and Lundstrom Lake (Shackleton Range), Antarctica. Microbial. Ecol. *62*, 399–413.

Ravagnani, A., Finan, C.L., and Young, M. (2005). A novel firmicute protein family related to the actinobacterial resuscitation-promoting factors by non-orthologous domain displacement. BMC Genomics *6*, 39.

Reddy, G.S., Prakash, J.S., Matsumoto, G.I., Stackebrandt, E., and Shivaji, S. (2002). *Arthrobacter roseus* sp. nov., a psychrophilic bacterium isolated from an antarctic cyanobacterial mat sample. Int. J. Syst. Evol. Microbiol. *52*, 1017–1021.

Reichardt, W., and Morita, R.Y. (1982). Influence of temperature adaptation on glucose metabolism in a psychrotrophic strain of *Cytophaga johnsonae*. Appl. Environ. Microbiol. *44*, 1282–1288.

Rosak, D.B., Grimes, D.J., and Colwell, R.R. (1984). Viable but nonrecoverble stage of *Salmonella enteritidis* in aquatic system. Can. J. Microbiol. *30*, 334–338.

Ruggiero, A., Tizzano, B., Pedone, E., Pedone, C., Wilmanns, M., and Berisio, R. (2009). Crystal structure of the resuscitation-promoting factor (Delta DUF) RpfB from *M. tuberculosis*. J. Mol. Biol. *385*, 153–162.

Shivaji, S., and Prakash, J.S. (2010). How do bacteria sense and respond to low temperature? Arch. Microbiol. *192*, 85–95.

Shleeva, M., Mukamolova, G.V., Young, M., Williams, H.D., and Kaprelyants, A.S. (2004). Formation of 'non-culturable' cells of *Mycobacterium smegmatis* in stationary phase in response to growth under suboptimal conditions and their Rpf-mediated resuscitation. Microbiology *150*, 1687–1697.

Steven, B., Briggs, G., McKay, C.P., Pollard, W.H., Greer, C.W., and Whyte, L.G. (2007). Characterization of the microbial diversity in a permafrost sample from the Canadian high Arctic using culture-dependent and culture-independent methods. FEMS Microbiol. Ecol. *59*, 513–523.

Steven, B., Leveille, R., Pollard, W.H., and Whyte, L.G. (2006). Microbial ecology and biodiversity in permafrost. Extremophiles *10*, 259–267.

Steven, B., Niederberger, T.D., and Whyte, L.G. (2009). Bacterial and archaeal diversity in permafrost. In Permafrost Soils, R. Margesin, ed. (Berlin, Germany: Springer-Verlag), pp. 59–72.

Vishnivetskaya, T., Kathariou, S., McGrath, J., Gilichinsky, D., and Tiedje, J.M. (2000). Low-temperature recovery strategies for the isolation of bacteria from ancient permafrost sediments. Extremophiles *4*, 165–173.

Vishnivetskaya, T.A., Petrova, M.A., Urbance, J., Ponder, M., Moyer, C.L., Gilichinsky, D.A., and Tiedje, J.M. (2006). Bacterial community in ancient Siberian permafrost as characterized by culture and culture-independent methods. Astrobiology *6*, 400–414.

Wai, S.N., Morita, T., Kondo, K., Misumi, H., and Amato, K. (1996). Resuscitation of *Vibrio cholerae* O1 strain TSI-4 from a viable but non-culturable state by heat shock. FEMS Microbiol. Lett. *136*, 187–191.

Wang, Q., Garrity, G.M., Tiedje, J.M., and Cole, J.R. (2007). Naive Bayesian classifier for rapid assignment of rRNA sequences into the new bacterial taxonomy. Appl. Environ. Microbiol. *73*, 5261–5267.

Wilhelm, R.C., Niederberger, T.D., Greer, C., and Whyte, L.G. (2011). Microbial diversity of active layer and permafrost in an acidic wetland from the Canadian high Arctic. Can. J. Microbiol. *57*, 303–315.

Xiang, S., Yao, T., An, L., Xu, B., and Wang, J. (2005). 16S rRNA sequences and differences in bacteria isolated from the Muztag Ata glacier at increasing depths. Appl. Environ. Microbiol. *71*, 4619–4627.

Xu, H.S., Roberts, N., Singleton, F.L., Attwell, R.W., Grimes, D.J., and Colwell, R.R. (1982). Survival and viability of nonculturable *Escherichia coli* and *Vibrio cholerae* in the estuarine and marine environment. Microbal. Ecol. *8*, 313–323.

Lipids in Cold-adapted Microorganisms

11

Ahmad Iskandar Bin Haji Mohd Taha, Rifat Zubair Ahmed, Taro Motoigi, Kentaro Watanabe, Norio Kurosawa and Hidetoshi Okuyama

Abstract

Ever since *Escherichia coli,* which is a mesophilic bacterium, was found to adjust its membrane fluidity in a liquid crystalline state by modulating fatty acid composition and the designation of this process as homeoviscous or homeophasic adaptation, numerous analogous phenomena have been reported in cold-adapted bacteria [psychrophilic or psychrotrophic (psychrotolerant) bacteria]. Unsaturation, which includes the biosynthesis of monounsaturated or long-chain polyunsaturated fatty acids, and branched-chain formation are the most important types of fatty acid modulation in psychrophilic and psychrotrophic (psychrotolerant) bacteria. The distribution of these fatty acids is not restricted to cold-adapted microorganisms: rather, it appears to depend on bacterial diversity (Gram- positive or negative) and/or habitat (terrestrial or marine environment) than on temperature. Eicosapentaenoic acid, which has been detected only in marine Gram-negative bacteria, had been regarded to confer significant membrane fluidity in bacteria, but it is now considered that it may also have a function in antioxidation or membrane modulation by constraining membrane fluidity. The mode of fatty acid modulation is unlikely to differ between psychrophilic and psychrotrophic bacteria, which had narrower and wider growth temperature ranges, respectively. The sole difference seems to be higher capacity to modulate fatty acid composition in psychrotrophic bacteria than in psychrophilic bacteria.

Introduction

Lipids in cold-adapted microorganisms, particularly psychrophilic and psychrotrophic (psychrotolerant) bacteria are characterized primarily by an abundance of fatty acids with a low melting temperature. Polar lipids are common structural components of biomembranes in living cells. The biomembrane is an amphiphilic lipid matrix, comprising a bilayer that exists in stable association with various hydrophobic components including proteins. Fatty acids are a major hydrophobic constituent of various membrane lipids, which commonly include polar (complex) lipids, mainly glycerophospholipids and glyceroglycolipids, as well as sphingolipids in some organisms. The fatty acid composition of these lipids can greatly affect the membrane properties by influencing (1) the physical state (fluidity) and (2) biological functions, such as permeability to solutes and nutrient uptake, which consequently affect motility, osmotic pressure and energy production through the transmembrane electrochemical gradient.

Microorganisms must maintain physiological functions of their biomembranes under all environmental conditions. The proper membrane functions depend on the liquid crystalline state of the lipid structure. When a microorganism encounters a change in ambient temperature, immediate and proper adjustment of the physical state of its membrane is indispensable for its survival, particularly for poikilothermic microorganisms, which do not have the capacity of thermal control. This type of response is called homeoviscous or homeophasic adaptation (Sinensky, 1974). Homeoviscous adaptation enables the microorganisms to cope with both low and high temperatures by maintaining the fluidity of the cell membrane. This process is important, because it is directly related to membrane structure and functions, accordingly, to the survival of bacteria including cold-adapted microorganisms (psychrophiles and psychrotrophs).

The categorization of cold-adapted microorganisms as psychrophiles and psychrotrophs has been challenged recently because of its ambiguity (D'Amico *et al.*, 2006; Zhang *et al.*, 2011a). Psychrophiles are defined here as cold-loving bacteria or archaea with an optimum temperature for growth of about 15°C or lower, maximal temperature for growth of about 20°C and a minimal temperature for growth at 0°C or lower. Psychrotrophs are defined as cold-loving bacteria or archaea whose optimal and maximal temperatures for growth are above 15°C and 20°C, respectively (Moyer and Morita, 2007). These definitions are based on the classic definition by Morita (1975) and are used in this chapter for convenience as there has been widespread use in scientific papers. In Moyer and Morita (2007), the maximum growth temperature was set at 20°C because the laboratory temperature in the USA is around 21–22°C, which is not considered cold.

Microorganisms achieve homeoviscous adaptation to low temperature through one or a combination of five main lipid structural or compositional strategies involving (1) increase in unsaturation of the fatty acid acyl chain, (2) increase in methyl branching, (3) increase in the anteiso to iso-branching ratio, (4) decrease in chain length, and (5) decrease in the sterol to phospholipid ratio. Bacteria are the most diverse in using all five strategies, whereas eukaryotes are limited to fatty acid unsaturation, change in chain length and changes in sterol to phospholipid ratio (Russell, 2008). In addition to these parameters, changes in the ratio of polar to nonpolar carotenoid and compositional changes in membrane lipid classes (polar group composition) can affect membrane fluidity in psychrophilic and psychrotrophic bacteria (Chintalapati *et al.*, 2004). Among these parameters, the change in the fatty acid composition of the membrane polar lipid is the most significant and most widely distributed in poikilothermic organisms including psychrophilic and psychrotrophic bacteria. The various modes of fatty acid modulation in bacteria in response to decreased ambient temperature (summarized in Table 11.1) can be classified into two processes: one is achieved by modification of fatty acids (unsaturation, *cis*-to-*trans* isomerization, cyclization and hydroxylation), and the other involves *de novo* synthesis of fatty acids. The immediate response is achieved by fatty acid modifications. The process involving the *de novo* synthesis of fatty acids is responsible for a microorganism's ability to adapt to low temperature over a long term or for the ability to inhabit permanently cold environments such as deep-sea, high-mountains, glacier or polar regions.

Bacteria synthesize unsaturated fatty acids by both anaerobic and aerobic mechanisms. In the anaerobic mechanism, a double bond is introduced through the process of *de novo* synthesis of fatty acids catalysed by fatty acid synthetase; only monounsaturated fatty acids are produced anaerobically. The polyketide synthetase-type enzymes (Pfa enzymes; Okuyama *et al.*, 2007) are involved in the *de novo* synthesis of long-chain polyunsaturated fatty

Table 11.1 Modes of fatty acid modulation in bacteria in response to decreased growth temperature

Reaction/property/conditions		Fatty acid products and compositional changes[a]	Mode of fatty acid modulation	Distribution	Reference
Unsaturation	Anaerobic	16:1(9c), 18:1(11c)	*De novo* synthesis	Most anaerobic or facultatively anaerobic bacteria	Marr and Ingraham (1962)
		20:4, 20:5, 22:6	*De novo* synthesis	Exclusively Gram-negative bacteria	DeLong and Yayanos (1986)
	Aerobic	16:1(5c), 16:1(9c)	Modification	Aerobic or facultatively anaerobic bacteria	Aguilar and de Mendoza (2006), Mansilla and de Mendoza (2005)
Chain length	Elongating	16:0/18:1(11c)	*De novo* synthesis	Most bacteria	Marr and Ingraham (1962)
	Shortening	18:1/16:1	*De novo* synthesis	Mainly Gram-negative bacteria	Russell (1971)
		i17:0/i15:0	*De novo* synthesis	Mainly Gram-positive bacteria	Kaneda (1991)
Branching	Iso/anteiso-branching	i17:0/a17:0; i15:0/a15:0	*De novo* synthesis	Most bacteria	Kaneda (1991)
Isomerization	*cis*-to-*trans* isomerization	16:1 (9c)/16:1 (9t)[b], 18:1(11c)/18:1 (11t)[b]	Modification	Mainly Gram-negative bacteria	Okuyama *et al*. (1990)
Cyclization	Cyclopropanation	18:1(11c)/18:0(cyclo)[c,d]	Modification	Most bacteria	Grogan and Cronan (1997)
Hydroxylation	Hydroxylation	2-OH 14:0, 3-OH i15:0[e]	Modification[f]	Most bacteria	Matsunaga *et al*. (1998), Okuyama *et al*. (1991)

[a]Representative products and changes are shown. Fatty acids, the number before the colon indicate the length of the carbon chain and the number after the colon indicate the number of double bonds in the chain; a and i designate anteiso and iso branching, respectively; the distance of the double bond from the carboxylic end of the fatty acid and the *cis* and *trans* configuration in the parentheses and c and t, respectively. X/Y indicates an increase in Y in accompanying with a decrease in X. 2-OH 14:0 and 3-OH i15:0 are 2-hydroxyl 14:0 and 3-hydroxyl i15:0, respectively.

[b]Increases in *trans* fatty acids are observed at higher temperatures.

[c,d]*cis*-11,12-Methylene octadecanoic acid, a 19-carbon cyclopropane analogue of 18:1(11c). Increases in 18:0(cyclo) are observed at higher temperatures. *cis*-11,12-Methylene octadecanoic acid, a 19-carbon cyclopropane analogue of 18:1(11c).

[e]Increases in 2-OH fatty acids are observed at higher temperatures.

[f]Direct hydroxylation and partial β-oxidation of fatty acid are involved.

acids such as eicosapentaenoic acid (EPA) and docosahexaenoic acid (DHA). This process is also an anaerobic process for unsaturated fatty acid production. In the aerobic mechanism, monounsaturated fatty acids can be synthesized by molecular oxygen-dependent fatty acid unsaturation of the existing saturated fatty acids as the substrate in bacteria, except for cyanobacteria, in which mono- and polyunsaturated fatty acids (PUFAs) are produced by various types of fatty acid desaturases (Wada and Murata, 2010). Branched fatty acids are synthesized *de novo* by fatty acid synthetase using branched amino acids (Kaneda, 1991; Zhu *et al.*, 2005). The *cis*-to-*trans* isomerization of monounsaturated fatty acids without a shift of the position of the double bond is observed only in a few species of psychrophilic and psychrotrophic bacteria. This process is used only to decrease the membrane fluidity by the *cis*-to-*trans* isomerization of monounsaturated fatty acids such as *cis*-16:1 and *cis*-18:1 after the temperature has been upshifted. Hydroxylation and cyclization of fatty acids are regarded as modification processes of fatty acids.

Psychrophilic bacteria can adapt to low-temperature environments, such as terrestrial or marine polar regions, deep seas, and high mountains, which can be under permanently low temperature, whereas psychrotrophic bacteria are tolerant to low temperature and have a relatively wider range of temperature for growth. It has not been clarified whether psychrophilic and psychrotrophic bacteria possess lipid modulation modes specific to these bacterial groups. In this chapter, a summary of the characteristics and distribution of fatty acids and polar lipid classes, habitat, and diversity (Gram staining, characteristics and habitat) is presented focusing mainly on recently isolated psychrophilic and psychrotrophic bacteria. The functions of lipids and fatty acids in modulation of the membrane fluidity, with a particular emphasis on the involvement of branched fatty acids and long-chain polyunsaturated fatty acids (LC-PUFAs) are discussed, followed by the potential biotechnological applications of lipids (LC-PUFAs and polyunsaturated hydrocarbons) derived from psychrophilic and psychrotrophic bacteria.

Polar lipid classes

Numerous taxonomic reports of newly isolated psychrophilic or psychrotrophic bacteria have been published recently, which regard the cellular fatty acid composition as a crucial chemotaxonomic importance. These taxonomic studies have included relatively little information on the lipid components, in contrast to the fatty acid information. However, recent reports, such as those published in Margesin *et al.* (2004), Reddy *et al.* (2010), Margesin *et al.* (2011a), and Zhang *et al.* (2011d), also include information on the types of polar lipids, such as phospholipids and glycolipids, and their composition in newly described bacterial isolates. When characterizing psychrophilic, psychrotrophic, and other types of bacteria, it is therefore preferable to include such information.

In general, bacterial cells have a simpler lipid composition than eukaryotic cells. The most common polar lipids comprising the bacterial cell membrane (plasma membrane) are the glycerophospholipids: phosphatidylethanolamine (PE), phosphatidylglycerol (PG), and diphosphatidylglycerol (DPG; cardiolipin), as in the case of *Escherichia coli*. These three types of glycerophospholipids are also common in most psychrophilic and psychrotrophic bacteria (Okuyama *et al.*, 1986; Margesin *et al.*, 2004).

Table 11.2 summarizes the tentative polar lipid compositions, which are derived from recently isolated psychrophilic and psychrotrophic bacteria. Although data from psychrophiles are available for only one species, it can be seen that PE, PG, and DPG are commonly detected in psychrotrophs but probably not in psychrophiles. However, PE, PG, and DPG have been detected in the psychrophilic marine *Vibrio* sp. strain ABE-1 (Okuyama *et al.*, 1986; presently *Colwellia maris* strain ABE-1, Yumoto *et al.*, 1998) and *Vibrio marinus* strain MP-1 (H. Okuyama, unpublished result; presently *Moritella marina* strain MP-1). In addition to these phospholipids, phosphatidylinositol (PI) has been detected in some terrestrial psychrotrophs, such as *Arthrobacter psychrophenolicus* strain AG31 (Margesin *et al.*, 2004) and *Sphingomonas alpina* (Margesin *et al.*, 2011b). No PE has been detected in psychrotrophic *Arthrobacter cryoconiti* (Margesin *et al.*, 2011a), or *Devosia psychrophila* (Zhang *et al.*, 2011d) or in psychrophilic *Cryobacterium roopkundense* (Reddy *et al.*, 2010), or *Arthrobacter psychrophenolicus* (Margesin *et al.*, 2004).

Although not studied thoroughly, glycolipids have been detected in terrestrial psychrophiles (Zhang *et al.*, 2011a; Margesin *et al.*, 2004, 2011b) or psychrotrophs, such as *Cryobacterium roopkundense* (Reddy *et al.*, 2010) and *Deinococcus alpinitundrae* (Callegan *et al.*, 2008). An unidentified sphingoglycolipid was detected in psychrotrophs *Sphingomonas alpina* (Margesin *et al.*, 2011b) and *Sphingomonas molluscorum* (Romanenko *et al.*, 2007). Interestingly, psychrotrophic *Deinococcus* species contain glycophospholipid similar to 2′-*O*-(1,2-diacyl-*sn*-glycero-3-phospho)–3′-*O*-(α-galactosyl)-*N*-D-glyceroyl alkylamine as a primary polar lipid (Callegan *et al.*, 2008). Since the size and charge of the head group of polar lipids affect the membrane fluidity, the presence of such glycophospholipid in psychrophiles and psychrotrophs may be involved in their cold adaptation. This type of lipid has been detected in gamma and UV radiation-resistant Gram-positive terrestrial mesophilic bacteria (Asker *et al.*, 2011) but not in marine bacteria, suggesting that it may also be involved in the resistance to radiation.

For the moment being, no general conclusion can be drawn whether membrane polar lipid composition is involved in the cold adaptation in psychrophilic and psychrotrophic bacteria. Thin-layer chromatography has been used to separate types of glycolipids in some psychrophilic bacteria, such as *Devosia psychrophila* strain Cr7-05 and *Devosia glacialis* strain Cr4-44 (Zhang *et al.*, 2011d), and *Sphingomonas alpina* S8-3, but these glycolipids have not been characterized. The phase transition temperatures of such glycophospholipids and glycolipids need to be determined.

Fatty acids

Fatty acid compositions of various types of bacteria have been summarized by Padley *et al.* (1986), O'Leary and Wilkinson (1988) and Wilkinson (1988). Bacterial lipids generally comprise C_{12} to C_{20} straight-chain saturated or monounsaturated fatty acids with an even number of carbons, although occasionally those with an odd number of carbon atoms are also detected. In this chapter, these fatty acids are designated 'normal' fatty acids. The major saturated fatty acids are tetradecanoic (14:0), hexadecanoic (16:0), and octadecanoic (18:0) acids, while the major unsaturated fatty acids are hexadecenoic (16:1) and/or octadecenoic (18:1) acids. Moyer and Morita (2007) emphasized that psychrophiles are endowed with a higher proportion of 16:1 and 18:1 than are mesophiles. Iso-branched fatty acids, such as iso-pentadecanoic acid (i15:0) and iso-heptadecanoic acid (i17:0) and anteiso-branched fatty acids, such as anteiso-pentadecanoic acid (a15:0) and anteiso-heptadecanoic acid (a17:0) are detected in psychrophiles, psychrotrophs, mesophiles and thermophiles. Branched

Table 11.2 Distribution of various polar lipid classes in psychrophilic and psychrotrophic bacteria

Organism	Temperature (°C)[a]	Site[b]	Gram[c]	Polar lipid classes[d]															Reference
				PG	DPG	PC	PI	PE	PME	PDE	*PL*	*AL*	*APL*	*GPL*	*GL*	*AGPL*	SGL	*L*	
Psychrophilic sp.																			
Cryobacterium roopkundense	0–18	S	+	l	l										l				Reddy *et al*. (2010)
Psychrotrophic sp.																			
Deinococcus alpinitundrae	5–25	S	+										l	l	l	m			Callegan *et al*. (2008)
Salinibacterium xinjiangense	4–23	G	+	l	l														Zhang *et al*. (2008)
Leifsonia pindariensis	4–35	G	+	l	l														Reddy *et al*. (2008)
Arthrobacter cryoconiti	1–25	G	+	m	m		l								l			l	Margesin *et al*. (2011a)
Arthrobacter psychrophenolicus	1–25	Si	+	l	l		l								l				Margesin *et al*. (2004)
Deinococcus radiomollis	5–25	S	+								m		l	l	l				Callegan *et al*. (2008)
Deinococcus caudionis	5–25	S	+										l	l	m				Callegan *et al*. (2008)
Deinococcus altitudinis	5–25	S	+										l	l	l	m			Callegan *et al*. (2008)
Leifsonia psychrotolerans	–6–28	S	+	l	l														Ganzert *et al*. (2011b)
Leifsonia rubra	0–22	L	+	l	l														Reddy *et al*. (2003a)
Leifsonia aurea	0–30	L	+	l	l														Reddy *et al*. (2003a)

Arthrobacter flavus	3–30	Lp	+	l	l		l										Reddy *et al*. (2000)
Paenibacillus algorifonticola	10–37	Ls	+		l		l						l				Tang *et al*. (2011)
Hymenobacter psychrophilus	1–25	S	–				l				m	m		m		m	Zhang *et al*. (2011a)
Pseudomonas bauzanensis	5–30	S	–	l	l		l					m					Zhang *et al*. (2011c)
Candidimonas bauzanensis	1–37	S	–	l	l		l	l			m	m				m	Zhang *et al*. (2011b)
Sphingomonas alpina	1–30	S	–	l	l		l	l		l		m		l	l	l	Margesin *et al*. (2011b)
Sphingomonas molluscorum	3–37	Mb	–	l	l	l	l		m		m				l	m	Romanenko *et al*. (2007)
Pseudochrobactrum glaciei	5–40	Mi	–	l	l	l	l	l		l	l						Romanenko *et al*. (2008)
Roseovarius indicus	10–39	D	–	l	l	l	l									m	Lai *et al*. (2011).

[a]Growth temperature range.
[b]The site where bacterial sample was collected: G, glacier; S, soil; Si, icecave; L, lake; Lp, pond; Ls, cold spring; Lf, low-salinity fjord; Mb, bay (bivalve); M, sea ice; D, deep sea.
[c]Gram staining.
[d]Polar lipid classes: PG, phosphatidylglycerol; DPG, diphosphatidylglycerol; PC, phosphatidylcholine; PI, phosphatidylinositol; PE, phosphatidylethanolamine; PME, phosphatidylmonomethylethanolamine; PDE, phosphatidyldimethylethanolamine; PL, phospholipid; AL, aminolipid; APL, aminophospholipid; GPL, glycophospholipid; GL, glycolipid; AGPL, aminoglycophospholipid; SGL, sphingolipid; L, polar lipid; l, major lipid; m, minor lipid. Lipid classes in italics are unidentified or unknown lipid.
*Minor and major based on author's definition.

fatty acids are probably more common in Gram-positive bacteria than in Gram-negative bacteria (Kaneda, 1991). Although C_{20} and C_{22} LC-PUFAs are uncommon particularly in bacteria, some marine Gram-negative psychrophilic bacteria contain these fatty acids such as EPA (20:5) or DHA (22:6) (see below). The bacterial fatty acid composition is affected by growth conditions such as the media, growth phase, temperature, and concentration of dissolved oxygen. These variations may lead to misinterpretation of reports from different laboratories (Russell, 2008). Chemostat cultivation has been recommended to eliminate the influence of various factors on the fatty acid composition and thus to allow comparisons between different laboratories (Wilkinson, 1988).

The fatty acid compositions of some recently discovered psychrotrophic and psychrophilic bacteria isolated from different environments are presented in Table 11.3. Although it was not confirmed for all bacterial species in this table, growth ability at 0°C is expected. Normal saturated and monounsaturated fatty acids, such as 16:0, 16:1, and 18:1, are detected most commonly in psychrotrophic and psychrophilic bacteria. When psychrophilic and psychrotrophic bacteria with normal fatty acids are grown at low temperature, there is a trend for 16:1 to be more abundant in psychrophiles and psychrotrophs than in mesophiles and vice versa for 18:1 (Moyer and Morita, 2007).

Branched fatty acids

Iso-branched (such as i15:0 and i17:0) and anteiso-branched (such as a15:0 and a17:0) have been detected in both psychrotrophic and psychrophilic bacteria. Anteiso-branched fatty acids are more abundant in terrestrial Gram-positive bacteria than in terrestrial Gram-negative bacteria (from glacier, soil, and lake environments; Table 11.3). Although branched fatty acids were scarcely detected in the marine Gram-negative bacteria (from sea, coastal sediment, sea ice, and deep-sea environments) shown in Table 11.3, psychrotrophic *Gelidibacter algens* from a sea ice core contains i15:1 and a15:0 (Bowman *et al.*, 1997). Gram-positive psychrophilic and psychrotrophic bacterial species isolated from marine environments are not included in this table. Some marine Gram-positive *Bacillus marinus* strains, whose growth temperature range is from 1°C to 4°C, are considered psychrophilic (Rüger *et al.*, 1983), but information on their fatty acid composition is not available.

The involvement of fatty acids in bacterial cold adaptation is summarized by Rose (1988) and Russell (1988, 2008). Biosynthesis of unsaturated fatty acids is necessary to increase the membrane fluidity during transient and permanent low-temperature adaptation in most psychrophilic and psychrotrophic bacteria. In bacteria, monounsaturated fatty acids can be produced via both anaerobic and aerobic pathways. Investigation on these two mechanisms has been conducted only in mesophilic bacteria in *E. coli* (Cronan, 2003) for the anaerobic pathway and in *Bacillus subtilis* (Aguilar and de Mendoza, 2006) and *Pseudomonas aeruginosa* for the aerobic mechanism (Zhu *et al.*, 2006). It has been generally accepted that anaerobic and facultatively anaerobic bacteria possess only an anaerobic pathway and that some aerobic bacteria possess only an aerobic pathway. However, there is physiological evidence that both pathways operate under aerobic conditions in the same psychrophilic or psychrotrophic bacterial species as in the marine psychrophile, *Vibrio* sp. strain ABE-1 (Morita *et al.*, 1992), and the terrestrial psychrotroph *Pseudomonas* sp. strain E-3 (Wada *et al.*, 1989; presently *Pseudomonas psychrophila* strain E-3, Yumoto *et al.*, 2001). In the mesophilic bacterium *Pseudomonas aeruginosa, fabA* (accession no. AAG04999 for strain PA01) encoding 3-hydroxydecanoyl-acyl carrier protein (ACP) dehydratase/isomerase and *desA*

Table 11.3 Distribution of fatty acids in psychrophilic and psychrotrophic bacteria

Organism	Temperature (°C)[a]	Site[b]	Gram[c]	Types of fatty acids[d] Straight	Iso	Anteiso	OH	LCPUFA	Others	Major and minor fatty acid[e]	References
Psychrophilic sp.											
Alpinimonas psychrophila	1–20	G	+		l	l				**a15:0**, **i16:0**	Schumann *et al*. (2012)
Cryobacterium roopkundense	0–18	S	+			l				**a15:0**, a15:1	Reddy *et al*. (2010)
Alkalilactibacillus ikkensis	0–10	Lf	+	l	m	l				14:0, i14:0, **a15:0**, **16:0**, 16:1(11c)	Schmidt *et al*. (2012)
Devosia glacialis	1–20	G	–	l						16:0, **16:1(7c)**, **18:1(7c)**	Zhang *et al*. (2011d)
Glaciimonas immobilis	1–20	G	–	l						16:0, **16:1(7c)**, 18:1(7c)	Zhang *et al*. (2011e)
Psychroserpens burtonensis	0-≤20	L	–	m	m	m				15:0, i15:0, i15:1(10c), a15:0	Bowman *et al*. (1997)
Thiomicrospira arctica	–2–20.8	Ms	–	l						14:1, 16:0, **16:1**, **18:1(7)**	Knittel *et al*. (2005)
Thiomicrospira psychrophila	–2–20.8	Ms	–	l						**16:1**, **18:0**	Knittel *et al*. (2005)
Desulfuromonas svalbardensis	–2–20	Mc	–	l						16:0, 16:1 (7c)	Vandieken *et al*. (2006)
*Moritella dasanensis**	–3–18	M	–	l				l		14:0, 14:1, 16:0, 16:1, 22:6	Kim *et al*. (2008)
*Moritella profunda**	2–12	D	–	l				l		14:0, 14:1, 16:0, 16:1, 22:6	Xu *et al*. (2003a)
*Moritella abyssi**	2–14	D	–	l				l		14:0, 14:1, 16:0, 16:1, 22:6	Xu *et al*. (2003a)
Photobacterium frigidiphilum	6–20	D	–	l				(+)		**16:0**, **16:1**, 18:1	Seo *et al*. (2005)
Psychromonas kaikoae	4–15	D	–	l				(+)		14:1(7t), 16:0, **16:1(7c)**	Nogi *et al*. (2002)
Psychromonas profunda	2–14	D	–	l						14:1, **16:0**, **16:1**	Xu *et al*. (2003b)
Photobacterium profundum	4–18	D	–	l	m			m		**16:1**, i16:0, 20:5	Nogi *et al*. (1998a)
Shewanella violacea	4–15	D	–	m	m			m		i15:0, 16:0, 16:1, 20:5	Nogi *et al*. (1998b)

Table 11.3 (Continued)

Organism	Temperature (°C)[a]	Site[b]	Gram[c]	Types of fatty acids[d] Straight	Iso	Anteiso	OH	LCPUFA	Others	Major and minor fatty acid[e]	References
Psychrotrophic sp.											
Clostridium algoriphilum	–5–25	S	+	l					m	14:0, 16:1(9c), 16:1(7c)dma	Shcherbakova *et al.* (2005)
Deinococcus alpinitundrae	5–25	S	+	l					l	16:0, **16:1(7c)**, **C17:0**	Callegan *et al.* (2008)
Salinibacterium xinjiangense	4–23	G	+		l	l				i14:0, **a15:0**, **i16:0**	Zhang *et al.* (2008)
Leifsonia pindariensis	4–35	G	+			l				**a15:0**, **a17:0**	Reddy *et al.* (2008)
Arthrobacter cryoconiti	1–25	G	+		m	l				i15:0, **a15:0**	Margesin *et al.* (2011a)
Arthrobacter psychrophenolicus	1–25	Si	+			l				**a15:0**	Margesin *et al.* (2004)
Arthrobacter alpinus	1–25	S	+		m	l				**a15:0**, i16:0, a17:0	D.C. Zhang *et al.* (2010)
Arthrobacter livingstonensis	–6–28	S	+	l		l				**a15:0**, 16:0, **a17:0**, **18:0**	Ganzert *et al.* (2011a)
Arthrobacter cryotolerans	–6–24	S	+	l		l				**a15:0**, 16:0, 18:0	Ganzert *et al.* (2011a)
Acetobacterium tundrae	1–30	S	+	m					l	16:1(9c), 16:1(11c), **16:0(al)**	Simankova *et al.* (2000)
Deinococcus radiomollis	5–25	S	+	l	m					**16:1(7c)**, i17:0	Callegan *et al.* (2008)
Deinococcus claudionis	5–25	S	+	l	m					**16:1(7c)**, i17:0	Callegan *et al.* (2008)
Deinococcus altitudinis	5–25	S	+	l	m					15:1(6c), **16:1(7c)**, i17:0	Callegan *et al.* (2008)
Leifsonia psychrotolerans	–6–28	S	+	m	m	l				i15:0, **a15:0**, 18:0	Ganzert *et al.* (2011b)
Paenibacillus algorifonticola	10–37	Ls	+	m		l				**a15:0**, 16:0	Tang *et al.* (2011)
Arthrobacter flavus	5–30	Lp	+	m		l				a15:0, 17:0	Reddy *et al.* (2000)
*Clostridium schirmacherense**	5–35	Lo	+	l						15:0, 16:0, 17:0	Alam *et al.* (2006)
Leifsonia rubra	0–22	Lp	+		m	l				a15:0, i16:0, **a17:0**	Reddy *et al.* (2003a)
Leifsonia aurea	0–30	Lp	+			l				**a15:0**, **a17:0**	Reddy *et al.* (2003a)

Devosia psychrophila	1–25	G	–	l					**16:1(7c)**, **18:1(7c)**	Zhang *et al*. (2011d)
Pseudomonas bauzanensis	5–30	S	–	l					16:1(7c), **18:1(7c)**	Zhang *et al*. (2011c)
Candidimonas bauzanensis	1–37	S	–	l				l	**16:0**, **16:1(7c)**, **C17**	Zhang *et al*. (2011b)
Sphingomonas alpina	1–30	S	–	l			m		**18:1(7c)**, 14:0 2-OH	Margesin *et al*. (2011b)
Hymenobacter psychrophilus	1–25	S	–	l	l	l			**i15:0**, **a15:0**, **16:1(7c)**	Zhang *et al*. (2011a)
Psychroflexus lacisalsi	10–28	L	–			l	m		**a15:0**, a15:1, 16:0 3-OH	H. Zhang *et al*. (2010)
Sphingomonas molluscorum	3–37	Mb	–	l			m		16:0, 18:1(7c), 18:1(9c), 14:0 2-OH	Romanenko *et al*. (2007)
Pseudoalteromonas arctica	4–25	M	–	l					16:0, **16:1(7c)**	Khudary *et al*. (2008)
Marinomonas polaris	4–37	M	–	l	m				**16:1(7c)**, i16:0, **18:1(7c)**	Gupta *et al*. (2006)
Psychrobacter okhotskensis	0–35	M	–	l					**17:1**, 18:1(9c)	Yumoto *et al*. (2003)
Colwellia chukchiensis	0–30	M	–	m			l		16:0, 16:(7c) 2-OH, 17:1	Yu *et al.* (2011)
Desulfuromusa ferrireducens	–2–23	Mc	–	l					**16:0**, **16:1(7c)**	Vandieken *et al*. (2006)
Pseudochrobactrum glaciei	5–40	Mi	–	l				l	**18:1(7c)**, **C19:0**	Romanenko *et al*. (2008)
Halomonas glaciei	4–30	Mi	–	l				m	16:0, 16:1(9c), **18:1**, Δ, ■	Reddy *et al*. (2003b)
Gelidibacter algens	0–≤25	Mi	–		m	m			i15:1(10c), a15:0	Bowman *et al*. (1997)
Roseovarius indicus	10–39	D	–	l					**18:1(7c)**	Lai *et al*. (2011)

[a]Growth temperature range.
[b]The site where bacterial sample was collected: G, glacier; S, soil; Si, icecave; L, lake; Lo, oasis; Lp, pond; Ls, cold spring; Lf, low-salinity fjord; Mb, bay (bivalve); Mi, sea ice; Mc, coastal sediment; Ms, sea sediment; D, deep sea.
[c]Gram staining.
[d]Types of fatty acids: straight, straight chain fatty acid; iso, iso branched fatty acid; anteiso, anteiso branched fatty acid; OH, hydroxylated fatty acid; LCPUFA, long-chain polyunsaturated fatty acid; l, major component (≥20% total fatty acid); m, minor component with 10–20% of total fatty acid; (+), presence of LCPUFA but less than 10% of total fatty acid; (al), aldehyde; cyclopropane; methoxymethyl; cyclo 2-octyl methyl; Δ, 18:0 (11-methoxymethyl); ■, 18:0 (cyclo 2-octyl methyl); dma, dimethyl acetal.
[e]Individual major fatty acids are in bold.

(ABJ14485) and *desB* (AAG08273), both of which encode fatty acid desaturases, have been identified (Zhu *et al.*, 2006). *fabA* and *desA* and *desB* are responsible for the anaerobic and aerobic synthesis, respectively, of monounsaturated fatty acids. These results imply that the coexistence of anaerobic and aerobic pathways in the same bacterium may be more common in aerobic bacteria than previously considered, in particular in aerobic psychrophilic and psychrotrophic bacteria. This may be because of the cooperative involvement of the two pathways in the synthesis of unsaturated fatty acids. This would be advantageous to cold-adapted bacteria by increasing the membrane fluidity at low temperature and particularly in psychrotrophic bacteria with a wider temperature range for growth, as the aerobic pathway is useful for adjusting the membrane solidification induced by an abrupt decrease in ambient temperature. This speculation is supported by the findings that putative *fabA* and *des* genes are both present in the genome of psychrotrophic *Shewanella oneidensis* MR-1 (AAN54908 and AAN53284, respectively) and *Shewanella baltica* OS185 (ABS08740 and ABS06345, respectively), and of psychrophilic *Shewanella piezotolerans* WP3 (ACJ29528 and ACJ27047, respectively), *Photobacterium profundum* SS9 273-4 (CAG20180 and CAG22614, respectively), and *Psychromonas ingrahamii* 37 (ABM03750 and ABM03309, respectively).

Function and biosynthesis of branched fatty acids

Iso-branched and anteiso-branched fatty acids seem to be regarded as fatty acids characteristic of Gram-positive bacteria (Kaneda, 1991), although they are also detected in Gram-negative bacteria. These branched fatty acids have a function similar to that of *cis*-unsaturated fatty acid (Kaneda, 1991) decreasing the phase-transition temperature of the membrane phospholipids that are acylated with these fatty acids. For example, the phase transition temperatures of phosphatidylcholine (PC) acylated with two molecules of n15:0, i15:0, or a15:0 is 34.2°C, 16.5°C, and 6.5°C, respectively, and those of PC acylated with n17:0, i17:0, or a17:0 are 48.8°C, 27.0°C, and 7.6°C, respectively (Silvius, 1982; Kaneda, 1991). Although PC is not commonly present in bacteria, the same trend is observed for other phospholipid classes in bacteria, such as PG, DPG, and PE (Silvius, 1982), and probably for glycolipids. Iso- and anteiso-branched monounsaturated fatty acids, such as a15:1 in *Psychroflexus lacisalsi* (H. Zhang *et al.*, 2010) and *Cryobacterium roopkundense* (Reddy *et al.*, 2010) are also found in both psychrophilic and psychrotrophic bacteria (Table 11.3), suggesting the mode of low-temperature adaptation involves increasing the polar lipid composition by increasing the amount of (1) anteiso-branched fatty acids, (2) branched fatty acids with shorter chain length, and 3) branched and monounsaturated fatty acids. However, these trends are not restricted to psychrotrophic and psychrophilic bacteria and may be partially applicable to mesophilic and even to thermophilic bacteria (Suutari and Laakso, 1994).

The biosynthetic mechanisms of branched-fatty acids in bacteria have been well documented (Kaneda, 1991; Zhu *et al.*, 2005). In these mechanisms, fatty acid synthetase is involved in the synthesis of iso- and anteiso-branched fatty acids using leucine and isoleucine, respectively, as an initiation primer. Branched fatty acids with a double bond, such as i15:1 and a15:1, are also considered to be synthesized by fatty acid synthetase via the so-called anaerobic pathway. The final product in the branched-chain fatty acid biosynthesis is primarily dependent on the activity of branched-chain amino acid transaminase using isoleucine or leucine as the substrate and the availability of the substrate. In mesophilic

Listeria monocytogenes, the addition of exogenous isoleucine increases the content of a15:0 at a lower growth temperature; it was suggested that the cold-regulated chain shortening increased a15:0 at the expense of a17:0 (Zhu *et al.*, 2005). Although no molecular analysis of branched fatty acid biosynthesis has been performed in psychrophiles and psychrotrophs, an antiparallel change in the contents of a15:0 and a17:0 was observed in *Pseudomonas* sp., a terrestrial psychrotroph from Antarctica, when grown under changed temperature (Fukunaga and Russell, 1990). Interestingly, a15:0 and a17:0 combined accounted for more than 90% of the total fatty acids in this psychrotrophic *Pseudomonas* grown in rich medium.

Recently, three types of cold-adapted bacteria were isolated from Antarctic soil associated with moss vegetation (N. Kurosawa, unpublished data): psychrophilic *Frigoribacterium* sp. and *Devosia* sp.; and psychrotrophic *Aurantimonas* sp. These bacteria grew in temperature ranges from −10°C to around 15°C with the optimal growth temperature around 10°C, from −10°C to around 20°C with the optimal growth temperature around 15°C, and from −10°C to around 27°C with the optimal growth temperature around 25°C, respectively (Fig. 11.1). Significant levels of branched fatty acids (a15:1, a15:0, i16:0 and a17:0) were detected only in *Frigoribacterium* sp. Interestingly, their combined levels (approximately 93% of total fatty acids) and their composition were scarcely changed by altered growth temperature. When *Aurantimonas* sp., whose major fatty acids included 18:1 in cells initially grown at 20°C, was grown at 10–15°C, the amount of shorter 16:1 increased. When it was grown at −10°C instead, the amounts of much shorter 14:0 and a15:0 increased while the amount of 16:1 decreased significantly. These results suggest that psychrophilic bacteria may be regarded

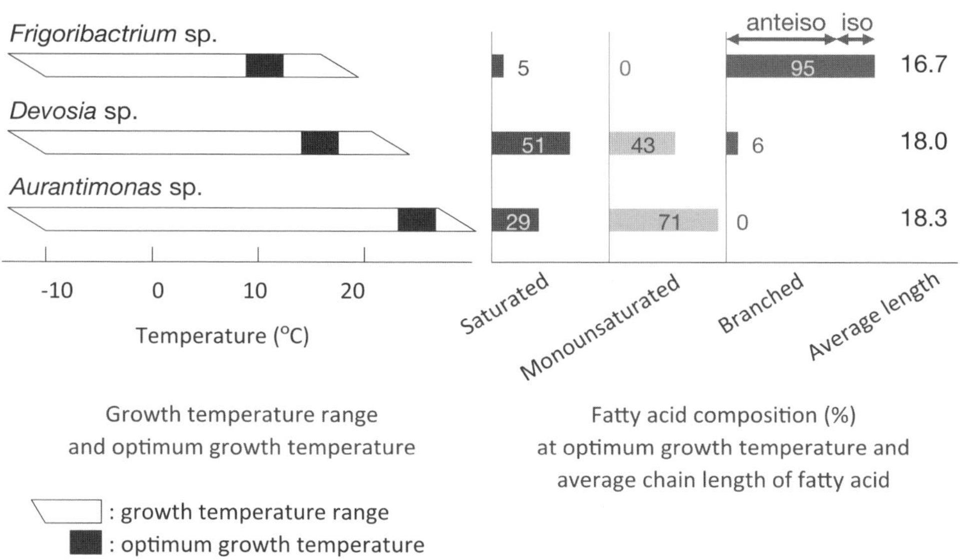

Figure 11.1 Growth temperature range, optimal temperature for growth, and fatty acid composition of psychrophilic *Frigoribacterium* sp. and *Devosia* sp. and psychrotrophic *Aurantimonas* sp. isolated from Antarctic soil. Fatty acid composition was determined using cells grown in modified Brock's basal salts agar (Kurosawa *et al*., 1998) at optimal temperature. Antarctic soil associated with moss vegetation was used to isolate individual strains (N. Kurosawa, unpublished). A colour version of this figure is located in the plate section at the back of the book.

as microorganisms that possess fatty acids with much lower melting temperature, such as anteiso-branched fatty acids in terrestrial bacteria, which would enable these bacterial membranes to be fluid even at temperatures as low as −10°C. Some bacteria may have limited capacity to adjust their membrane fluidity by changing fatty acid composition, which may explain why psychrophilic bacteria have much lower and narrower temperature ranges than psychrotrophic bacteria. In Fig. 11.2A (left), the possible functions of branched fatty acids in bacteria in cold terrestrial regions are illustrated. Iso- and anteiso-branched fatty acids are considered to be responsible for the maintenance of membrane fluidity at low temperature. Almost all colonies of psychrophilic and psychrotrophic bacteria from the Antarctic soil had colour, irrespective of growth temperature (Fig. 2A, right). Although no direct evidence

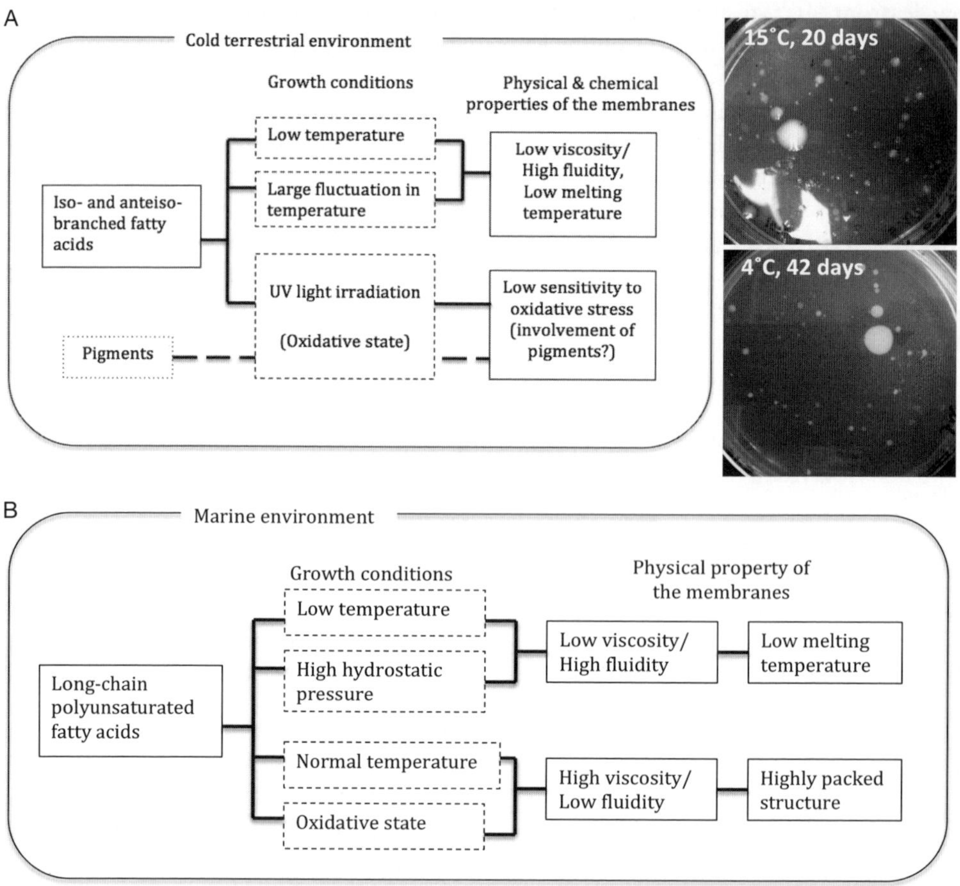

Figure 11.2 Schematic presentation of possible physiological roles of branched fatty acids in bacteria in cold terrestrial environments (A, left panel) and long-chain polyunsaturated fatty acids (LC-PUFAs) in Gram-negative psychrophilic and psychrotrophic bacteria in marine environments (B). The right panel of A shows agar plates of bacteria grown at 4°C and 15°C for 42 days and 20 days, respectively. In this experiment, Antarctic soil associated with moss vegetation was used (Y. Tsuboi and N. Kurosawa, unpublished). Environmental conditions are enclosed in boxes with dashed line. Although it has not been certified yet, possible involvement of pigments in antioxidation is shown with a dotted box and a broken line in the left panel of A. A colour version of this figure is located in the plate section at the back of the book.

is available at present, pigments may be involved in the antioxidative function in cold terrestrial bacterial cells.

As described above (see Tables 11.1 and 11.3), the modes of fatty acid adaptation probably do not differ between psychrophilic and psychrotrophic bacteria. Straight-chain and branched-chain fatty acids have been detected in both psychrophilic and psychrotrophic bacteria. Although branched-chain fatty acids are likely to be more abundant in Gram-positive bacteria than in Gram-negative and in terrestrial (non-marine) than in marine bacteria, their presence is probably not exclusive to one type of organism. By contrast, LC-PUFAs have been detected only in Gram-negative marine bacteria.

LC-PUFAs

Some psychrophilic and psychrotrophic bacteria (Nichols *et al.*, 1997; Yano *et al.*, 1994, 1998), and some mesophilic bacteria (Hosoya *et al.*, 2006; Sugihara *et al.*, 2010a, 2012) contain LC-PUFAs, such as EPA, DHA, and arachidonic acid (AA), which are considered to be acyl constituents of membrane phospholipids. Curiously, however, with the exception of cyanobacteria, bacteria generally do not possess C_{16} or C_{18} PUFAs, which are very common in cyanobacteria, plants, fungi, and animals. As described above, these LC-PUFAs are detected exclusively in Gram-negative bacteria, and most are from marine sources, particularly polar and deep marine sources.

Functions and biosynthesis of LC-PUFAs

Since LC-PUFAs were first found in bacteria, these fatty acids and other unsaturated fatty acids have been considered to be crucial for maintaining optimal membrane fluidity. This is because their acyl components are comprised of structural lipids under low-temperature or high-hydrostatic pressure conditions (DeLong and Yayanos, 1986). This expected role of LC-PUFAs is based on the following findings.

1. LC-PUFAs have lower melting temperatures than C_{16} or C_{18} monounsaturated fatty acids, which are the most common unsaturated fatty acids in bacteria.
2. Phospholipids with LC-PUFAs have lower gel-to-liquid crystalline phase transition temperatures than do phospholipids with C_{16} or C_{18} monounsaturated fatty acids.
3. The levels of cellular LC-PUFAs in growing cells increase with a decrease in temperature or an increase in hydrostatic pressure (DeLong and Yayanos, 1986; Nichols *et al.*, 1997). A psychrophilic bacterium strain 651 isolated from Antarctic sea ice produces significantly increased levels of EPA at temperatures over the range of 2–15°C (Nichols *et al.*, 1997). Many researchers shared the belief that LC-PUFAs are found only in psychrophiles and are involved in bacterial cold adaptation (Valentine and Valentine, 2004; Moyer and Morita, 2007). However, it has become apparent that this is not necessarily correct, as described below.

Recent studies have focused on the specific functions of bacterial LC-PUFAs using EPA-producing bacteria and their EPA-lacking mutants. In these studies, insertional inactivation mutagenesis is frequently used to generate EPA-deficient mutants by targeting any one of five *pfa* genes of EPA-producing bacteria. EPA-deficient mutants derived from marine psychrophilic and the piezotolerant bacterium *Photobacterium profundum* SS9 do not necessarily become cold sensitive or high hydrostatic pressure sensitive, and mono-unsaturated fatty acid but no EPA is needed for these bacteria to adapt to low temperature and high hydrostatic pressure (Allen *et al.*, 1999). These results suggest that membrane fluidity is maintained not by EPA under low-temperature or high-hydrostatic pressure conditions, even in EPA-producing bacteria, but rather by monounsaturated fatty acids. These findings imply that the membrane-fluidizing function of LC-PUFAs in psychrophilic and psychrotrophic bacteria is not universal and their function varies between organisms (Okuyama *et al.*, 2008). Interestingly, Kawamoto *et al.* (2009) reported that there is little difference in bulk membrane fluidity between *Shewanella livingstonensis* Ac10, whose EPA content is about 5% of the total fatty acid content in cells grown at 4°C, and its EPA-deficient mutant. Thus, the defect in EPA in the *Shewanella livingstonensis* Ac10 mutant at low temperature is explained by a specific association of EPA with a membrane function such as membrane transport or cell division at low temperature. A similar observation was reported for a psychrophilic deep-sea bacterium *Shewanella violacea* DSS12 and its EPA-deficient mutant (Kawamoto *et al.*, 2011). The piezophilic property of this EPA-producing bacterium was lost by disrupting the EPA biosynthesis gene and this property was restored by the addition of exogenous phospholipid containing EPA but not by the addition of exogenous phospholipid containing oleic acid (18:1). The loss of the piezophilic property in this mutant is accompanied by the formation of filamentous cells, suggesting that EPA has some specific function, for example in the step of Z-ring formation (Kawamoto *et al.*, 2011) in cell growth at low temperature. However, it is unclear whether it is involved in cold adaptation in this psychrophilic bacterium.

Nishida *et al.* (2006a,b, 2007) first reported that EPA has an antioxidative effect against extracellularly added hydrogen peroxide (H_2O_2) in *E. coli* recombinants carrying the *pfa* genes responsible for the biosynthesis of EPA that were derived from *Shewanella pneumatophori* SCRC-2738 and *Shewanella marinintestina* strain IK-1 and its EPA-deficient mutant. Although the molecular mechanism responsible for this function has not been elucidated, a membrane-shielding effect of the EPA-containing membrane against membrane-diffusible hydrophilic compounds including reactive oxygen species (ROS), such as H_2O_2 and *tert*-butyl hydroperoxide (*tert*-BHP), has been proposed (Nishida *et al.*, 2006b, 2007; Okuyama *et al.*, 2008; Hori *et al.*, 2011). The molecular structure of phospholipids having EPA and hexadecanoic or hexadecenoic acids is thought to be involved in the proposed mechanism. Phospholipids with LC-PUFAs would have a more highly packed structure than fatty acids that are less unsaturated, and lipid membranes comprising phospholipids with saturated fatty acids, such as 16:0, and LC-PUFAs, such as DHA and probably EPA, may form more hydrophobic interfaces between the phospholipid bilayers (Rajamoorthi *et al.*, 2005; Okuyama *et al.*, 2008). This hydrophobic interface of the cell membrane might prevent the entry of the hydrophilic H_2O_2 molecule. The *in vivo* antioxidative function of EPA probably arises through a combination of EPA and its counterpart fatty acid of the phospholipid molecules to shield membranes against the effects of exogenous ROS (Okuyama *et al.*, 2008). Similar hydrophobic interfaces between the phospholipid bilayers can be formed using phospholipid comprising equimolar medium straight-chain saturated and monounsaturated fatty acids such as 16:0 or 16:1 and AA, EPA, or DHA. The membrane-shielding effect of LC-PUFAs is applicable for H_2O_2 and other hydrophilic compounds (Nishida *et al.*, 2010; Hori *et al.*, 2011) and can be regarded as a nonspecific function in bacteria including psychrophilic and psychrotrophic bacteria.

Usui *et al.* (2012) investigated the membrane fluidity of *Shewanella violacea* DSS12 and its EPA-deficient mutant by labelling these cells with a fluorescent probe. Interestingly, they showed that EPA is involved in stabilizing but not fluidizing the membrane structure. The results of this study suggest that EPA prevents the membrane from becoming hyperfluid under high hydrostatic pressure. Because the motion of membrane components is considered to be more restricted under higher hydrostatic pressure, the finding by Usui *et al.* (2012) conflicts with the generally accepted concept that greater membrane fluidity is a characteristic common in bacteria inhabiting high-hydrostatic-pressure environments (see above and DeLong and Yayanos, 1986).

Five *pfa* genes, *pfaA–E*, are responsible for the biosynthesis of EPA and DHA (Orikasa *et al.*, 2004; Okuyama *et al.*, 2007). Of the five *pfa* genes *pfaA*, *pfaB*, *pfaC*, and *pfaD* are located in tandem in this order in the chromosome in most EPA or/DHA-producing bacteria. Although EPA and DHA have been considered to exclusively be synthesized in one bacterium, this is not the case for DHA-synthesizing bacteria, which can produce, in addition to DHA, small amounts of EPA: however, bacteria possessing the EPA biosynthesis genes produce only EPA. PfaB has been identified to be the key enzyme determining the final product in bacterial EPA or DHA biosynthesis (Orikasa *et al.*, 2009). Generally, in EPA-producing bacteria *pfaE* encoding phosphopantetheinyl transferase is located in the vicinity of *pfaA–D*, whereas *pfaE* gene is located far from clustered *pfaA–D* in the DHA biosynthesis gene (Okuyama *et al.*, 2007). The function of *pfaE* is not specific for LC-PUFA species (Sugihara *et al.*, 2008) and *pfaE* in the biosynthesis of DHA can be replaced by *E. coli entD* (Sugihara *et al.*, 2010b), which encodes the phosphopantetheinyl transferase responsible for the biosynthesis of siderophore. All of these findings suggest that only the *pfaA–D* genes responsible for biosynthesis of EPA or DHA move horizontally as a clustered gene in the restricted groups of Gram-negative marine bacteria belonging to *Shewanella*, *Photobacterium*, *Colwellia*, *Psychromonas*, *Psychroflexus*, and *Moritella*. Although presently there is no information on genes for AA biosynthesis, AA may be also synthesized by *pfa* genes because its biosynthesis is accompanied by EPA biosynthesis in psychrophilic *Psychroflexus torquis* from Antarctic sea ice (Bowman *et al.*, 1997). Although it is neither psychrophilic nor psychrotrophic, *Aureispira marina* would be a material appropriate for cloning genes responsible for the biosynthesis of AA because of its easy cultivation, abundance of cellular AA, and relatively high optimum temperature for growth (Hosoya *et al.*, 2006).

The preferential distribution of LC-PUFAs in marine Gram-negative bacteria has been shown clearly. However, the assumption that they are distributed in only cold-adapted bacteria is not valid for the following reasons:

1 Highly AA-producing mesophilic *Aureispira marina* bacterium was isolated from the southern coastline of Thailand (Hosoya *et al.*, 2006).
2 Mesophilic EPA-producing *Shewanella* strains were isolated from Hokkaido (43.19°N, 141.00°E; Sugihara *et al.*, 2012) and Okinawa (26.36°N, 128.84°E; S. Sugihara and H. Okuyama, unpublished data), Japan.
3 EPA-producing *Vibrio* was isolated from Okinawa (26.35°N, 127.59°E; S. Sugihara and H. Okuyama, unpublished data).

No clear explanation has been given why LC-PUFAs are distributed preferentially or exclusively in Gram-negative bacteria inhabiting marine environments. LC-PUFAs may have diverse functions in marine environments given the following observations: 71% of the earth's surface is marine and 90% by volume of the ocean is 5°C or lower (Moyer and Morita, 2007), and almost all organisms in addition to bacteria possess LC-PUFAs (Bergé and Barnathan, 2005). The role of these fatty acids as a modulator of membrane fluidity in cold-adapted marine bacteria is thought to fulfil at least one of these functions. Okuyama *et al.* (2008) regarded the membrane-shielding effects of LC-PUFAs as a protective 'breakwater' against ROS in marine environments, because the production of ROS such as H_2O_2 is prevalent in the world's oceans (Lesser, 2006). Fig. 11.2B illustrates the possible functions of LC-PUFAs in bacteria in marine environments.

Practical applications of lipids

As described by Perveen *et al.* (2006) LC-PUFAs are important for human nutrition. DHA, EPA, and AA are the building block of membrane phospholipids of nervous, visual and reproductive tissue cells and have physiologically important functions in humans (Gill and Valivety, 1997; Kroes *et al.*, 2003; Ratledge, 2004) and animals (Bezard *et al.*, 1994). Marine fish oil and eukaryotic microbial oil are the major sources of DHA and EPA, and AA, respectively, and are used in pharmaceuticals and food supplements. EPA-, DHA-, and AA-producing psychrophilic, psychrotrophic, and mesophilic bacteria are potential candidates for the commercial production of these LC-PUFAs. *Shewanella* sp. strain SCRC 2748 (presently *Shewanella pneumatophori* SCRC 2738, Hirota *et al.*, 2005) is an EPA-producing marine bacterium that was isolated for such a purpose (Yazawa *et al.*, 1988). However, bacteria, particularly psychrophilic and psychrotrophic bacteria, have serious limitations in terms of their low cell yields and long cultivation times compared with eukaryotic microorganisms (Perveen *et al.*, 2006).

Since the first cloning of *pfa* genes responsible for the biosynthesis of EPA (Yazawa, 1996), similar genes have been obtained from various bacteria (Tanaka *et al.*, 1999). Success in the heterologous expression of *pfa* genes, in *E. coli* has been limited, and besides bacteria, yeast, and plant cells have been also tried as models to express these genes (Takeyama *et al.*, 1997; Yu *et al.*, 2000; Vrinten *et al.*, 2007; Bin Haji Mohd Taha *et al.*, 2011).

A unique long straight-chain polyunsaturated hydrocarbon with an odd number of carbon atoms, hentriacontanonaene ($C_{31:9}$), was first discovered in marine Antarctic bacteria (Nichols *et al.*, 1995). These $C_{31:9}$-producing bacteria are psychrophilic, and this hydrocarbon was contained in bacteria with eicosapentaenoic acid (EPA) or/and docosahexaenoic acid (DHA) (Nichols *et al.*, 1995). The geometry and position of the double bonds in this compound were determined physicochemically to be *cis* at positions 3, 6, 9, 12, 15, 19, 22, 25 and 28 (Sugihara *et al.*, 2010a) in an EPA-producing marine mesophilic bacterium that was identified as *Shewanella oshoroensis* osh08 (Sugihara *et al.*, 2012). $C_{31:9}$ is thought to be synthesized by the head-to-head condensation reaction of two molecules of Δ4,7,10,13-all-*cis*-hexatetradecenoic acid (16:4; Sugihara *et al.*, 2010a; Sukovich *et al.*, 2010). Involvement of $C_{31:9}$ in cold adaptation has been shown in *Shewanella oneidensis* MR-1 (Sukovich *et al.*, 2010). However its physiological role has yet to be known, because levels of $C_{31:9}$ increased at the expense of EPA when growth temperature was raised in *Shewanella oshoroensis* osh08 (Sugihara *et al.*, 2010a). Sukovich *et al.* (2010) identified *oleA–D* genes involved in the biosynthesis of $C_{31:9}$ in EPA-producing *Shewanella oneidensis* MR-1. In *S. oneidensis* MR-1, genetic knockout of the *oleABCD* gene cluster caused a loss of the reported hydrocarbon. Sukovich *et al.* (2010) also found that *pfa* genes are involved in the biosynthesis of $C_{31:9}$ in

S. oneidensis MR-1 because the knockout of *pfaA* caused a loss of the ability to produce any hydrocarbon and this was restored by plasmid-encoded *pfaA*. Thus, a possible intermediate 16:4 can be synthesized by Pfa proteins. Investigation of 3558 genomes of *Bacteria*, *Archaea*, and *Eukaryota* for *ole* gene organization showed the presence of *ole* genes only in 69 bacterial genomes (Sukovich *et al.*, 2010). These 69 genomes were phylogenetically and ecologically diversified, but they all showed the presence of *ole* genes and the phenotypic expression in the form of long-chain olefins, although the type of olefin synthesized differed greatly with respect to the number of carbon–carbon double bonds.

Conclusions

As with other microorganisms, psychrophilic and psychrotrophic bacteria respond to low temperature by optimizing their membrane fluidity through a combination of lipid structural and compositional strategies, of which fatty acid branching and unsaturation are considered the most significant modulation modes for their cold adaptation. However, the distribution of branched and unsaturated (monounsaturated and long-chain polyunsaturated) fatty acids is not restricted to psychrophilic and psychrotrophic bacteria. Monounsaturated fatty acids are used most frequently as a membrane modulator at low temperature in microorganisms. Iso- and anteiso-branched fatty acids have been detected in psychrophilic to thermophilic bacteria, but LC-PUFAs have been found only in psychrophilic, psychrotrophic, and mesophilic bacteria. Branched fatty acids are distributed preferentially in Gram-positive terrestrial bacteria but are also distributed in Gram-negative marine bacteria. By contrast, LC-PUFAs are distributed exclusively in Gram-negative marine bacteria. LC-PUFAs such as EPA and DHA are more easily oxidized by oxygen or UV light, which may be a physiological and ecological reason why terrestrial bacteria preferentially possess branched fatty acids but not LC-PUFAs (Tables 11.1 and 11.3). The function of LC-PUFAs in psychrophilic, piezophilic and psychrotrophic marine bacteria has been explored recently using LC-PUFA-deficient mutants. The results of these studies suggest that the role of LC-PUFAs in increasing the membrane fluidity at low temperature is only one of their possible functions, which is supported by the recent culture-independent studies showing that *pfa* genes are widely distributed in marine environments (Shulse and Allen, 2011). Interestingly, EPA is thought to decrease the hyperfluid state of the membrane in piezophiles grown under high-hydrostatic-pressure conditions (Usui *et al.*, 2012). *pfa* genes responsible for the biosynthesis of EPA and DHA are available from various psychrophilic and psychrotrophic bacteria, and these have been used to identify the functions of EPA and DHA in bacteria. *pfa* genes are also involved in the biosynthesis of a unique very long-chain polyunsaturated hydrocarbon, hentriacontanonaene ($C_{31:9}$), in a manner coordinated with *ole* genes (Sukovich *et al.*, 2010). $C_{31:9}$ is distributed mainly in psychrophilic and psychrotrophic marine Gram-negative bacteria, but its physiological function has not been clarified. A clue to its role in cold-adapted microorganisms can be obtained by studying the combined expression of heterologous *pfa* and *ole* genes, for example in *Escherichia coli.*

Future trends

The likely presumption is that terrestrial and marine psychrophilic bacteria have less capacity to modulate their fatty acid composition and that these bacteria have a narrower growth range compared with psychrotrophic bacteria (see Fig. 11.1). This presumption can be confirmed using molecular methods to generate, anteiso-branched fatty acid- or LC-PUFA-deficient mutants. Physiological role(s) of branched fatty acids and LC-PUFAs in relation to their bulk lipid fraction and that in their specificities must be clarified. Various types of *pfa* gene-transformed *E. coli* K-12 mutants lacking either membrane transport proteins or pumping proteins were generated. There was no difference in the susceptibility of EPA-producing *E. coli* K-12 parent and mutant cells against *tert*-butyl hydroperoxide, indicating that EPA is not involved in the import or the export of *tert*-butyl hydroperoxide (R. Hori and H. Okuyama, unpublished results). By contrast, EPA-producing *E. coli* K-12 parent and mutants lacking *acrA, acrB,* or *tolC* became more resistant to antibiotics than did their recombinants that did not produce EPA. These results suggest that the function of EPA is specific to the membrane transport systems. However, for further understanding of the function of the branched fatty acids we need the cloning of genes, such as a gene for branched-chain amino acid aminotransferase (Madsen *et al.*, 2002), involved in the synthesis of branched fatty acids and the generation of mutants of psychrophilic bacteria without the capacity to synthesize branched fatty acids. The availability of genome sequence information and biotechnologies such the insertional mutagenesis method would facilitate further research.

Our preliminary unpublished results indicate that coexpression of *pfa* (*pfaA–E*) and *ole* (*oleA–D*) genes in *E. coli* cells does not lead to the synthesis of $C_{31:9}$ (S. Sugihara and H. Okuyama, unpublished results). However, many unidentified olefinic compounds have been detected by gas–liquid chromatography and gas chromatography–mass spectrometry, suggesting that one or more genes may be necessary for microbial production of $C_{31:9}$. Hydrocarbons are one of the most important and versatile groups of compounds, and that the need to develop renewable petroleum or sustainable biofuels emphasizes the importance of fatty acids to the modern world. Better understanding of the processes and genetic control of the biosynthesis of hydrocarbons in microorganisms may show that cold-adapted microorganisms have potential use in the microbial production of biofuels.

Acknowledgements

Various types of *pfa* gene clones responsible for biosynthesis of EPA in our laboratory were kindly provided by Sagami Chemical Research Centre, Ayase 252-1193, Japan. A.I.B.H.M.T. is a recipient of the scholarship of the Ministry of Education, Science, Sports, and Culture (MEXT) of Japan. This work was partly supported by Grant-in-Aid for Scientific Research ((C) no. 22570130) from MEXT and by the National Institute of Polar Research, Japan, through General Collaboration Projects (no. 22-15).

References

Aguilar, P.S., and de Mendoza, D. (2006). Control of fatty acid desaturation: a mechanism conserved from bacteria to humans. Mol. Microbiol. *62*, 1501–1514.

Alam, S.I., Dixit, A., Reddy, G.S.N., Dube, S., Palit, M., Shivaji, S., and Singh, L. (2006). *Clostridium schirmacherense* sp. nov., an obligately anaerobic, proteolytic, psychrophilic bacterium isolated from lake sediment of Schirmacher Oasis, Antarctica. Int. J. Syst. Evol. Microbiol. *56*, 715–720.

Allen, E.E., Facciotti, D., and Bartlett, D.H. (1999). Monounsaturated but not polyunsaturated fatty acids are required for growth of the deep-sea bacterium *Photobacterium profundum* SS9 at high pressure and low temperature. Appl. Environ. Microbiol. *65*, 1710–1720.

Asker, D., Award, T.S., McLandsborough, L., Beppu, T., and Ueda, K. (2011). *Deinococcus depolymerans* sp. nov., a gamma- and UV-radiation-resistant bacterium, isolated from a naturally radioactive site. Int. J. Syst. Evol. Microbiol. *61*, 1448–1453.

Bergé, J.P., and Barnathan, G. (2005). Fatty acids from lipids of marine organisms: molecular biodiversity, roles as biomarkers, biologically active compounds, and economical aspects. Adv. Biochem. Eng. Biotechnol. *96*, 49–125.

Bézard, J., Blond, J.P., Bernard, A., and Clouet, P. (1994). The metabolism and availability of essential fatty acids in animal and human tissues. Reprod. Nutr. Dev. *34*, 539–568.

Bin Haji Mohd Taha, A.I., Okuyama, H., Ohwada, T., Yumoto, I., and Orikasa, Y. (2011). Exogenous catalase gene expression as a tool for enhancing metabolic activity and production of biomaterials in host microorganisms. In Innovations in Biotechnology, E.C. Agbo, eds. (Rijeka, Croatia: InTech – Open Access Publisher), pp. 251–274.

Bowman, J.P., McCammon, S.A., Brown, J.L., Nichols, P.D., and McMeekin, T.A. (1997). *Psychroserpens burtonensis* gen. nov., sp. nov., and *Gelidibacter algens* gen. nov., sp. nov., psychrophilic bacteria isolated from Antarctic lacustrine and sea ice habitats. Int. J. Syst. Bacteriol. *47*, 670–677.

Callegan, R.P., Nobre, M.F., McTernan, P.M., Battista, J.R., Navarro-González, R., McKay, C.P., da Costa, M.S., and Rainey, F.A. (2008). Description of four novel psychrophilic, ionizing radiation-sensitive *Deinococcus* species from alpine environments. Int. J. Syst. Evol. Microbiol. *58*, 1252–1258.

Chintalapati, S., Kiran, M.D., and Shivaji, S. (2004). Role of membrane lipid fatty acids in cold adaptation. Cell. Mol. Biol. *50*, 631–642.

Cronan, J.E. (2003). Bacterial membrane lipids: Where do we stand? Annu. Rev. Microbiol. *57*, 203–224.

D'Amico, S., Collins, T., Marx, J.C., Feller, G., and Gerday, C. (2006). Psychrophilic microorganisms: challenges for life. EMBO Rep. *7*, 385–389.

DeLong, E.F., and Yananos, A.A. (1986). Biochemical function and ecological significance of novel bacterial lipids in deep-sea prokaryotes. Appl. Environ. Microbiol. *51*, 730–737.

Fukunaga, N., and Russell, N.J. (1990). Membrane lipid composition and glucose takeup in two psychrotolerant bacteria from Antarctica. J. Gen. Microbiol. *136*, 1669–1673.

Ganzert, L., Bajerski, F., Mangelsdorf, K., Lipski, A., and Wagner, D. (2011a). *Arthrobacter livingstonensis* sp. nov. and *Arthrobacter cryotolerans* sp. nov., salt-tolerant and psychrotolerant species from Antarctic soil. Int. J. Syst. Evol. Microbiol. *61*, 979–984.

Ganzert, L., Bajerski, F., Mangelsdorf, K., Lipski, A., and Wagner, D. (2011b). *Leifsonia psychrotolerans* sp. nov., a psychrotolerant species of the family Microbacteriaceae from Livingston Island, Antarctica. Int. J. Syst. Evol. Microbiol. *61*, 1938–1943.

Gill, I., and Valivety, R. (1997). Polyunsaturated fatty acids, part 1: Occurrence, biological activities and applications. Trends Biotechnol. *15*, 401–409.

Grogan D.W., and Cronan, J.E. Jr. (1997). Cyclopropane ring formation in membrane lipids of bacteria. Microbiol. Mol. Biol. Rev. *61*, 429–441.

Gupta, P., Chaturvedi, P., Pradhan, S., Delille, D., and Shivaji, S. (2006). *Marinomonas polaris* sp. nov., a psychrohalotolerant strain isolated from coastal sea water off the subantarctic Kerguelen islands. Int. J. Syst. Evol. Microbiol. *56*, 361–364.

Hirota, K., Nodasaka, Y., Orikasa, Y., Okuyama, H., and Yumoto, I. (2005). *Shewanella pneumatophori* sp. nov., eicosapentanoic-acid-producing marine bacterium isolated from the intestines of pacific mackerel (*Pneumatophorus japonicus*) intestine. Int. J. Syst. Evol. Microbiol. *55*, 2355–2359.

Hori, R., Nishida, T., and Okuyama, H. (2011). Hydrophilic and hydrophobic compounds antithetically affect the growth of eicosapentaenoic acid-synthesizing *Escherichia coli* recombinants. Open Microbiol. J. *5*, 114–118.

Hosoya, S., Arunpairojana, V., Suwannachart, C., Kanjana-Opas A., and Yokota, A. (2006). *Aureispira marina* gen. nov., sp. nov., a gliding, arachidonic acid-containing bacterium isolated from the southern coastline of Thailand. Int. J. Syst. Evol. Microbiol. *56*, 2931–2935.

Kaneda, T. (1991). Iso- and anteiso-fatty acids in bacteria: biosynthesis, function and taxonomic significance. Microbiol. Rev. *55*, 288–302.

Kawamoto, J., Kurihara, T., Yamamoto, K., Nagayasu, M., Tani, Y., Mihara, H., Hosokawa, M., Baba, T., Sato, S.B., and Esaki. N. (2009). Eicosapentaenoic acid plays a beneficial role in membrane organization and cell division of a cold-adapted bacterium, *Shewanella livingstonensis* Ac10. J. Bacteriol. *191*, 632–640.

Kawamoto, J., Sato, T., Nakasone, K., Kato, C., Mihara, H., Esaki, N., and Kurihara, T. (2011). Favourable effects of eicosapentaenoic acid on the late step of the cell division in a piezophilic bacterium, *Shewanella violacea* DSS12, at high-hydrostatic pressures. Environ. Microbiol. *13*, 2293–2298.

Khudary, R.A., Stösser, N.I., Qoura, F., and Antranikian, G. (2008). *Pseudoalteromonas arctica* sp. nov., an aerobic, psychrotolerant, marine bacterium isolated from Spitzbergen. Int. J. Syst. Evol. Microbiol. *58*, 2018–2024.

Kim, H.J., Park, S., Lee, J.M., Park, S., Jung, W., Kang, J.S., Joo, H.M., Seo, K.W., and Kang, S.H. (2008). *Moritella dasanensis* sp. nov., a psychrophilic bacterium isolated from the Arctic ocean. Int. J. Syst. Evol. Microbiol. *58*, 817–820.

Knittel, K., Kuever, J., Meyerdierks, A., Meinke, R., Amann, R., and Brinkhoff, T. (2005). *Thiomicrospira arctica* sp. nov. and *Thiomicrospira psychrophila* sp. nov., psychrophilic, obligately chemolithoautotrophic, sulfur-oxidizing bacteria isolated from marine Arctic sediments. Int. J. Syst. Evol. Microbiol. *55*, 781–786.

Kroes, R., Schaefer, E.J., Squire, R.A., and Williams, G.M. (2003). A review of the safety of DHA45 oil. Food Chem. Toxicol. *41*, 1433–1446.

Kurosawa, N., Itoh, Y.H., Iwai, T., Sugai, A., Uda, I., Kimura, N., Horiuchi, T., and Itoh, T. (1998). *Sulfurisphaera ohwakuensis* gen. nov., sp. nov., a novel extremely thermophilic acidophile of the order *Sulfolobales*. Int. J. Syst. Bacteriol. *48*, 451–456.

Lai, Q., Zhong, H., Wang, J., Yuan, J., Sun, F., Wang, L., Zheng, T., and Shao, Z. (2011). *Roseovarius indicus* sp. nov., isolated from deep-sea water of the Indian Ocean. Int. J. Syst. Evol. Microbiol. *61*, 2040–2044.

Lesser, M.P. (2006). Oxidative stress in marine environments: biochemistry and physiological ecology. Annu. Rev. Physiol. *68*, 253–278.

Mansilla, M.C., and de Mendoza, D. (2005). The *Bacillus subtilis* desaturase: a model to understand phospholipid modification and temperature sensing. Arch. Microbiol. *183*, 229–235.

Madsen, S.M., Beck, H.C., Ravn, P., Vrang, A., Hansen, A.M., and Israelsen, H. (2002). Cloning and inactivation of a branched-chain-amino-acid aminotransaminase gene from *Staphylococcus carnosus* and characterization of the enzyme. Appl. Environ. Microbiol. *68*, 4007–4014.

Margesin, R., Schumann, P., Spröer, C., and Gounot, A.M. (2004). *Arthrobacter psychrophenolicus* sp. nov., isolated from an alpine ice cave. Int. J. Syst. Evol. Microbiol. *54*, 2067–2072.

Margesin, R., Schumann, P., Zhang, D.C., Redzic, M., Zhou, Y.G., Liu, H.C., and Schinner, F. (2011a). *Arthrobacter cryoconiti* sp. nov., a psychrophilic bacterium isolated from alpine glacier cryoconite. Int. J. Syst. Evol. Microbiol. *62*, 394–402.

Margesin, R., Zhang, D.C., and Busse, H.J. (2011b). *Sphingomonas alpina* sp. nov., a psychrophilic bacterium isolated from alpine soil. Int. J. Syst. Evol. Microbiol. *62*, 1558–1563.

Morita N., Gotoh, M., Okajima, N., Okuyama, H., Hayashi, H., Higashi, S., and Murata, N. (1992). Both the anaerobic pathway and aerobic desaturation are involved in the synthesis of unsaturated fatty acids in *Vibrio* sp. strain ABE-1. FEBS Lett. *297*, 9–12.

Morita, R.Y. (1975). Psychrophilic bacteria. Bacteriol. Rev. *39*, 144–167.

Marr, A.G., and Ingraham, J.L. (1962). Effect of temperature on composition of fatty acids in *Escherichia coli*. J. Bacteriol. *84*, 1260–1267.

Matsunaga, I., Yamada, M., Kusunose, E., Miki, T., and Ichihara, K. (1998). Further characterization of hydrogen peroxide-dependent fatty acid alpha-hydroxylase from *Sphingomonas paucimobilis*. J. Biochem. Tokyo*124*, 105–110.

Moyer G.L., and Morita, R.Y. (2007). Psychrophiles and psychrotrophs. In: eLS (Chichester, UK: John Wiley & Sons Ltd). http://www.els.net [doi: 10.1002/9780470015902.a0000402.pub2].

Nichols, D.S., Brown, J.L., Nichols, P.D., and McMeekin, T.A. (1997). Production of eicosapentaenoic and arachidonic acids by an Antarctic bacterium: response to growth temperature. FEMS Microbiol. Lett. *152*, 349–354.

Nichols, D.S., Nichols, P.D., and McMeekin, T.A. (1995). A new n-$C_{31:9}$ polyene hydrocarbon from Antarctic bacteria. FEMS Microbiol. Lett. *125*, 281–285.

Nishida, T., Hori, R., Morita, N., and Okuyama, H. (2010). Membrane eicosapentaenoic acid is involved in the hydrophobicity of bacterial cells and affects the entry of hydrophilic and hydrophobic compounds. FEMS Microbiol Lett. *306*, 91–96.

Nishida, T., Orikasa, Y., Ito, Y., Yu, R., Yamada, A., Watanabe, K., and Okuyama, H. (2006a). *Escherichia coli* engineered to produce eicosapentaenoic acid becomes resistant against oxidative damages. FEBS Lett. *580*, 2731–2735.

Nishida, T., Orikasa, Y., Watanabe, K., Morita, N., Yano, Y., Orikasa, Y., and Okuyama, H. (2006b). The cell membrane-shielding function of eicosapentaenoic acid for *Escherichia coli* against exogenously added hydrogen peroxide. FEBS Lett. *580*, 6690–6694.

Nishida, T., Yano, Y., Morita, N., and Okuyama, H. (2007). The antioxidative function of eicosapentaenoic acid in a marine bacterium, *Shewanella marinintestina* IK-1. FEBS Lett. *581*, 4212–4216.

Nogi, Y., Kato, C., and Horikoshi, K. (1998b). Taxonomic studies of deep-sea barophilic *Shewanella* strains and description of *Shewanella violacea* sp. nov. Arch. Microbiol. *170*, 331–338.
Nogi, Y., Kato, C., and Horikoshi, K. (2002). *Psychromonas kaikoae* sp. nov., a novel piezophilic bacterium from the deepest cold-seep sediments in the Japan Trench. Int. J. Syst. Evol. Microbiol. *52*, 1527–1532.
Nogi, Y., Masui, N., and Kato, C. (1998a). *Photobacterium profundum* sp. nov., a new, moderately barophilic bacterial species isolated from a deep sea sediment. Extremophiles 2, 1–7.
Okuyama, H., Fukunaga, N., and Sasaki, S. (1986). Homeoviscous adaptation in a psychrophilic bacterium, *Vibrio* sp. strain ABE-1. J. Gen. Appl. Microbiol. *32*, 437–482.
Okuyama, H., Kawazoe, R., Morita, N., and Reichardt, W. (1991). The structure and functions of lipoamino acids in *Cytophaga johnsonae*. In Frontiers and new horizons in amino acid research, K. Takai, ed. (Amsterdam, The Netherlands: Elsevier), pp. 371–374.
Okuyama, H., Orikasa, Y., and Nishida, T. (2008). Significance of antioxidative functions of eicosapentaenoic and docosahexaenoic acids in marine microorganisms. Appl. Environ. Microbiol. *74*, 570–574.
Okuyama, H., Orikasa, Y., Nishida, T., and Morita, N. (2007). Bacterial genes responsible for the biosynthesis of eicosapentaenoic and docosahexaenoic acids and their heterologous expression. Appl. Environ. Microbiol. *73*, 665–670.
Okuyama, H., Sasaki, S., Higashi, S., and Murata, N. (1990). A *trans*-unsaturated fatty acid in a psychrophilic bacterium, *Vibrio* sp. strain ABE-1. J. Bacteriol. *172*, 3515–3518.
O'Leary, W.M., and Wilkinson, S.G. (1988). Gram-positive bacteria. In Microbial Lipids, Vol. 1, C. Ratledge and S.G. Wilkinson, eds. (Oxford, UK: Academic Press), pp. 117–202.
Orikasa, Y., Tanaka, M., Sugihara, S., Hori, R., Nishida, T., Ueno, A., Morita, N., Yano, Y., Yamamoto, K., Shibahara, A., Hayashi, H., Yamada, A., Yu, R., Watanabe, K., and Okuyama, H. (2009). *pfaB* products determine the molecular species produced in bacterial polyunsaturated fatty acid biosynthesis. FEMS Microbiol. Lett. *295*,170–176.
Orikasa, Y., Yamada, A., Yu, R., Ito, Y., Nishida, T., Yumoto, I., Watanabe, K., and Okuyama, H. (2004). Characterization of the eicosapentaenoic acid biosynthesis gene cluster from *Shewanella* sp. strain SCRC-2738. Cell. Mol. Biol. *50*, 625–630.
Padley, F.B., Gunstone, F.D., and Harwood, J.L. (1986). Occurrence and characteristics of oils and fats. In The Lipid Handbook, F.D. Gunstone, J.L. Harwood and F.B. Padley, eds. (Cambridge, UK: Chapman & Hall), pp. 47–223.
Perveen, Z., Ando, H., Ueno, A., Ito, Y., Yamamoto, Y., Yamada, Y., Takagi, T., Kaneko, T., Kogame, K., and Okuyama, H. (2006). Isolation and characterization of a novel thraustochytrid–like microorganism that efficiently produces docosahexaenoic acid. Biotechnol. Lett. *28*, 197–202.
Rajamoorthi, K., Petrache, H.I., McIntosh, T.J., and Brown, M.F. (2005). Packing and viscoelasticity of polyunsaturated omega-3 and omega-6 lipid bilayers as seen by ^{2}H NMR and X-ray diffraction. J. Am. Chem. Soc. *127*, 1576–1588.
Ratledge, C. (2004). Fatty acid biosynthesis in microorganisms being used for single cell oil production. Biochimie. *86*, 807–815.
Reddy, G.S.N., Aggarwal, R.K., Matsumoto, G.I., and Shivaji, S. (2000). *Arthrobacter flavus* sp. nov., a psychrophilic bacterium isolated from a pond in McMurdo Dry Valley, Antarctica. Int. J. Syst. Evol. Microbiol. *50*, 1553–1561.
Reddy, G.S.N., Prabagaran, S.R., and Shivaji, S. (2008). *Leifsonia pindariensis* sp. nov., isolated from the Pindari glacier of the Indian Himalayas, and emended description of the genus *Leifsonia*. Int. J. Syst. Evol. Microbiol. *58*, 2229–2234.
Reddy, G.S.N., Pradhan, S., Manorama, R., and Shivaji, S. (2010). *Cryobacterium roopkundense* sp. nov., a psychrophilic bacterium isolated from glacial soil. Int. J. Syst. Evol. Microbiol. *60*, 866–870.
Reddy, G.S.N., Prakash, J.S.S., Srinivas, R., Matsumoto, G.I., and Shivaji, S. (2003a). *Leifsonia rubra* sp. nov. and *Leifsonia aurea* sp. nov., psychrophiles from a pond in Antarctica. Int. J. Syst. Evol. Microbiol. *53*, 977–984.
Reddy, G.S.N., Raghavan, P.U.M., Sarita, N.B., Prakash, J.S.S., Nagesh, N., Delille, D., and Shivaji, S. (2003b). *Halomonas glaciei* sp. nov. isolated from fast ice of Adelie Land, Antarctica. Extremophiles *7*, 55–61.
Romanenko, L.A., Tanaka, N., Frolova, G.M., and Mikhailov, V.V. (2008). *Pseudochrobactrum glaciei* sp. nov., isolated from sea ice collected from Peter the Great Bay of the Sea of Japan. Int. J. Syst. Evol. Microbiol. *58*, 2454–2458.
Romanenko, L.A., Uchino, M., Frolova, G.M., Tanaka, N., Kalinovskaya, N.I., Latyshev, N., and Mikhailov, V.V. (2007). *Sphingomonas molluscorum* sp. nov., a novel marine isolate with antimicrobial activity. Int. J. Syst. Evol. Microbiol. *57*, 358–363.

Rose, A.H. (1988). Influence of the environment on microbial lipid composition. In Microbial Lipids, Vol. 2, C. Ratledge and S.G. Wilkinson, eds. (Oxford, UK: Academic Press), pp. 255–279.

Rüger, H.J. (1983). Differentiation of *Bacillus globisporus, Bacillus marinus* comb. nov., *Bacilus aminovorans,* and *Bacillus insolitus.* Int. J. Syst. Bacteriol. *33,* 157–161.

Russell, N.J. (1971). Alteration in fatty acid chain length in *Micrococcus cryophilus* grown at different temperatures. Biochim. Biophys. Acta *231,* 254–256.

Russell, N.J. (1988). Functions of lipids: structural roles and membrane functions. In Microbial Lipids, Vol. 2, C. Ratledge and S.G. Wilkinson, eds. (Oxford, UK: Academic Press), pp. 279–366.

Russell, N.J. (2008). Membrane components and cold sensing. In Psychrophiles From Biodiversity to Biotechnology, R. Margesin, F. Schinner, J-C. Marx, C. Gerday, eds. (Berlin, Germany: Springer), pp. 177–190.

Schmidt, M., Priemé, A., Johansen, A., and Stougaard, P. (2012). *Alkalilactibacillus ikkensis,* gen. nov., sp. nov., a novel enzyme-producing bacterium from a cold and alkaline environment in Greenland. Extremophiles *16,* 297–305.

Schumann, P., Zhang, D.C., Redzic, M., and Margesin, R. (2012). *Alpinimonas psychrophila* gen. nov., sp. nov., a novel actinobacterium of the family Microbacteriaceae isolated from alpine glacier cryoconite. Int. J. Syst. Evol. Microbiol. In press doi:10.1099/ijs.0.036160–0.

Seo, H.J., Bae, S.S., Lee, J.H., and Kim, S.J. (2005). *Photobacterium frigidiphilum* sp. nov., a psychrophilic, lipolytic bacterium isolated from deep-sea sediments of Edison Seamount. Int. J. Syst. Evol. Microbiol. *55,* 1661–1666.

Shcherbakova, V.A., Chuvilskaya, N.A., Rivkina, E.M., Pecheritsyna, S.A., Laurinavichius, K.S., Suzina, N.E., Osipov, G.A., Lysenko, A.M., Gilichinsky, D.A., and Akimenko, V.K. (2005). Novel psychrophilic anaerobic spore-forming bacterium from the overcooled water brine in permafrost: description *Clostridium algoriphilum* sp. nov. Extremophiles *9,* 239–246.

Shulse, C.N., and Allen, E.E. (2011). Diversity and distribution of microbial long-chain fatty acid biosynthetic genes in the marine environment. Environ. Microbiol. *13,* 684–695.

Silvius, J.R. (1982). Thermotropic phase transitions of pure lipids in model membranes and their modification by membrane protein. In Lipid–Protein Interactions, P. Jost and O.H. Griffith, eds. (New York, USA: John Wiley and Sons), vol. 2, pp. 239–281.

Simankova, M.V., Kotsyurbenko, O.R., Stackebrandt, E., Kostrikina, N.A., Lysenko, A.M., Osipov, G.A., and Nozhevnikova, A.N. (2000). *Acetobacterium tundrae* sp. nov., a new psychrophilic acetogenic bacterium from tundra soil. Arch. Microbiol. *174,* 440–447.

Sinensky, M. (1974). Homeoviscous adaptation – a homeoviscous process that regulates the viscosity of membrane lipids in *Escherichia coli.* Proc. Nat. Acad. Sci. USA. *71,* 522–525.

Sugihara, S., Orikasa, Y., and Okuyama, H. (2008). An EntD-like phosphopantetheinyl transferase gene from *Photobacterium profundum* SS9 complements *pfa* genes of *Moritella marina* strain MP-1 involved in biosynthesis of docosahexaenoic acid. Biotechnol. Lett. *30,* 411–414.

Sugihara, S., Hori, R., Nakanowatari, H., Takada, Y., Yumoto, I., Morita, N., Yano, Y., Watanabe, K., and Okuyama, H. (2010a). Possible biosynthetic pathways for all *cis*-3,6,9,12,15,19,22,25,28-hentriacontanoaene in bacteria. Lipids *45,* 167–177.

Sugihara, S., Orikasa, Y., and Okuyama, H. (2010b). The highly expressed *Escherichia coli entD* gene complements the *pfaE* deficiency in a *pfa* gene clone responsible for the biosynthesis of long-chain *n*-3 polyunsatuarted fatty acids. FEMS Microbiol. Lett. *307,* 207–211.

Sugihara, S., Bin Haji Mohd Taha, I.A., Motoigi, T., Ueno, A., Shimizu, S., Sato, M., Kaneko, T., Watanabe, K., Yumoto, I., and Okuyama, H. (2012). *Shewanella oshoroensis* sp. nov., mesophilic eicosapentaenoic acid and hentriacontanonaene-producing bacterium. Res. J. Microbiol. *7,* 131–138.

Sukovich, D.J., Seffernick, J.L., Richman, J.E., Gralnick, J.A., and Wackett, L.P. (2010). Widespread head-to-head hydrocarbon biosynthesis in bacteria and role of OleA. Appl. Environ. Microbiol. *76,* 3850–3862.

Suutari M., and Laakso, S. (1994). Microbial fatty acids and thermal adaptation. Crit. Rev. Microbiol. *20,* 285–328.

Takeyama, H., Takeda, D., Yazawa, K., Yamada, A., and Matsunaga, T. (1997). Expression of the eicosapentaenoic acid synthesis gene cluster from *Shewanella* sp. in a transgenic marine cyanobacterium, *Synechococcus* sp. Microbiology *143,* 2725–2731.

Tanaka, M., Ueno, A., Kawasaki, K., Yumoto, I., Ohgiya, S., Hoshino, T., Ishizaki, K., Okuyama, H., and Morita, N. (1999). Isolation of clustered genes that are notably homologuos to the eicosapentaenoic acid biosynthesis gene cluster from the docosahexaenoic acid-producing bacterium *Vibrio marinus* strain MP-1. Biotechnol. Lett. *21,* 939–945.

Tang, Q.Y., Yang, N., Wang, J., Xie, Y.Q., Ren, B., Zhou, Y.G., Gu, M.Y., Mao, J., Li, W.J., Shi, Y.H., and Zhang, L.X. (2011). *Paenibacillus algorifonticola* sp. nov., isolated from a cold spring. Int. J. Syst. Evol. Microbiol. *61*, 2167–2172.

Usui, K., Hiraki, T., Kawamoto, J., Kurihara, T., Nogi, Y., Kato, C., and Abe, F. (2012). Eicosapentaenoic acid plays a role in stabilizing dynamic membrane structure in the deep-sea piezophile *Shewanella violacea*: A study employing high-pressure time-resolved fluorescence anisotropy measurement. Biochim. Biophys. Acta. *1818*, 574–583.

Valentine, R.C., and Valentine, D.L. (2004). *Omega*-3 fatty acids in cellular membranes: a unified concept. Prog. Lipid Res. *43*, 383–402.

Vandieken, V., Muβmann, M., Niemann, H., and Jørgensen, B.B. (2006). *Desulfuromonas svalbardensis* sp. nov. and *Desulfuromusa ferrireducens* sp. nov., psychrophilic, Fe(III)-reducing bacteria isolated from Arctic sediments, Svalbard. Int. J. Syst. Evol. Microbiol. *56*, 1133–1139.

Vrinten, P., Wu, G., Truska, M., and Qiu, X. (2007). Production of polyunsaturated fatty acids in transgenic plants. Biotechnol. Genet. Eng. Rev. *24*, 263–279.

Wada, M., Fukunaga, N., and Sasaki, S. (1989). Mechanism of biosynthesis of unsaturated fatty acids in *Pseudomonas* sp. strain E-3, a psychrotrophic bacterium. J. Bacteriol. *171*, 4267–4271.

Wada H., and Murata, N. (2010). Lipids in Thylakoid Membranes and Photosynthesizing Cells. In Lipids in Photosynthesis: Essential and Regulatory Functions (Advances in Photosynthesis and Respiration, vol. 30), H. Wada and N. Murata, eds. (The Netherlands: Springer), pp. 1–9.

Wilkinson, S.G. (1988). Gram-negative bacteria. In Microbial Lipids, Vol. 1, C. Ratledge and S.G. Wilkinson, eds. (Oxford, UK: Academic Press), pp. 299–488.

Xu, Y., Nogi, Y., Kato, C., Liang, Z., Rüger, H.J., Kegel, D.D., and Glansdorff, N. (2003a). *Moritella profunda* sp. nov. and *Moritella abyssi* sp. nov., two psychropiezophilic organisms isolated from deep Atlantic sediments. Int. J. Syst. Evol. Microbiol. *53*, 533–538.

Xu, Y., Nogi, Y., Kato, C., Liang, Z., Rüger, H.J., Kegel, D.D., and Glansdorff, N. (2003b). *Psychromonas profunda* sp. nov., a psychropiezophilic bacterium from deep Atlantic sediments. Int. J. Syst. Evol. Microbiol. *53*, 527–532.

Yano, Y., Nakayama, A., Saito, H., and Ishihara, K. (1994). Production of docosahexaenoic acid by marine bacteria isolated from deep sea fish. Lipids *29*, 527–528.

Yano, Y., Nakayama, V., Ishihara, K., and Saito, H. (1998). Adaptive changes in membrane lipids of barophilic bacteria in response to changes in growth pressure. Appl. Environ. Microbiol. *64*, 479–485.

Yazawa, K. (1996). Production of eicosapentaenoic acid from marine bacteria. Lipids *31*, S297–S300.

Yazawa, K., Araki, K., Okazaki, N., Watanabe, K., Ishikawa, C., Inoue, A., Numao, N., and Kodo, K. (1988). Production of eicosapentaenoic acid by marine bacteria. J. Biochem. Tokyo *103*, 5–7.

Yu, R., Yamada, A., Watanabe, K., Yazawa, K., Takeyama, H., Matsunaga, T., and Kurane, R. (2000). Production of eicosapentaenoic acid by a recombinant marine cyanobacterium, *Synechococcus* sp. Lipids 35, 1061–1064.

Yu, Y., Li, H.R., and Zeng, Y.X. (2011). *Colwellia chukchiensis* sp. nov., a psychrotolerant bacterium isolated from the Arctic Ocean. Int. J. Syst. Evol. Microbiol. *61*, 850–853.

Yumoto, I., Kawasaki, K., Iwata, H., Matsuyama, H and Okuyama, H. (1998). Assignment of *Vibrio* sp. strain ABE-1 to *Colwellia maris* sp. nov., a new psychrophilic bacterium. Int. J. Syst. Bacteriol. *48*, 1357–1362.

Yumoto, I., Kusano, T., Shingyo, T., Nodasaka, Y., Matsuyama, H., and Okuyama, H. (2001). Assignment of *Pseudomonas* sp. strain E-3 to *Pseudomonas psychrophila* sp. nov., a new facultatively psychrophilic bacterium. Extremophiles 5, 343–349.

Yumoto, I., Hirota, K., Sogabe, Y., Nodasaka, Y., Yokota, Y., and Hoshino, T. (2003). *Psychrobacter okhotskensis* sp. nov., a lipase-producing facultative psychrophile isolated from the coast of the Okhotsk Sea. Int. J. Syst. Evol. Microbiol. *53*, 1985–1989.

Zhang, D.C., Liu, H.C., Xin, Y.H., Yu, Y., Zhou, P.J., and Zhou, Y.G. (2008). *Salinibacterium xinjiangense* sp. nov., a psychrophilic bacterium isolated from the China No. 1 glacier. Int. J. Syst. Evol. Microbiol. *58*, 2739–2742.

Zhang, D.C., Schumann, P., Liu, H.C., Xin, Y.H., Zhou, Y.G., Schinner, F., and Margesin, R. (2010). *Arthrobacter alpinus* sp. nov., a psychrophilic bacterium isolated from alpine soil. Int. J. Syst. Evol. Microbiol. *60*, 2149–2153.

Zhang, D.C., Busse, H.J., Liu, H.C., Zhou, Y.G., Schinner, F., and Margesin, R. (2011a). *Hymenobacter psychrophilus* sp. nov., a psychrophilic bacterium isolated from soil. Int. J. Syst. Evol. Microbiol. *61*, 859–863.

Zhang, D.C., Busse, H.J., Wieser, C., Liu, H.C., Zhou, Y.G., Schinner, F., and Margesin, R. (2011b). *Candidimonas bauzanensis* sp. nov., isolated from soil and emended description of the genus *Candidimonas* Vas-Moreira *et al.* 2011, emend Zhang *et al.* 2011. Int. J. Syst. Evol. Microbiol. *62*, 2084–2088.

Zhang, D.C., Liu, H.C., Zhou, Y.G., Schinner, F., and Margesin, R. (2011c). *Pseudomonas bauzanensis* sp. nov., isolated from soil. Int. J. Syst. Evol. Microbiol. *61*, 2333–2337.

Zhang, D.C., Redzic, M., Liu, H.C., Zhou, Y.G., Schinner, F., and Margesin, R. (2011d). *Devosia psychrophila* sp. nov. and *Devosia glacialis* sp. nov., two novel bacteria from alpine glacier cryoconite. Int. J. Syst. Evol. Microbiol. *62*, 710–715.

Zhang, D.C., Redzic, M., Schinner, F., and Margesin, R. (2011e). *Glaciimonas immobilis* gen. nov., sp. nov., a member of the family Oxalobacteraceae isolated from alpine glacier cryoconite. Int. J. Syst. Evol. Microbiol. *61*, 2186–2190.

Zhang, H., Tanabe, S.H., Nagata, S., Ban, S., and Imura, S. (2010). *Psychroflexus lacisalsi* sp. nov., a moderate halophilic bacterium isolated from a hypersaline lake (Hunazoko-Ike) in Antarctica. J. Microbiol. *48*, 160–164.

Zhu, K., Choi, K.H., Schweizer, H.P., Rock, C.O., and Zhang1 Y.M. (2006). Two aerobic pathways for the formation of unsaturated fatty acids in *Pseudomonas aeruginosa*. Mol. Microbiol. *60*, 260–273.

Zhu, K., Ding, X., Julotok, M., and Wilkinson, B.J. (2005). Exogenous isoleucine and fatty acid shortening ensure the high content of anteiso-$C_{15:0}$ fatty acid required for low-temperature growth of *Listeria monocytogenes*. Appl. Environ. Microbiol. *71*, 8002–8007.

Index

A

C

D

E

F

N

O

P

R

S

W

X

Y

Z

Figure 2.1 Salmon *Izushi. Izushi* is a traditional fermented food in the northern part of Japan. It is prepared from raw and salted sockeye salmon (*Oncorhynchus nerka*) fillets, rice and rice malt and vegetables (cabbage, carrot, ginger and red pepper). *Izushi* has been produced as a preserved food in the winter season. It is considered that the presence of lactic acid bacteria and the preservation at low temperature prevent its spoilage in a short period. Bar= 3cm.

Figure 2.2 Squid *Shiokara. Shiokara* is a traditional fermented food in Japan. The original materials are shredded squid meat, squid liver and salt. *Shiokara* has been produced as a preserved food in a wide area of Japan and in all seasons. The high salt concentration and squid liver prevent its spoilage. Bar= 2cm.

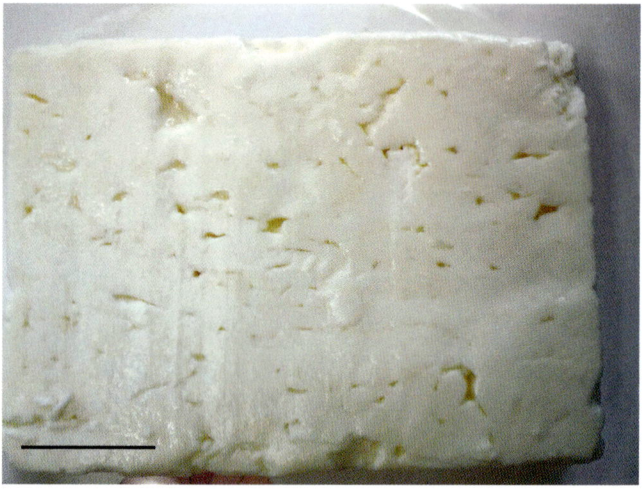

Figure 2.3 Feta cheese. Feta cheese is one of the most important and popular dairy products in Greece. It tastes slightly acidic and salty and is usually ripened in brine. Its characteristic flavour and taste are developed by the action of natural lactic acid microbiota of milk. It is considered that the presence of lactic acid and ripening at low temperature prevent its spoilage. Bar= 3cm.

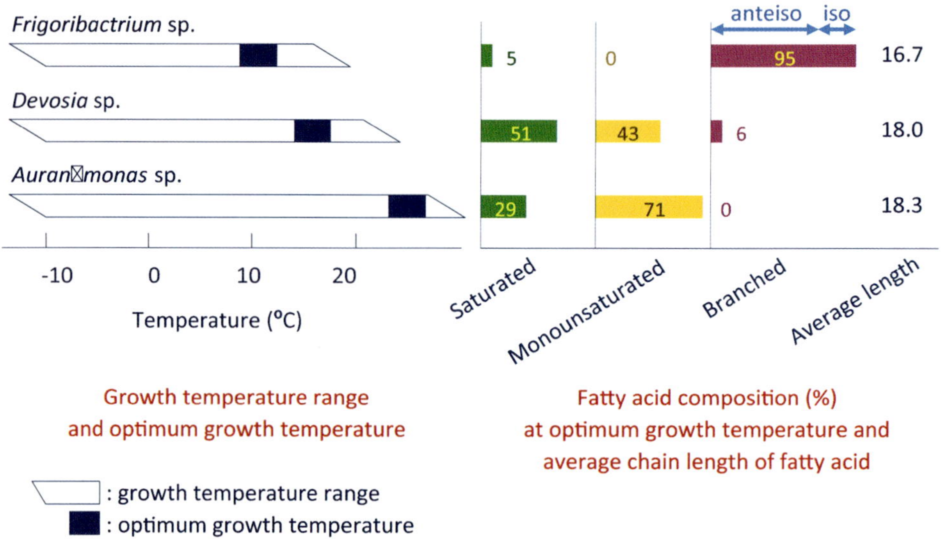

Figure 11.1 Growth temperature range, optimal temperature for growth, and fatty acid composition of psychrophilic *Frigoribacterium* sp. and *Devosia* sp. and psychrotrophic *Aurantimonas* sp. isolated from Antarctic soil. Fatty acid composition was determined using cells grown in modified Brock's basal salts agar (Kurosawa *et al.*, 1998) at optimal temperature. Antarctic soil associated with moss vegetation was used to isolate individual strains (N. Kurosawa, unpublished).

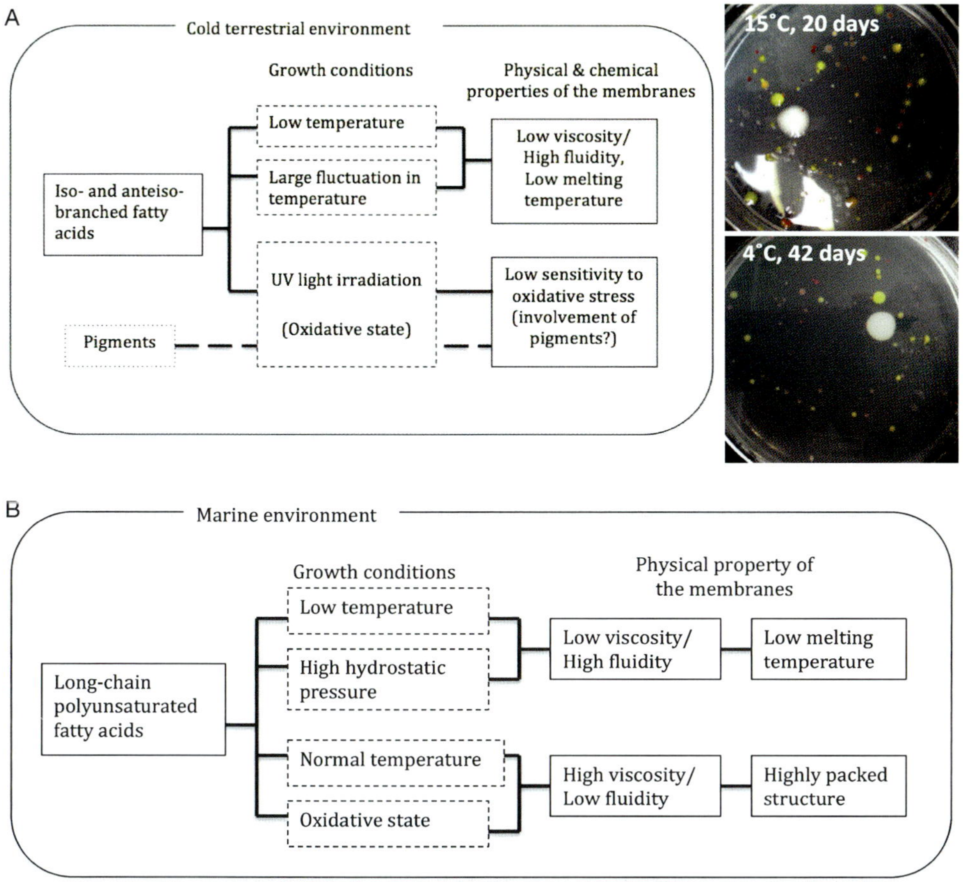

Figure 11.2 Schematic presentation of possible physiological roles of branched fatty acids in bacteria in cold terrestrial environments (A, left panel) and long-chain polyunsaturated fatty acids (LC-PUFAs) in Gram-negative psychrophilic and psychrotrophic bacteria in marine environments (B). The right panel of A shows agar plates of bacteria grown at 4°C and 15°C for 42 days and 20 days, respectively. In this experiment, Antarctic soil associated with moss vegetation was used (Y. Tsuboi and N. Kurosawa, unpublished). Environmental conditions are enclosed in boxes with dashed line. Although it has not been certified yet, possible involvement of pigments in antioxidation is shown with a dotted box and a broken line in the left panel of B.